P9-DHT-359

SCHAUM'S OUTLINE OF

THEORY AND PROBLEMS

OF

FLUID DYNAMICS

Second Edition

•

WILLIAM F. HUGHES, Ph.D.

Professor of Mechanical Engineering
Carnegie-Mellon University

JOHN A. BRIGHTON, Ph.D.

Dean of Engineering
Pennsylvania State University

SCHAUM'S OUTLINE SERIES

McGRAW-HILL, INC.

New York   St. Louis   San Francisco   Auckland   Bogotá   Caracas
Hamburg   Lisbon   London   Madrid   Mexico   Milan   Montreal
New Delhi   Paris   San Juan   São Paulo   Singapore
Sydney   Tokyo   Toronto

WILLIAM F. HUGHES holds the B.S., M.S., and Ph.D. degrees from Carnegie Institute of Technology. He was a National Science Foundation Postdoctoral Fellow at Cambridge University in England and a Fulbright Fellow at the University of Sydney, Australia. He is the author of *An Introduction to Viscous Flow*, coauthor of *The Electromagnetodynamics of Fluids* and coauthor of *Schaum's Outline of Basic Equations of Engineering Science*. He is also the author of numerous scientific and technical publications in the fields of fluid mechanics and magnetohydrodynamics. Presently, he is Professor of Mechanical Engineering at Carnegie Mellon University.

JOHN A. BRIGHTON, P.E., holds the B.S., M.S., and Ph.D. degrees from Purdue University. He is Dean of the College of Engineering at the Pennsylvania State University and has taught at Purdue, Carnegie Mellon, Penn State, Michigan State and Georgia Tech Universities. He is a Fellow of ASME and ASEE. His research in fluid mechanics includes turbulent mixing and blood flow.

Schaum's Outline of Theory and Problems of
FLUID DYNAMICS

1 2 3 4 5 6 7 8 9 10 11 12 13 14 15 16 17 18 19 20 SHP SHP 9 2 1 0

ISBN 0-07-031117-X

Sponsoring Editor, John Aliano
Production Supervisor, Denise Puryear
Editing Supervisors, Meg Tobin, Maureen Walker

Library of Congress Cataloging-in-Publication Data
Hughes, William F. (William Frank)
    Schaum's outline of theory and problems of fluid dynamics /
William F. Hughes, John A. Brighton. — 2nd ed.
        p.   cm. — (Schaum's outline series)
    ISBN 0-07-031117-X
    1. Fluid dynamics—Outlines, syllabi, etc.   2. Fluid dynamics—
Problems, exercises, etc.   I. Brighton, John A. II. Title.
QA911.H84   1991
532'.05—dc20
                                                                    90-34320
                                                                        CIP

Preface

If one casually glances around, most things seem to be solids, but when one thinks of the oceans, the atmosphere, and on out into space it becomes rather obvious that a good portion of the earth's surface and of the entire universe is in the fluid state.

Aside from the scientists' interest in the nature of the universe which is mostly gas, the engineer's interest in devices useful to mankind can seldom drift far from fluids. It is indeed difficult to think of any machine, device, or tool which doesn't have some fluid hidden in it somewhere and some fluid mechanics behind its design. Pumps, fans, blowers, jet engines, rockets, gas turbines are primarily fluid machines. Aircraft and ships move through fluids. The atmosphere and the weather are governed by the dynamics of fluids. All machines must be lubricated, and the lubricant is a fluid. Even the vacuum tube relies on an electron gas for its operation. And, no matter how complex or esoteric the device, the basic concepts of fluid dynamics still apply.

Fluid dynamics occupies an important place in modern science and engineering. It forms one of the foundations of aeronautics and astronautics, mechanical engineering, meteorology, marine engineering, civil engineering, bio-engineering and, in fact, just about every scientific or engineering field.

This book may be used as either a text or supplementary text for a first undergraduate course in fluid mechanics. However, one of the unique features is the treatment of a broad spectrum of fluid mechanics topics and a few specialized topics such as hypersonic flow, magnetohydrodynamics and non-Newtonian fluids. The coverage of this material makes this book useful as a reference and supplementary text for either an intermediate or first year graduate course.

The first few chapters are written primarily for the beginning student, with considerable emphasis on basic ideas of fluid motion. The first three chapters contain rather complete derivations of the conservation equations both in integral and differential form. Many examples are presented in order to convey the very important ideas of a control volume, Bernoulli's equation, and the motion of fluids in general. A convenient summary of important equations and a general discussion of problem solving techniques, which will be helpful to the beginning student, are provided in Chapter 3.

The level of the book changes from chapter to chapter. Chapters 1 through 5 and Chapter 7 serve as a first introduction to fluid mechanics at the undergraduate level. Chapters 6 and 8 extend to the aerodynamics of subsonic and supersonic flow and are pitched at a more advanced undergraduate level.

The last part of the book deals with topics which are of current research interest. For example, much of the fluid mechanics literature and research efforts of today are in the areas of turbulence, hypersonic flow, magnetohydrodynamics, and non-Newtonian fluids. These chapters are written in such a way that one who is not familiar with these particular fields may obtain an introduction to the form of the mathematical models, the simplifications and techniques, and the present state of the art. If one is interested in a more in-depth study, the references at the end of the chapters may be pursued. These, along with this book, could serve as the basis for an individual study program.

In a book of this type the topics and applications which can be covered are obviously limited and many important areas of current interest are not discussed. For example, the topics of wave motion, stability, applications to fluid machinery, and perhaps most importantly the rapidly growing field of CFD (Computational Fluid Dynamics) are not covered here. Much of the research in modern fluid dynamics is centered on computational techniques, but it was thought that an emphasis on the physical aspects of fluid dynamics would be more appropriate for a book at this introductory level. A thorough understanding of the material presented in this book should prepare the reader to approach the current literature in a variety of specialized topics.

In this second edition we have added a substantial amount of material in the first eight chapters and in the appendix in order to make the book more self-contained and readable. The references have been updated and the later chapters have been revised to reflect recent research trends.

The authors wish to thank Dr. E. W. Gaylord for his valuable help with Chapter 12, Mrs. Dorothy C. Wakefield and Mrs. Mary Bathurst for typing the original manuscript, and Ms. Ruth Davis for word processing the second edition.

William F. Hughes
John A. Brighton

Contents

Chapter 1

Introduction

1.1 WHAT IS A FLUID

Dynamics, the study of motion of matter, may be divided into two parts—dynamics of rigid bodies and dynamics of non-rigid bodies. The latter is usually further divided into two general classifications—elasticity (solid elastic bodies) and fluid mechanics. Since a large portion of the Earth is in the fluid state, it is rather obvious that engineers and scientists have to know something about fluids. But first, what is a fluid? How does a fluid differ from what we call a solid elastic substance such as a bar of steel?

In simple terms a fluid is a substance which cannot resist a shear force or stress without moving as can a solid. Fluids are usually classified as liquids or gases. A liquid has intermolecular forces which hold it together so that it possesses volume but no definite shape. A liquid poured into a container will fill the container up to the volume of the liquid regardless of the container's shape. Liquids have but slight compressibility and the density varies little with temperature or pressure. A gas, on the other hand, consists of molecules in motion which collide with each other tending to disperse it so that a gas has no set volume or shape. A gas will fill any container into which it is placed. For a given mass or system of gas, the pressure, temperature and enclosing volume are related by the appropriate equation of state of the gas.

The many applications of fluid mechanics make it one of the most vital and fundamental of all engineering and applied scientific studies. The flow of fluids in pipes and channels is of importance to civil engineers. The study of fluid machinery such as pumps, compressors, heat exchangers, jet and rocket engines, and the like, makes fluid mechanics of importance to mechanical engineers. The flow of air over objects, aerodynamics, is of fundamental interest to aeronautical and space engineers in the design of aircraft, missiles and rockets. In meteorology, hydrology and oceanography the study of fluids is basic since the atmosphere and the ocean are fluids. And today in modern engineering many new disciplines combine fluid mechanics with classical disciplines. For example, fluid mechanics and electromagnetic theory are studied together as magnetohydrodynamics. In new types of energy conversion devices and in the study of stellar and ionospheric phenomena, magnetohydrodynamics is vital.

We see that a good familiarity with fluid mechanics is essential to the modern engineer and scientist, and it is probably obvious that fluid mechanics and its applications is a broad subject with far-flung fields of specialization. What we will do in this book is to present most of the fundamental ideas with many of the applications. Once these fundamentals are mastered more advanced books and research literature may be studied to increase one's understanding of more specialized aspects of fluid mechanics. However, it is essential to be well grounded in the fundamentals which do not change with age, and which can always be relied upon, before going on to more advanced topics.

We will provide an introduction to the subject of fluid mechanics by presenting fundamental ideas, from the basic mathematical formulation to modern hypersonic flow theory and magnetohydrodynamics.

1.2 THE MATHEMATICAL MODEL

In solid rigid body mechanics, we usually ask the question: what is the position in space of a particle as a function of time? From this information all other questions, such as what are the velocity and acceleration, may be answered. If the position vector $\mathbf{r}$ denotes the position of a particle, then $\mathbf{r}(t)$ is the important parameter. Velocity and acceleration are simply $d\mathbf{r}/dt$ and $d^2\mathbf{r}/dt^2$.

However, in a fluid we are not dealing with a single particle. We are concerned with a *continuum*. In fact we don't have to keep track of individual particles or even little blobs of fluid. Rather, it is convenient to ask the question: at some point in space (relative to an arbitrary fixed coordinate system) what is the velocity, acceleration and thermodynamic properties at that point in space as a function of time? As time proceeds the fluid at that point in space is being constantly exchanged for new fluid as the fluid flows past, so that we keep track not of any individual particle of fluid but of history at some point in space regardless of what bit of fluid happens to be there at any particular time. Such a description of the fluid is called an Eulerian description as opposed to a Lagrangian description which is used to keep track of an individual particle as in rigid body dynamics. We will discuss these problems in detail in Chapter 3.

We have mentioned the word continuum. What does this mean in a fluid sense? We assume that the distance between fluid particles (or molecules), or more precisely the mean free path, is very small. By small we mean small compared to any physical dimensions of the problem to which we are applying the principles of fluid mechanics. In aerodynamics, the thickness, say, of a wing is many orders of magnitude greater than the mean free path of the air flowing over the wing. Hence we assume that all mathematical limiting processes (of the calculus) can be taken in a meaningful sense and that any volume of fluid can be continuously subdivided into smaller and smaller volumes while retaining the fluid continuum character. Obviously this division would break down eventually, but we assume that by the time it does the dimensions are so small that they are microscopic and of no concern to us.

However, if such is not the case, as in a rarefied gas flow where the mean free path may be of the same order of magnitude as the physical dimensions of the problem, then the fluid continuum assumption breaks down and we must use a strictly microscopic approach such as free molecule flow theory. We will not be concerned with these rarefied gas theories here and will always be concerned with a homogeneous isotropic continuum which can be treated strictly by macroscopic methods.

The number of basic variables in fluid mechanics is five: three velocity components and two thermodynamic properties. Any two of the thermodynamic properties, such as pressure, temperature, density, enthalpy, entropy, etc., suffice to determine the state and hence all the other properties. The fluid flow field is completely determined once we specify the velocity vector **V** and two thermodynamic properties as a function of space and time. Hence we need five independent equations. These are usually the three components of the equation of motion, a continuity equation and an energy equation. Often an equation of state is also introduced in order to allow the writing of the energy equation in terms of three variables (temperature, density and pressure) instead of just two. In that case we have six variables and six equations. In turbulent flow additional unknowns appear for the same number of equations, which prevents a complete theoretical formulation of the problem.

In an incompressible fluid the energy equation is not needed since density is taken as known and only pressure along with velocity need be found in order to completely describe the fluid flow. The temperature is uncoupled then, but an energy equation must be used to find it if this information is needed.

1.3 SOME DEFINITIONS

The flow of fluids may be classified in many different ways. Let us look at some of the different types of flow and relate them to some ordinary everyday experiences and observations. But first a few terms must be defined so that we will have a language in which to talk about fluid flow. Later we will provide more rigorous definitions, but for now, let us give a few simple and somewhat heuristic definitions.

Pressure: The pressure in a static fluid (one at rest) is defined as the normal compressive force per unit area (normal stress) acting on a surface immersed in the fluid. The pressure might be measured by the force on the face of a unit cube (one with unit dimensions) inserted in the fluid. We have to imagine that the cube does not disturb the fluid, so that the actual pressure at a point in the fluid would be the force acting on a face of the cube divided by the area of that face in the limit as the area becomes infinitesimally small. The pressure at a point is isotropic in a fluid at rest, that is, the force would be the same on all faces of the cube and would be the same regardless of the orientation of the cube in space. Such an isotropic pressure is called hydrostatic pressure. This is the pressure used in thermodynamics (gas laws) and is a thermodynamic property. If the pressure varies from place to place in the fluid a net pressure force would exist on any fixed volume of fluid and must be balanced by a body force such as gravity, or else the fluid will move, the pressure force generating an acceleration in the fluid.

In a dynamical situation (when the fluid moves) there may exist not only pressure forces in the fluid, but shear forces or stresses as well. However, the pressure is still isotropic and defined in the same way as above but must be measured as the normal stress on an area which moves along locally with the fluid. There are sometimes difficulties in moving gases where the normal stresses on a cube are not quite the same in all directions. We can still define a hydrostatic isotropic pressure, but small additional forces in certain directions come about because of fluid viscosity effects. These concepts will be discussed in more detail in the next two chapters.

Viscosity, Friction, and Ideal Flow: All fluids have viscosity which causes friction. The importance of this friction in physical situations depends on the type of fluid and the physical configuration or flow pattern. If the friction is negligible, we say the flow is ideal. Friction may arise because of viscosity or turbulence.

Roughly speaking, viscosity is a measure of the fluid's resistance to shear when the fluid is in motion (remember a fluid cannot resist shear without moving as can a solid). Imagine two parallel plates of large size, Fig. 1-1, in steady relative motion. Fluid between the plates has a linear velocity profile as shown (if no pressure gradient exists along the plates in the direction of motion). There is no slip between the fluid and plates; that is, at an interface between a fluid and solid, the velocity of the fluid must be the same as the solid. If we consider a small element of the fluid, as shown in Fig. 1-1, the shear stress τ on the top (which is numerically the same as on the bottom in this case) may be written:

$$\tau = \mu \, \frac{\partial u}{\partial y} \qquad\qquad (1.1)$$

where μ, the viscosity, is the proportionality constant between shear stress and velocity gradient. The units of viscosity are clearly lb_f-sec/ft^2 in English engineering units, and N-s/m^2 in SI units. Also, viscosity is often measured in a unit called the poise. One poise is one (dyne-sec)/cm^2. The ratio of viscosity to mass density ρ is called kinematic viscosity and is usually denoted as v.

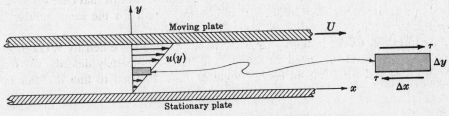

Fig. 1-1. Flow between parallel plates to illustrate viscosity. The velocity u is linear across the channel, zero at the bottom and U at the top. A small element shows the shear stress.

The viscosity of a liquid decreases with increasing temperature (as one knows from trying to start an automobile on a cold morning) but gases increase their viscosity with increasing temperature. The viscosity of fluids also depends on pressure, but this dependence is usually of little importance compared to the temperature variation in engineering problems.

Such a simple relationship between shear stress and velocity gradient is known as the Newtonian relationship. In general, fluids which obey such a relationship are known as Newtonian fluids. (A more general expression for Newtonian fluids which allows velocity gradients in three dimensions will be developed later.)

Although the linear Newtonian relationship is only an approximation, it is surprisingly good for a wide class of fluids. For some substances, however, the shear stress may be a function not only of velocity gradient (which is the same as shear strain rate) but also ordinary strain as well. Such substances are known as viscous-elastic. And even for simple viscous fluids, in which the shear stress depends only on velocity gradient, the fluid may not be Newtonian and in fact there may exist a rather complicated nonlinear relationship between shear stress and strain rate. If the fluid stress-strain rate relationship depends on prior working or straining, the fluid is said to be thixotropic (such a substance is printer's ink).

Another type of fluid is one with plastic behavior which is characterized by an apparent yield stress; that is, the fluid behaves as a solid until it yields, then behaves like a viscous fluid. Some greases and sludges behave in this manner. On the other extreme from plastic fluids are fluids called dilatants which flow easily with a low viscosity at low strain rates, but become more like a solid as the strain rate increases (such a substance is quicksand). Fig. 1-2 shows the behavior of these fluids graphically.

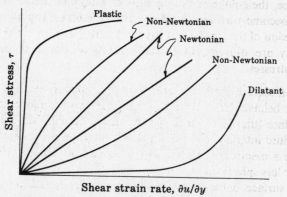

Fig. 1-2.   Types of viscous and plastic fluids.

Water, air and gases are essentially Newtonian, but considerations of fluids which are visco-elastic or non-Newtonian are very important in fluid mechanics, although generally not so well understood or appreciated.

Internal friction or stresses can be generated, then, by the viscosity in a fluid. In addition, turbulence may generate shear stresses in a fluid, and we will mention turbulence in the next section.

If a fluid has no viscosity and does not flow in a turbulent manner, the fluid is said to be an ideal fluid, or more correctly the flow is said to be ideal. An ideal flow then has no internal friction and hence no internal dissipation or losses. Actually, of course no fluid is ever really ideal, but some fluids, at least in certain regions of flow under certain circumstances, approach ideal conditions very closely and are considered such for analysis. For instance, the flow of air over objects (aerodynamics) is considered ideal flow except in a thin layer called the boundary layer just next to the wing or surface. As we will see, it is convenient to divide real fluid flows into different regions each of which may be considered ideal, viscous, or turbulent.

Laminar Flow and Turbulent Flow: The terms laminar flow and purely viscous flow are used synonymously to mean a fluid flow which flows in laminas or layers, as opposed to turbulent flow in which the velocity components have random turbulent fluctuations imposed upon their mean values (Fig. 1-3). A stream of dye or ink inserted in a laminar flow will streak out a thin line and always be composed of the same fluid particles. However, in turbulent flow the dye line would quickly become tangled up and mixed in with the fluid as it flows along, and we would see myriads of threads and clouds ever widening and dispersing as the fluid flows along. A vivid example of laminar flow is a heavy molasses being poured from a bottle.

What determines whether a flow is laminar or turbulent? For a given fluid the velocity and channel configuration or size determine it. As the velocity increases the flow will change from laminar to turbu-

Fig. 1-3. Laminar and turbulent flow. The lines indicate the
paths of particles.

lent, passing through a transition regime. Both laminar and turbulent flow occur in nature, but turbulent flow seems to be the more natural state of affairs.

A simple example of this transition may be seen by observing the smoke rising from a cigarette or smoke stack. For a distance, the smoke rises in a smooth laminar manner. Then, rather abruptly, the smoke begins to mix up, become turbulent, and the smoke column rapidly widens and diffuses. The turbulence aids in the diffusion of the smoke and causes it to diffuse into a widening chaotic stream.

The effects of viscosity are still present in turbulent flow, but they are usually masked by the dominant turbulent shear stresses.

Surface Tension: The term surface tension is used loosely to identify the apparent stress in the surface layer of a liquid. This layer behaves like a stretched membrane and can give rise to a pressure difference across a curved liquid surface (that is an air-liquid interface). Actually the surface tension is an energy associated with any fluid-fluid interface and the liquid-air interface is the most common type. Since the liquid surface behaves like a membrane, we see why a liquid may form a miniscus in a capillary, and why raindrops are more or less spherical.

Across an interfacial surface between two fluids, the pressure difference Δp is balanced by the surface tension T (measured in force per unit length). At any point on the surface, the surface may be characterized by the two radii R_1 and R_2 of curvature of the surface (in any two mutually orthogonal planes, both perpendicular to the surface). The result is $T(1/R_1 + 1/R_2) = \Delta p$. Well-known geometrical relationships can be used to express the radii of curvature in terms of the equation of the surface.

Compressible and Incompressible Flow: It is customary to divide fluids into two groups—gases and liquids. Gases are compressible and their density changes readily with temperature and pressure. Liquids, on the other hand, are rather difficult to compress and for most problems one might consider them incompressible. Only in such situations as sound propagation in liquids does one need to consider their compressibility.

The density of a fluid is a thermodynamic property which depends on the state of the fluid. It is convenient to express the density ρ as a function of pressure and temperature. Such a relationship, which may be phenomenological or derived from microscopic considerations, is known as an equation of state. For an ideal gas the equation of state may be expressed as $p = \rho RT$ where R is the gas constant. More complicated equations may be used for real gases whose behavior departs from the ideal gas approximation. For a liquid, the density is related to the temperature by a coefficient of expansion just as for a solid, and the pressure dependence may be written as

$$dp = \beta \, d\rho/\rho \tag{1.2}$$

where β is the bulk compression modulus. For water β is about 3×10^5 psi, so that it requires enormous pressures to effect a slight density change. (For adiabatic compression of atmospheric air β is about 20 psi.) For most practical purposes then, liquids are incompressible. And, in fact, under certain flow conditions involving slight pressure changes, gases may be considered incompressible. (This is the case of subsonic aerodynamics where the air is assumed incompressible for low Mach numbers.)

Subsonic and Supersonic Flow: In compressible flow there is a great distinction between flow involving velocities less than that of sound (subsonic flow) and flow involving velocities greater than that of sound (supersonic). (Sonic speed in air at STP is about 1080 ft/sec or 810 mi/hr.) These differences between subsonic and supersonic flow will be pointed out later, but it is useful to remember that shock waves can only occur in flows which are supersonic.

The Mach number, M, is a measure of this relative speed and is defined as the ratio of the fluid speed to the local speed of sound:

$$M = V/a \qquad (1.3)$$

where V is the fluid speed and a the local sonic speed. When $M < 1$ we have subsonic flow, and when $M > 1$ we have supersonic flow. For flows around objects, where M is less than about 0.3, the flow may be treated as approximately incompressible. Transonic flow occurs when part of a body (airplane or missile, say) has fluid flowing over it at $M < 1$ and another part of the same body has fluid flowing over it at $M > 1$, so that at some point on the body $M = 1$. How can $M < 1$ and $M > 1$ on the same body at the same time? The answer is that the sonic speed and the fluid speed vary over the body. The temperature varies over a body in general and hence the local sonic speed must vary.

Steady Flow: By a steady flow we mean one in which the velocity components and thermodynamic properties at every point in space do not change with time. Actually if we were to follow an individual fluid particle its properties and velocity might change as it flows along. However, this does not matter. In fluid mechanics we always ask: what is happening at a particular point in space regardless of what fluid particle happens to be there at any particular time? In this sense, then, steady flow means steady in that nothing changes with time at any point in space. A motion picture or snapshot would look the same no matter what time it was made. An important point to understand is that a fluid may have an acceleration at a point in space even in steady flow. A fluid particle may be moving along, but at any particular point in space it behaves just as any other fluid particle when it was at that place.

Types or Classes of Flow: We have discussed some basic definitions as applied to types of fluid flow. Now we can see what types of flow occur in real physical situations and classify them accordingly. We will see that we may have such classifications as incompressible-laminar, compressible-ideal-supersonic, compressible-laminar, incompressible-turbulent, etc.

1.4 PHYSICAL CLASSIFICATIONS AND TYPES OF FLOW

There are many ways to classify flow, according to the structure of the flow mentioned above, or according to the physical situation or configuration which also allows classification into one of these groups. Let us mention a few of these classifications.

Basically there are two types of fluid configurations, or spatial regions of flow: external and internal flow. By internal flow we mean pipe and channel flow and the like, where the fluid flows within a confining structure. External flow is the flow of a fluid over an object, such as in aerodynamics. Let us examine each of these flow types in more detail.

External Flow: The flow region around an object may be divided into three regions. Far from the body the flow is essentially ideal, with friction unimportant. Near the body the fluid develops a shear layer (since the velocity must be zero relative to the body on the surface of the body) where viscosity and/or turbulence is important. This frictional layer is called a boundary layer. The boundary layer may be laminar or turbulent. Behind the body a wake (a third distinct region) develops and is generally a region of high turbulence and low pressure (hence the drag due to the wake). Fig. 1-4 shows the flow over a cylinder with the wake. The wake is due to the separation of the boundary layer from the surface of the body. In fact, the ideal flow region behind the body, but outside the wake, and the wake region are rather well delineated by a shear layer (Fig. 1-4a).

The boundary layer brought about because viscosity in the fluid is the cause of the wake. If there were absolutely no viscosity so that the fluid were frictionless in an absolute sense, then there would be no separation and no wake. If there were no wake the flow pattern (which would be ideal) would be symmetrical from front to back of the cylinder and the pressure would be the same on the front as on

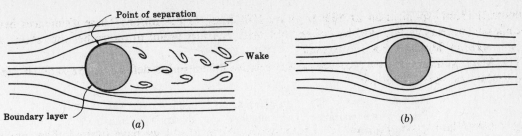

Fig. 1-4. (a) Boundary layer and separation over a cylinder. (b) Ideal flow if no separation occurred.

the back and there would be no drag on an object inserted into a flowing fluid. This absence of drag is contrary to experience and we realize that all fluids must have some internal friction. In the early days of the development of fluid mechanics as a science, it was thought that viscosity was rather negligible and mathematically the flow would be ideal everywhere and hence there would be no drag predicted by theory (Fig. 1-4b). Since this conclusion was contrary to experience it was known as d'Alembert's paradox. It was not until the boundary layer concept was introduced by Prandtl (a German fluid dynamicist) during the early part of this century that it became apparent that there was no paradox at all, but that any amount of viscosity, no matter how small, could bring about a wake and consequent drag.

If a body is streamlined (Fig. 1-5), that is, the trailing edge is faired into a gradual smooth contour, no separation will occur and the boundary layer will hug the body all the way around. Streamlining substantially reduces the drag and most aerodynamic objects, wings, etc., are streamlined. In such cases the flow is ideal completely around the body except for the boundary layer and a thin wake. As we shall see, the boundary layer in such cases may be rather thin and the flow pattern is well described by ideal flow except for the friction drag calculations. In subsonic aerodynamics the lift is determined by the potential (ideal) flow, and the drag essentially by the boundary layer.

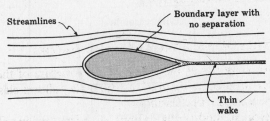

Fig. 1-5. Streamline flow about an object. The gradual taper of the trailing edge prevents separation of the boundary layer.

The boundary layer itself may be laminar or turbulent, depending on the parameters involved. In most practical cases the boundary layer changes from laminar to turbulent along the body. The transition to turbulence usually retards separation, but in partially streamlined bodies separation may not be distinct and the turbulent boundary layer merges into the wake region as shown in Fig. 1-6.

If the velocity is small, then the density variations are small and the flow may be considered incompressible. (This concept will be demonstrated in Chapter 7.) The flow then is as shown in Fig. 1-5 or Fig. 1-6.

If the velocity is increased until the Mach number is greater than about 0.3, then the density variations become important, but the general flow picture is still as shown above. However, when the Mach number is increased to a value greater than one, a shock wave will occur and the flow will be as shown in Fig. 1-7.

If the Mach number is increased to a value greater than about six, then dissociation and ionization will occur.

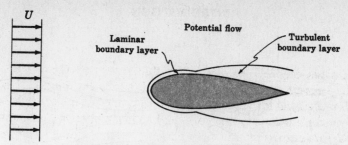

Fig. 1-6. Subsonic flow with a transition boundary layer with no separation. The boundary layer thickness is exaggerated.

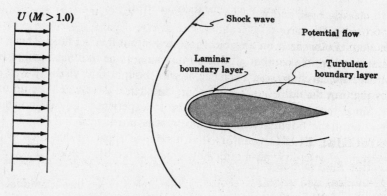

Fig. 1-7. Supersonic flow around an object.

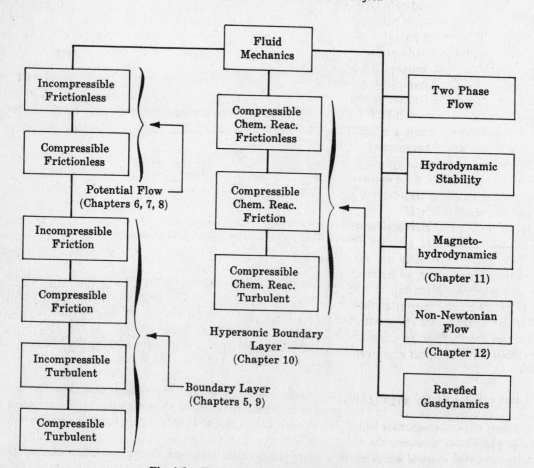

Fig. 1-8. Topical divisions of fluid mechanics.

Internal Flow: Inside pipes, channels, nozzles, and in fluid machinery the flow of the fluid is confined by walls and is usually referred to as internal flow. Such flow in the main part of a channel may be considered as approximately ideal for gases, but even so a boundary layer (usually turbulent) develops on the walls. In both viscous and turbulent flow the boundary layer thickens downstream so that it eventually extends all the way across the channel or pipe.

Now we are in a position to make topical divisions. These are shown in Fig. 1-8 with the left column generally going from the least to the most complicated. These topical divisions are purely arbitrary and represent different types of flow which are generally characterized by particular mathematical methods.

Another way to classify flow, and perhaps a bit easier to understand, is by the physical situation where the flow is occurring. We have tried to apply the results of each chapter to a variety of practical applications, rather than divide the book according to applications. Quite often, in standard textbooks there is a mix of classification. We could classify chapters accordingly as, for example: turbomachinery, open channel flow, aerodynamics, and waves and acoustics, to mention a few. Many such applications will be discussed throughout the text of the book and in the context of the solved example problems.

1.5 HOW WE DESCRIBE FLUID MOTION

In order to describe fluid motion we usually ask the question: What are the fluid properties, thermodynamic and mechanical, and what is the velocity and acceleration vector *at a particular location in space at a particular time*? We generally do not try to keep track of individual fluid particles. This scheme is based on a coordinate system which locates a point in space, just like coordinates on a map. We can then describe what is happening at that point as a function of time. Different fluid particles may be continuously sweeping past that point but if they all do the same thing and have the same properties we say that the flow is steady at that point. (The fluid particles may have a velocity and an acceleration but these quantities are constant in time for steady flow.)

Hence, in fluid mechanics we generally use spatial coordinates to denote "field coordinates" or position $\mathbf{r}$ in space. Derivatives of the coordinate with time such as dx/dt are meaningless. The main problem in fluid mechanics then is to find the velocity vector (and other properties) as a function of position $\mathbf{r}$ and time t. Such a coordinate system is known as an "Eulerian" system after the famous eighteenth century mathematician L. Euler.

This type of coordinate system is in contrast to the Lagrangian system used in rigid body mechanics (and occasionally in fluid mechanics for special purposes). There, the position of a particle may be represented as $\mathbf{r}(t)$. The coordinates locate the position and are functions of time. Velocity is simply $d\mathbf{r}/dt$ and acceleration $d^2\mathbf{r}/dt^2$.

Clearly, in the Eulerian system derivatives of $\mathbf{r}$ are meaningless and special means must be used to describe acceleration and velocity. We will discuss this procedure in Chapter 3. It should be remembered, however, that velocity and acceleration are vectors, and regardless of how they are represented they are the same in either an Eulerian or Lagrangian representation.

It is sometimes confusing to beginning students that in a "steady flow" the fluid may have an acceleration. We must remember that steady flow refers to what happens at a fixed point in space (with time). If one were to ride along with a fluid particle that is accelerating, one would observe changes in time even in steady flow. In summary, in steady flow each particle behaves the same way and has the same properties as it passes any given position in the flow field (i.e., any given point in space).

1.6 UNITS IN FLUID MECHANICS

The laws of fluid mechanics will be developed and expressed in the form of mathematical equations in this book. These equations do not depend on the system of units used. Currently, throughout the world, two systems of units are in widespread use: the English engineering system and the metric SI (Le Système International d'Unités). The SI system has been adopted in most of the world, replacing the

English system, but in engineering practice the English system is still in widespread use and will probably continue to be so for many years to come. It is imperative that one become familiar with both systems. In the Appendix we give a more complete discussion of units, but a brief introduction is in order.

In any given system of units there are generally two methods by which a consistent set of units in fluid mechanics may be obtained. Fundamental dimensions of either M, L, T, θ (mass, length, time, and temperature) or F, L, T, θ (force, length, time, and temperature) may be used and all other dimensions then expressed in terms of these fundamental dimensions by means of definitions and laws. (If electrical phenomena are to be considered, one further arbitrary electrical dimension must be chosen, say charge.)

The choice of either M, L, T, θ or F, L, T, θ is arbitrary, but they must be related in a precise manner dictated by Newton's second law of motion, Force = mass × acceleration ($\mathbf{F} = m\mathbf{a}$). For example, if we choose the unit of force as pound force (lb_f), then the unit of mass is the amount that is accelerated one ft/sec² when acted upon by one lb_f. This unit of mass is called a slug. In the SI system the unit of force is the newton (N) and the corresponding unit of mass is the kilogram (kg). One N will accelerate one kg one m/s².

One slug, acted upon by gravity, will accelerate g ft/sec² (where g is the acceleration due to gravity). Hence the gravitational force (at the Earth's surface at nominal sea level) on one slug is 32.174 lb_f. One slug "weighs" 32.174 pounds on the Earth. One kilogram mass is acted on by a force of g newtons on the surface of the Earth, where g is 9.807 m/s² at nominal sea level.

There are other ways to relate mass and force independently of Newton's law. The pound mass (lb_m) is defined as the quantity of mass that would (if suspended at the nominal Earth surface) be attracted to the Earth by one lb_f. Similarly, in SI units a kg_m may be defined as the mass that would be attracted to the Earth by one kg_f. In order to satisfy Newton's second law and dimensional homogeneity, a conversion factor g_c (which has units) must be used as $\mathbf{F} = m\mathbf{a}/g_c$ (with F in lb_f and m in lb_m, or F in kg_f and m in kg_m), where g_c is numerically the nominal Earth sea level value of g. Hence $g_c = 32.174 \dfrac{lb_m}{lb_f} \dfrac{ft}{sec^2}$ in English units and $9.807 \dfrac{kg_m}{kg_f} \dfrac{m}{s^2}$ in SI units.

In this book we use the English engineering system and SI system, which may be summarized in the following table. It should be noted that if we were to use the system where mass is defined in terms of lb_m and force in lb_f a g_c conversion factor would have to be used in all dynamical equations. This system is often confusing when first encountered; however, by defining mass in terms of slugs and kg we avoid the introduction of g_c and it never appears. In the system used throughout this book, g is always understood to be the local value of the acceleration due to gravity. The conversion from lb_m to slug is simply: one slug equals 32.174 lb_m so that the numerical value of g_c is the conversion factor.

Dimension	English engineering	SI
Force	pound (lb)	newton (N)
Mass	slug	kilogram (kg)
Time	second (sec)	second (s)
Length	foot (ft)	meter (m)
Pressure	lb/ft²	pascal = N/m² = Pa
Velocity	ft/sec	m/s
Acceleration	ft/sec²	m/s²
Viscosity	lb-sec/ft²	N-s/m² = Pa-s
Absolute temperature	degree Rankin (°R)	degree Kelvin (K)
Density	slug/ft³	kg/m³
Specific weight	lb/ft³	N/m³

Supplementary Problems

Below are listed a few simple experiments which illustrate some of the flow behavior of fluids.

1.1. Pour a bit of Karo syrup or molasses into a dish. Pour a little milk over the syrup and try to mix it. What happens? The viscosity is so great that the mixing caused by stirring is laminar and the mixing is rather slow. Turbulent mixing is much faster than laminar mixing. Why? Molecular diffusion also occurs but is rather unimportant in a heavy syrup, compared to the mixing effect.

1.2. Observe the smoke rising from a cigarette. Notice how it begins in a laminar fashion and suddenly becomes turbulent. As the smoke rises the flow becomes unstable and develops into turbulent flow.

1.3. Turn on a water faucet and adjust the flow so that there is a very small laminar trickle. What happens to the water a few inches down the stream? Again the flow becomes unstable and turbulent.

1.4. Increase the flow rate, in the above example. What happens? The water becomes turbulent in the pipe and is turbulent as it leaves the faucet.

1.5. Observe a low lying cloud system on a windy day, especially a heavy rain cloud system. Is the flow turbulent? Is it on a different scale from the water faucet flow?

1.6. Either with a burning cigarette or pipe create a rising column of smoke. Take a pencil and hold it (horizontal) in the stream of smoke. Observe, by looking at the pencil on end, the flow of smoke over the pencil and the separation of the boundary layer from the sides of the pencil. It is best to generate a heavy column of smoke by blowing into the pipe or cigarette.

1.7. Why does the smoke in the preceding problem rise?

Questions for thought

1.8. A mass of one slug weighs how much on the Earth?   *Ans.*   32.2 pounds.

1.9. How much does 1 lb_m weigh on the Earth?   *Ans.*   1 pound.

1.10. The moon's gravity is about 1/6 that of the Earth. How much does 1 lb_m weigh on the moon?
Ans.   1/6 pound.

1.11. Weight is often given in kilos (kg_f). How much does one kg mass weigh on the Earth?   *Ans.*   one kilo.

1.12. A special scale weighs in Newtons. A mass weighs 1 N. What would the same mass weigh on a standard scale?   *Ans.*   0.102 kilo.

1.13. A spherical soap bubble floats through the air. What can you say about the air pressure inside the bubble: If you are given the diameter D and the effective surface tension T of the soap film, what is the pressure inside?   *Ans.* $\Delta p = 4T/D$ where Δp is the rise above atmospheric pressure.

1.14. An air table is a large flat surface used to support heavy objects above a thin layer of air which allows the objects to be moved about with low friction. Air is blown through closely spaced holes in the table surface. Consider a heavy plate 1 meter square suspended 1 mm above the table on a cushion of air. What force is necessary to move the plate at a speed of 1 m/s along the table? $\mu = 2.18 \times 10^{-5}$ N-s/m^2.   *Ans.* 2.18×10^{-2} N.

1.15. Repeat the above problem with motor oil which has a viscosity of $\mu = 0.10$ N-s/7. Contrast the frictional properties of air vs. oil. Convert the answers to pounds and compare.

Chapter 2

Fluid Statics

2.1 PRESSURE

Before beginning the actual study of fluid dynamics, it is useful to discuss fluid statics, mainly because it allows immediate application of some everyday ideas to real engineering problems without becoming involved in complex notions. The concept of pressure is of particular importance in fluid statics. For example, it is important in determining the force applied to a dam by a body of water or in determining forces on submerged bodies in general. The exploration, recovery of objects and man's survival at large depths in the oceans has received considerable interest in recent years.

Pressure is defined as stress, or surface force per unit area,

$$p \equiv \lim_{\Delta A \to 0} \frac{\Delta F}{\Delta A} = \frac{dF}{dA} \qquad (2.1)$$

When we imagine the area going to zero for an actual fluid, a difficulty arises in that we are no longer dealing with a continuum. Thus there is a conflict between the continuum assumption and the usual definition of derivatives and differential quantities. This difficulty or conflict of ideas is present in the development of the differential equations of fluid mechanics, but we resolve it by assuming that the fluid is a continuum to any macroscopic scale of interest.

Let us look at a container of liquid at rest. If we take an imaginary boundary around a small part of the total volume, the net force acting on the fluid mass must be zero since it is not being accelerated. In particular, let us select a small volume as shown in Fig. 2-1.

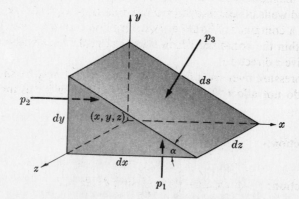

Fig. 2-1. Equilibrium of pressure forces on a fluid element.

Since there is no relative motion in the fluid, then the shear stress is everywhere zero. The only surface forces acting on the fluid element are normal (pressure). And the only body force is due to the earth's gravitational field which acts in the negative y direction. Thus summing the forces for the x direction we have

$$p_2 \, dy \, dz - p_3 \, dz \, ds \sin \alpha = 0$$

where $ds \sin \alpha = dy$ and thus $p_2 = p_3$. For the y direction[1],

$$p_1 \, dx \, dz - p_3 \, dz \, ds \cos \alpha - \tfrac{1}{2}\rho g \, dx \, dy \, dz = 0$$

where ρ is the mass density of the fluid and $ds \cos \alpha = dx$, giving

$$p_1 - p_3 - \tfrac{1}{2}\rho g \, dy = 0$$

Since the third term is small compared to the first two,

$$p_1 = p_2 = p_3$$

Since the volume element may be chosen with any orientation and the angle α was arbitrary, we have shown that the pressure at a point in a fluid at rest is the same in all directions, that is, isotropic.

In English engineering units pressure is given in lb_f/ft^2 or lb_f/in^2, written as psi. In SI units pressure is given in N/m^2 or pascals, written as Pa. One pascal is one N/m^2. Since one pascal is a rather small number, pressure is often expressed in kPa or MPa. As a general "feel" for the magnitude of pressure in the SI system, one atmosphere (14.7 psi) is about 101,000 Pa, which is 101 kPa or 0.101 MPa.

The pressure above atmospheric is referred to as gage pressure. In many problems only pressure differences are important, and the use of gage pressure is convenient. Care must be exercised, however, because the absolute pressure must often be used as, for example, in equations of state such as the ideal gas law. In English units the pressure expressed in lb_f/in^2 is written psia for absolute pressure and psig for gage pressure.

2.2 DIFFERENTIAL EQUATIONS OF FLUID STATICS

We will define equilibrium as a state in which each fluid particle is either at rest or has no relative motion with respect to other particles. The significant difference between these two conditions is that in the first there can be no acceleration of the total fluid system, and in the second there can be acceleration. We will consider both cases.

There are two kinds of forces to be considered: (1) body forces—forces acting on the fluid particles at a distance (e.g. gravity, magnetic field, etc.), and (2) surface forces—forces due to direct contact with other fluid particles or solid walls (forces due to pressure and tangential, that is, shear stress).

Again, we will look at a container of liquid at rest. This time, however, we shall choose an infinitesimal fluid particle from within the container which is cubical in shape as shown in Fig. 2-2. We assume gravity to act in the negative z direction.

We may assume the pressure over each face is constant since it may be shown that any variations are of second order and do not affect the final result. Summing the forces for the different directions gives

$$x \text{ direction:} \qquad p \, dy \, dz - \left(p + \frac{\partial p}{\partial x} \, dx\right) dy \, dz = 0$$

$$y \text{ direction:} \qquad p \, dx \, dz - \left(p + \frac{\partial p}{\partial y} \, dy\right) dx \, dz = 0$$

$$z \text{ direction:} \qquad p \, dx \, dy - \left(p + \frac{\partial p}{\partial z} \, dz\right) dx \, dy - \rho g \, dx \, dy \, dz = 0$$

---

[1] In this book the slug is used as the unit of mass in the English system, and kg as the unit of mass in the SI system. Then ρ has units of $slugs/ft^3$ or kg/m^3 and the specific weight γ has units lb_f/ft^3 or N/m^3. γ is given by $\gamma = \rho g$. We will generally not bother with the f subscript as in lb_f since lb is used here only as a force. For a detailed discussion of units and dimensions, the reader is referred to Appendix B.

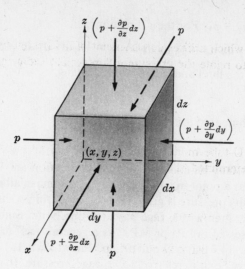

Fig. 2-2

Simplifying,
$$\frac{\partial p}{\partial x} = 0, \qquad \frac{\partial p}{\partial y} = 0, \qquad \frac{\partial p}{\partial z} = -\rho g \qquad (2.2)$$

By equations (2.2) we have

$$\frac{dp}{dz} = -\rho g \qquad (2.3)$$

which is an important equation of fluid statics. If the density of the fluid is constant, equation (2.3) may be integrated between two elevations z_1 and z_2 to give

$$\int_{p_1}^{p_2} dp = -\rho g \int_{z_1}^{z_2} dz$$

giving
$$p_2 - p_1 = -\rho g(z_2 - z_1) \qquad (2.4)$$

which describes the pressure change with elevation for a fluid in static equilibrium. An independent and simple derivation of equation (2.3) may be carried out by considering the forces on a right circular cylinder whose axis is vertical to the earth's surface.

Equation (2.4) is valid for an incompressible fluid. Let us consider briefly pressure variations in compressible fluids. Starting with equation (2.3) for an ideal gas, we have

$$dp = -\frac{pg}{RT}\, dz$$

where R = gas constant and T = absolute temperature. If we assume an isothermal atmosphere, we may integrate this equation to give

$$\ln \frac{p}{p_1} = -\frac{g}{RT}(z - z_1)$$

or
$$p = p_1 e^{-(g/RT)(z - z_1)} \qquad (2.5)$$

where p_1 is the pressure at some known elevation z_1. We may take $z_1 = 0$ so that p_1 corresponds to the atmospheric pressure at the Earth's surface and we see that the pressure would decrease exponentially in an isothermal atmosphere. In reality the temperature generally decreases with height above the surface of the Earth, although under unusual circumstances the temperature might actually rise for a short distance above the Earth as in a temperature inversion.

2.3 MANOMETRY

A manometer is a device which utilizes displacement of fluid columns to determine pressure differences. Equation (2.4) is used to relate the pressure differences to the heights of fluid columns. We may write this equation as

$$p_2 - p_1 = \gamma h = \rho g h \qquad (2.6)$$

where

$$h = -(z_2 - z_1) \qquad (2.7)$$

Fig. 2-3 shows a single U-tube manometer for measuring pressure difference. The difference in pressures p_A and p_B may be determined as follows. The pressure at point a is given by

$$p_a = h_1 \gamma_{H_2O} + (h_3 - h_1)\gamma_{air} + p_A$$

or

$$p_a = h_2 \gamma_{H_2O} + (h_3 - h_2)\gamma_{air} + p_B \qquad (2.8)$$

Subtracting,

$$p_A - p_B = (h_2 - h_1)(\gamma_{H_2O} - \gamma_{air}) \qquad (2.9)$$

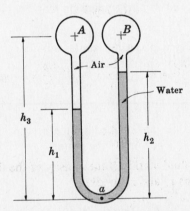

Fig. 2-3. The U-tube manometer.

The specific weight of air is small compared to water, which means that the pressure difference is approximately equal to the difference in column heights times the specific weight of water:

$$p_A - p_B = (h_2 - h_1)\gamma_{H_2O} \qquad (2.10)$$

Manometers may have different shapes, orientations and use different fluids depending on the application. For example, in order to attain an improved accuracy over the vertical manometer, an inclined manometer as shown in Fig. 2-4 may be used. Or, a two fluid manometer as shown in Fig. 2-5 could be

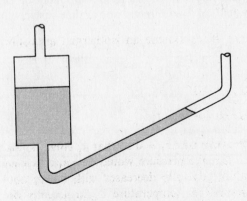

Fig. 2-4. An inclined manometer.

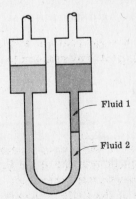

Fig. 2-5. The two fluid manometer.

used to achieve an improved accuracy. The method of relating pressure differences to fluid column deflections for these two examples is basically the same as that described for the U-tube manometer.

2.4 FLUID FORCES ON SUBMERGED BODIES

The total surface force on a body may be determined by taking the vector sum of the differential surface forces over the entire area,

$$\mathbf{F} = \int_A d\mathbf{F} \tag{2.11}$$

where

$$d\mathbf{F} = p \, d\mathbf{A}$$

and

$$p = p_0 + \gamma h \tag{2.12}$$

where p_0 = pressure at the free surface and h = depth below the free surface.

These equations are all that are needed for establishing the forces on submerged surfaces. In the specific applications then, all that is required is to write the depth h and differential area dA in terms of the same variable of integration. Also, before the equation may be integrated it must be written in scalar form. We shall consider some specific cases.

Horizontal Plane Surface: Consider a plane horizontal surface at distance h below the surface of a liquid as shown in Fig. 2-6. The force on the surface (one side) is

$$F = \int_A dF = \int_A (p_0 + \gamma h) \, dA = (p_0 + \gamma h)A \tag{2.13}$$

Inclined Plane Surface: Next, consider an inclined submerged surface as shown in Fig. 2-7. Here we have for the force on the plate (which is normal to the plate),

$$F = \int_A dF = \int_A (p_0 + \gamma h) \, dA = p_0 A + \gamma \int_{y_1}^{y_2} wy \sin \theta \, dy$$

$$= p_0 A + \tfrac{1}{2}\gamma w \sin \theta (y_2^2 - y_1^2) \tag{2.14}$$

Curved Surface: For a curved surface, such as shown in Fig. 2-8, the force components are

$$F_x = \int_A (p_0 + \gamma h) \, dA_x \tag{2.15}$$

$$F_y = \int_A (p_0 + \gamma h) \, dA_y \tag{2.16}$$

The integration of these equations may not always be the most convenient method for determining the forces on submerged bodies. For example, we may observe that the vertical component of force on any submerged surface is equal to the weight (or equivalent) of fluid directly above that surface plus the

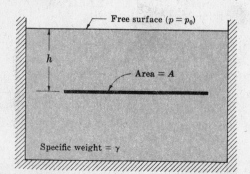

Fig. 2-6. A submerged horizontal plate.

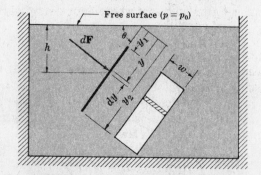

Fig. 2-7. An inclined submerged plate.

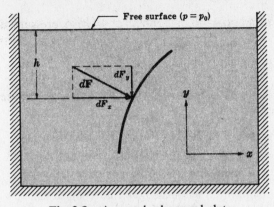

Fig. 2-8. A curved submerged plate.

force due to pressure at the free surface, and that the horizontal component of force may be determined by considering a plane vertical surface which is the projected area of the surface in question.

Buoyancy: Archimedes' principle states that a submerged body is subject to an upward force equal to the weight of the fluid displaced. Therefore this force (buoyant force) will be equal to the volume of the body which is submerged times the specific weight of the fluid.

It is important that we recognize that the buoyant force can also be computed by summing all of the vertical components of the pressure area products over the submerged surface of the body, i.e.,

$$F_B = \int_A p \, dA_y \qquad (2.17)$$

which as we have already noted leads to

$$F_B = \gamma \times \text{volume} \qquad (2.18)$$

2.5 ACCELERATING FLUIDS IN THE ABSENCE OF SHEAR STRESS

In the previous sections we have considered fluids at rest. Now we will consider fluids which have a constant acceleration (constant with time) but each fluid particle has no motion relative to its immediate neighbor, so that the fluid moves as a rigid body. For example, consider a container of liquid which has a constant acceleration **a** upward and to the right as shown in Fig. 2-9. Using the same method as in Section 2.2 it is easily shown that Newton's law for an infinitesimal particle in the container gives

$$\partial p/\partial x = -\rho a_x, \qquad \partial p/\partial y = -(\rho a_y + \gamma)$$

Integrating,
$$p = -[\rho a_x x + (\rho a_y + \gamma)y] + C \qquad (2.19a)$$

The shape of the free surface is determined by letting $p = p_0$, yielding a plane surface. Constant pressure surfaces are parallel planes whose slope with respect to the horizontal plane is given by

$$\theta = \tan^{-1} a_x/(a_y + g) \qquad (2.19b)$$

where θ is indicated in Fig. 2-9.

Another example of fluid particles undergoing a constant acceleration is encountered when a fluid body rotates uniformly as a whole without relative motion of its parts. Let ω be the constant angular velocity and ρ the density of the liquid. Choose polar coordinates with z and r as shown in Fig. 2-10. The angular coordinate is ϕ. The centripetal acceleration of a particle at radius r is $-\omega^2 r$ in the r direction. From Newton's law for the z direction,

$$-\rho g r \, dr \, d\phi \, dz + p r \, dr \, d\phi - \left(p + \frac{\partial p}{\partial z} \, dz\right) r \, dr \, d\phi = 0 \qquad \text{or} \qquad \frac{\partial p}{\partial z} = -\rho g$$

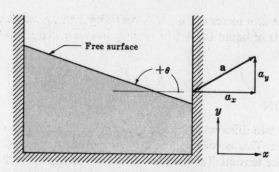

Fig. 2-9. An accelerating container of fluid. The direction of gravity is in the negative y direction so that a_x and a_y are respectively parallel and perpendicular to the Earth's surface. If a_x and a_y are positive, then the slope is negative as shown.

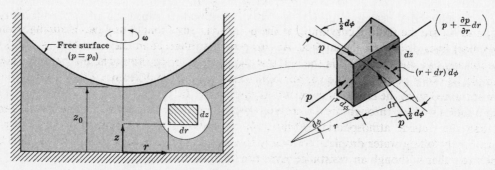

Fig. 2-10. A rotating container of fluid. Gravity acts in the negative z direction.

For the r direction, assuming $\sin \frac{1}{2}\,d\phi \approx \frac{1}{2}\,d\phi$,

$$p\,dz\,r\,d\phi - \left(p + \frac{\partial p}{\partial r}\,dr\right) dz(r + dr)\,d\phi + 2p\,dz\,dr(\tfrac{1}{2}\,d\phi) = -\rho\,dz\,r\,d\phi\,dr\omega^2 r$$

which becomes (neglecting higher order terms)

$$\partial p/\partial r = \rho\omega^2 r \tag{2.20}$$

We note that $p = p(r, z)$ and

$$dp = \frac{\partial p}{\partial r}\,dr + \frac{\partial p}{\partial z}\,dz$$

Substituting for $\partial p/\partial r$ and $\partial p/\partial z$,

$$dp = \rho\omega^2 r\,dr - \rho g\,dz$$

Integrating,
$$p = \tfrac{1}{2}\rho\omega^2 r^2 - \rho g z + C \tag{2.21}$$

At $r = 0$ and $z = z_0$ we have $p = p_0$. Then equation (2.21) becomes

$$p - p_0 = \tfrac{1}{2}\rho\omega^2 r^2 + \rho g(z_0 - z) \tag{2.22}$$

At the free surface, where $p = p_0$, we have

$$z = z_0 + \tfrac{1}{2}\omega^2 r^2/g \tag{2.23}$$

which is a paraboloid of revolution.

Note that for constant z, p increases as r^2. A centrifugal pump and a centrifuge make use of this principle. An enclosed mass of liquid is whirled rapidly to create a great difference in pressure between its center and its periphery.

2.6 SURFACE TENSION

The interface between two different immiscible fluids can sustain (i.e., balance) a pressure difference across it if it has curvature. This is because the interface itself acts like a membrane whose strength is characterized by the surface tension. This surface tension T measured in lb_f/ft or N/m is a property of the two fluids in contact. Most often we are concerned with a liquid-air interface, but the concept applies to any two fluids. The surface tension T for various fluid combinations may be found in reference tabulations.

If we consider a curved interfacial surface, the pressure difference Δp across the interface is given by

$$\Delta p = T\left(\frac{1}{R_1} + \frac{1}{R_2}\right) \qquad (2.24)$$

where R_1 and R_2 are the radii of curvature (at the point of intersection) of any two mutually orthogonal (perpendicular) lines drawn on the surface. At this point of intersection the sum $(1/R_1 + 1/R_2)$ will be the same for any two such lines. Δp is the value at the point of intersection where R_1 and R_2 are given. For example, at every position in the surface of a sphere a pressure difference $T(1/R_1 + 1/R_2) = 2T/R$ would be sustained since the radius of curvature is everywhere R.

Soap bubbles drifting through the air are examples of spherical surfaces where the pressure inside is greater than the outside atmospheric pressure, the tension in the soap film balancing the pressure difference. Freely falling water droplets are nearly spherical in shape because of surface tension holding the droplet together (although air resistance gives rise to some distortion).

2.7 SUMMARY

In this chapter we have looked at forces acting on fluids which (1) were in static equilibrium (fluids at rest) and (2) were undergoing a constant acceleration. In both cases the absence of relative motion among the fluid particles results in the tangential stresses being everywhere zero. Thus the only forces which were acting on the fluid particles were those due to (1) pressure (normal surface stress) and (2) gravity (a uniform body force).

It was shown that the pressure at a point in a fluid in static equilibrium is the same in all directions. Thus pressure is a scalar function of the spatial coordinates and time.

For a fluid in static equilibrium it was shown that the pressure change with elevation is given by

$$dp/dz = -\gamma = -\rho g$$

and for an incompressible fluid,

$$p_2 - p_1 = -\gamma(z_2 - z_1)$$

which was found useful in establishing pressure from various kinds of manometer readings.

Forces on submerged surfaces may be established by integrating the pressure area product over the entire area for the scalar components of the vector equation:

$$F_x = \int_A p\, dA_x, \qquad F_y = \int_A p\, dA_y$$

where $p = p_0 + \gamma h$. Thus in using these equations for a specific problem, the only remaining consideration is to relate the differential areas and the depth below the surface, h, to some common and convenient variable of integration.

The buoyant force may be calculated in two ways: (1) by integrating the vertical component of pressure area product over the submerged area of the body or (2) by multiplying the volume of fluid displaced by the body times the specific weight of the fluid.

For a fluid undergoing a constant acceleration but having no relative motion among the fluid particles, Newton's laws for an infinitesimal fluid particle gives

$$\partial p / \partial x = -\rho a_x, \qquad \partial p / \partial y = -(\rho a_y + \gamma)$$

where a_x and a_y are the components of the acceleration vector parallel and perpendicular to the earth's surface respectively.

Solved Problems

2.1.   The mercury manometer of Fig. 2-11 is connected to the inlet and outlet of a water pump (the left side to the inlet and the right side to the outlet). Assuming that the inlet and outlet are at the same elevation, determine the pressure increase for the pump.

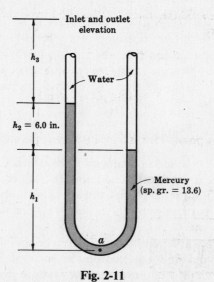

Fig. 2-11

We determine the pressure at point a by computing the pressure due to the column of liquid in the left side as

$$p_a = h_1 \gamma_{Hg} + h_2 \gamma_{Hg} + h_3 \gamma_{H_2O} + p_{in}$$

and in the right side as

$$p_a = h_1 \gamma_{Hg} + h_2 \gamma_{H_2O} + h_3 \gamma_{H_2O} + p_{out}$$

Subtracting,
$$p_{out} - p_{in} = h_2 (\gamma_{Hg} - \gamma_{H_2O})$$
$$= \tfrac{6}{12}[62.4(13.6 - 1)] = 393 \text{ psf}$$

We have used the fact that the specific weight of water is 62.4 lb_f/ft^3 and the specific gravity of mercury is 13.6.

2.2.   In Fig. 2-12 determine the gage pressure at point A.

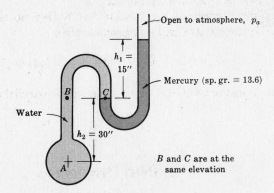

Fig. 2-12

The absolute pressure at C is the same as at B. The absolute pressure at B is $\gamma_1 h_1 + p_a$ and hence

$$p_A = (p_a + \gamma_1 h_1) + \gamma_2 h_2$$

For convenience we have denoted the specific weights of mercury and water as γ_1 and γ_2 respectively. Then by definition of gage pressure,

$$p_{A\,\text{gage}} = p_A - p_a = \gamma_1 h_1 + \gamma_2 h_2$$

$$= 62.4(13.6)(15/12) + 62.4(30/12)$$

$$= 1216 \text{ psf} = 8.45 \text{ psi}$$

2.3.   Determine the pressure at point A for the inclined manometer shown in Fig. 2-13.

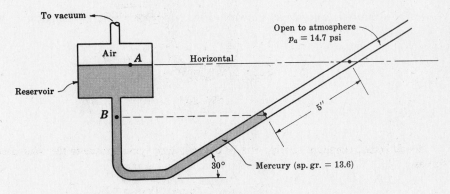

Fig. 2-13

$$p_B = p_A + \gamma h$$

p_B is atmospheric, so that

$$p_A = p_a - \gamma h = 14.7 - 62.4(13.6)(2.5/12)(1/144) = 13.47 \text{ psi}$$

Note that the level of mercury in the inclined tube is lower than that in the reservoir because a vacuum is being drawn on the reservoir.

2.4.   In Fig. 2-14 find the gage pressures at points A, B and C in the pipe filled with flowing water.

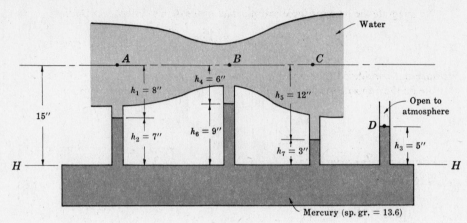

Fig. 2-14

Denoting the specific weight of water as γ_1 and the specific weight of mercury as γ_2 we start at point D, work down to level HH, and then up to various points in the tube.

$$p_A = p_a + \gamma_2 h_3 - \gamma_1 h_1 - \gamma_2 h_2$$

But $p_{\text{gage}} = p - p_a$, so that

$$p_{A\text{ gage}} = \frac{62.4(13.6)}{1728}(5) - \frac{62.4}{1728}(8) - \frac{62.4(13.6)}{1728}(7) = -1.28 \text{ psi}$$

(The number 1728 is the conversion factor to convert the specific weight from lb_f/in^3 to lb_f/ft^3.) Similarly,

$$p_{B\text{ gage}} = \gamma_2 h_3 - \gamma_1 h_4 - \gamma_2 h_6 = -2.17 \text{ psi}$$

$$p_{C\text{ gage}} = \gamma_2 h_3 - \gamma_1 h_5 - \gamma_2 h_7 = 0.55 \text{ psi}$$

2.5.   Water flows through a nozzle as shown in Fig. 2-15. Determine the height h if the gage pressure at A is 5 psi.

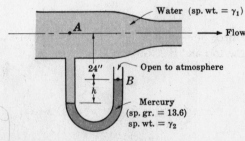

Fig. 2-15

Starting at the free surface B and working around to A, we have

$$p_A = p_a + \gamma_2 h - \gamma_1(h + 24)$$

$$p_{A\text{ gage}} = 5 = \frac{62.4(13.6h)}{1728} - \frac{62.4}{1728}(h + 24)$$

from which $h = 12.9$ in.

2.6. Determine the total force of the water acting on the surface of the dam as shown in Fig. 2-16. The dam is 5 ft wide and the water is 10 ft deep.

We integrate the pressure force over the face of the dam to find the total force.

$$F = \int_A dF = \int_A p \, dA = \int_0^{10} \left[5\gamma y \, dy = 5\gamma y^2/2 \right]_0^{10} = 15{,}600 \text{ lb}_f$$

We have not considered the atmospheric pressure at the water surface, since it is canceled out by the air pressure on the opposite side of the dam.

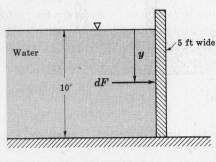

Fig. 2-16

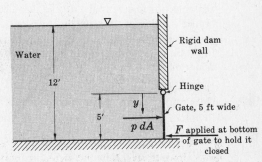

Fig. 2-17

2.7. Determine the total force F required to keep the dam gate, shown in Fig. 2-17, from opening. The gate is 5 ft wide.

The sum of moments about the hinge must be zero for static equilibrium. Taking moments,

$$5F - \int_A yp \, dA = 0$$

Thus

$$5F = \int_0^5 y\gamma(7 + y)5 \, dy = 5(62.4)\left[7y^2/2 + y^3/3 \right]_0^5$$

and $F = 8050 \text{ lb}_f$.

2.8. Determine the total force due to water pressure on the inclined surface shown in Fig. 2-18. The surface is hinged at the top and is 3 m wide.

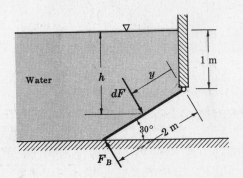

Fig. 2-18

The total force is normal to the gate and may be found by integrating the pressure over the surface.

$$F = \int_A dF = \int_A p\, dA = \int_A \gamma h\, dA$$

$$= \int_0^2 \gamma(1 + y \sin 30°)(3\ dy)$$

$$= 3(9810)\left[1y + \tfrac{1}{4}y^2 \right]_0^2 = 8.83 \times 10^4\ \text{N}$$

Again, since the right hand side of the surface is exposed to the atmosphere, we do not consider the atmospheric pressure in the calculation. The total force could be resolved into x and y components, and the proper location of the resultant force could be found so that the action of this single resultant force would be equivalent to the total pressure force in determining the reactions at the hinge and bottom of the plate. In Problem 2.9 we calculate the reaction force at the base of the plate. After following that calculation it is a simple matter to find the location of the resultant force.

2.9.   Determine the total reaction force of the bottom surface on the bottom end of the gate for Problem 2.8. Find the location of the single force F which is equivalent to the total water pressure force.

The sum of the moments about the hinge must be zero. Since the pressure can only exert a force normal to the gate, the reactions at the hinge and bottom will be normal to the plate if we neglect the weight of the plate. Neglecting the weight of the plate, we sum moments about the hinge to obtain

$$F_B \times 2 = \int_A y\, dF = \int_A y(\gamma h)\, dA$$

$$= \gamma \int_0^2 y(1 + y \sin 30°)(3\ dy) = (3)(9810)\left[\tfrac{1}{2}y^2 + y^3/6 \right]_0^2 = 9.81 \times 10^4\ \text{N-m}$$

and $F_B = 4.905 \times 10^4$ N.

If the equivalent resultant force of the pressure, F, is located at $y = y_0$, then $Fy_0 = F_B(2)$ so that

$$y_0 = 2F_B/F = 2(4.905 \times 10^4)/8.83 \times 10^4 = 1.11\ \text{m}$$

2.10.   Determine the total pressure force on the curved surface shown in Fig. 2-19. The width of the curved plate is 2 ft.

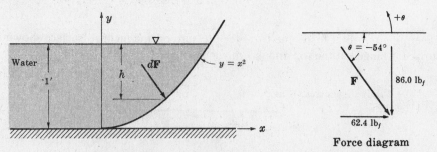

Fig. 2-19

We can write the general equation for the force of the water on the plate as $\mathbf{F} = \int_A d\mathbf{F}$, but this equation must be written in scalar form before it can be integrated. We write

$$F_x = \int_A dF_x = \int_A p\, dA_x = \int_A (\gamma h)2\ dy$$

Writing h in terms of y,

$$F_x = 2\gamma \int_0^1 (1-y)\,dy = 2(62.4)\left[y - y^2/2\right]_0^1 = 62.4\ \text{lb}_f$$

Similarly we can find F_y, but we must remember that F_y is defined positive in the positive y direction and the pressure force acts downward in the negative y direction. Hence

$$F_y = \int_A dF_y = -\int_A p\,dA_y = -2\gamma \int_0^1 (1-y)\,dx$$

Writing y in terms of x, we can integrate explicitly:

$$F_y = -2\gamma \int_0^1 (1-x^2)\,dx = -86.0\ \text{lb}_f \quad -83.2$$

The negative sign indicates that F_y acts downward. The magnitude of the total pressure force is

$$|\mathbf{F}| = (F_x^2 + F_y^2)^{1/2} = 106.4\ \text{lb}_f$$

and its direction is given by $\tan\theta = F_y/F_x = -1.38$ or $\theta = -54°$. θ is the inclination of $\mathbf{F}$ to the horizontal. As shown in Fig. 2-19, θ is negative and so the resultant force acts in a downward direction.

2.11. A U-tube accelerometer shown in Fig. 2-20 may be used for measuring the acceleration of an automobile. It is mounted in the vehicle so that the legs are vertical. The open U-tube is partially filled with a liquid of specific weight γ. Under constant acceleration a_x the liquid assumes the configuration shown. Relate the relevant parameters to the value of the acceleration.

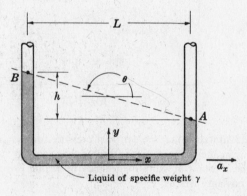

Liquid of specific weight γ

Fig. 2-20

The fluid behaves exactly as if it were in an open container of width L. The angle θ given by equation (2.19b) is then

$$\tan\theta = -h/L = -a_x/g$$

and we get simply $a_x = gh/L$.

2.12. What is the value of a_x at which the water just begins to run over the rear wall of the container shown in Fig. 2-21? When the container is at rest the water fills it to a depth of 5 ft.

As a_x increases the water rises against the rear wall. As it reaches the top edge, it has risen 5 ft and the front must have lowered by 5 ft. Hence at the onset of spilling, $\tan\alpha = 10/20 = a_x/g$ so that $a_x = g/2 = 16.1\ \text{ft/sec}^2$. Note that for convenience we have defined an angle α which is the negative of the angle θ of equation (2.19b).

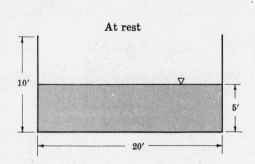 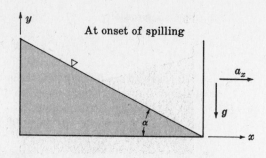

At rest   At onset of spilling

Fig. 2-21

2.13. Show that when a container of liquid is constantly accelerated at a value **a** in a gravitational field **g**, the pressure distribution in the container is the same as if the container were at rest in a fictitious gravitational field of value (**g** − **a**). That is, assuming gravity acts in the negative y direction, the fictitious field has magnitude $\sqrt{(a_y + g)^2 + a_x^2 + a_z^2}$ and is the vector sum of **g** and the negative of **a** as shown in Fig. 2-22. If a container were falling freely in a gravitational field, then of course the liquid would have zero gage pressure throughout. Explain.

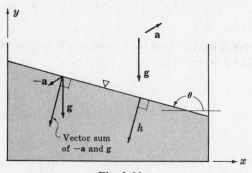

Fig. 2-22

From equation (*2.19a*) we see that the constant pressure surfaces in the fluid are planes. Assuming that there are no variations in the z direction and that gravity acts in the negative y direction, the pressure variation normal to the constant pressure planes may be written by using the rule for differentiation of a variable normal to a constant surface:

$$dp/dh = [(\partial p/\partial x)^2 + (\partial p/\partial y)^2]^{1/2}$$

Carrying out the calculation explicitly, we immediately arrive at

$$dp/dh = \rho\sqrt{(a_y + g)^2 + a_x^2}$$

which may be extended to three dimensions (with gravity acting in the negative y direction) to give

$$dp/dh = \rho\sqrt{(a_y + g)^2 + a_x^2 + a_z^2}$$

Integration yields
$$p = p_0 + \rho h\sqrt{(a_y + g)^2 + a_x^2 + a_z^2}$$

The free surface and constant pressure planes are inclined so that they are normal to the vector (**g** − **a**) which may be seen immediately from equation (*2.19b*).

2.14. Discuss the behavior of a tank filled with two immiscible fluids of different densities undergoing constant acceleration as shown in Fig. 2-23. Gravity acts in the negative y direction, and the vector **a** is assumed to have only x and y components.

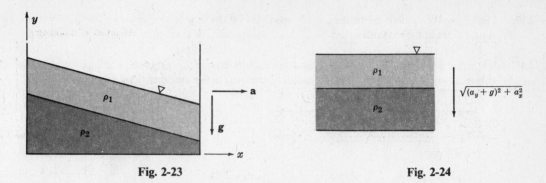

Fig. 2-23   Fig. 2-24

From the previous problem we know that the system behaves as if it were in a gravitational field of $(\mathbf{g} - \mathbf{a})$. Hence the constant pressure surfaces are planes and we can reduce the problem to the system shown in Fig. 2-24 with an equivalent gravitational field and treat it as a simple statics problem just as we did for a two fluid manometer. Alternatively we could solve the equilibrium equation (*2.19a*) and obtain the same result. The slope of the free surface and the interface is the same as for a single liquid.

Supplementary Problems

2.15. The pressure at the liquid surface in the tank shown in Fig. 2-25 is 4.0 psi larger than atmospheric pressure. Determine the height h if the liquid in the tank is (*a*) water, (*b*) mercury.

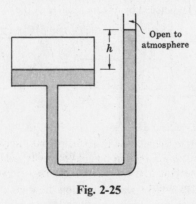

Fig. 2-25

2.16. Determine the pressure change (psi) between the points A and B for flow in the vertical pipe of Fig. 2-26.

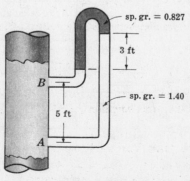

Fig. 2-26

2.17. Determine the manometer reading for Problem 2.16 if the flow is horizontal rather than vertical. The points *A* and *B* are at the same elevation but the manometer has the same orientation as in Problem 2.16.

2.18. An inclined manometer is used to measure the difference in air pressure in a pipe between two points as shown in Fig. 2-27. What is the difference in pressure (psi) for the conditions given in the figure?

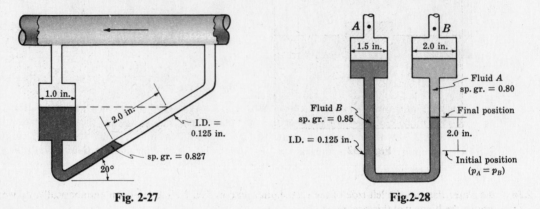

Fig. 2-27   Fig. 2-28

2.19. A bifluid manometer as shown in Fig. 2-28 may be used to determine small pressure differences with a better accuracy than a single fluid manometer. Find the pressure difference $p_A - p_B$ (psi) for a deflection of 2.0 in. of the boundary between the two fluids.

2.20. A tank as shown in Fig. 2-29 is partially filled with water and open to the atmosphere. A triangular gate is hinged at the bottom and held closed by a force applied at the top. Determine the force.

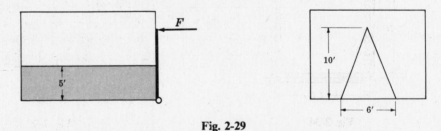

Fig. 2-29

2.21. Referring to Fig. 2-30, a skin diver wants to know how far below the water surface he can go with a container and still be able to open the access door. Assuming that the pressure inside is atmospheric, and the maximum pull the diver can exert is 150 lb, determine the maximum depth.

2.22. In Fig. 2-31 find the force at *O* on the solid gate in the shape of a quarter cylinder due to the water. The gate is 30 ft long and its weight may be neglected. Also, calculate the force F_N at the bottom of the gate.

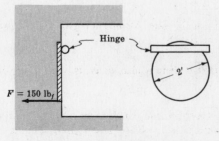   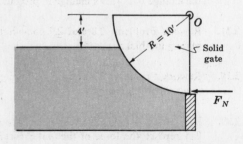

Fig. 2-30   Fig. 2-31

2.23. In Fig. 2-32 an object having specific gravity 0.5 is floating in a container of water. If the container and liquid are given an upward acceleration of 10 ft/sec², what will be the position of the object relative to the water surface?

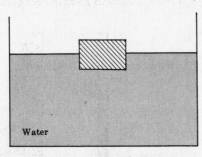

Fig. 2-32

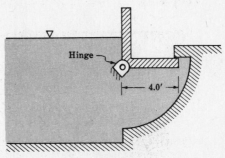

Fig. 2-33

2.24. As water rises on the left side of the rectangular gate in Fig. 2-33 it will open automatically. At what depth above the hinge will this occur?

2.25. A U-tube rotates about the line AB as shown in Fig. 2-34. The angular velocity is one revolution per second. Determine the pressures at C, D and E. The end at C is closed.

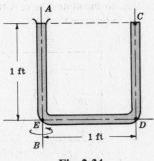

Fig. 2-34

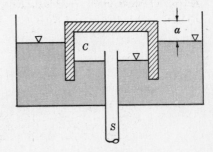

Fig. 2-35

2.26. A rather novel type manometer employing an inverted straight-sided thick-walled cup floating in liquid, as shown in Fig. 2-35, has been suggested as being much more sensitive than a regular U-tube manometer for the measurement of pressure. In this proposed manometer the pressure to be measured is communicated by a pipe S to the closed chamber C formed by the inverted cup and liquid. A container open at the top to the atmosphere holds the liquid in which the cup floats. The distance a from the outer water surface to the top of the cup would be used as a measure of the pressure. What would be the sensitivity of this manometer, i.e. the change in a with a change in pressure?

2.27. Rework Problems 2.8 and 2.9 considering the weight of the gate. Assume the gate to be a uniform steel plate $\frac{1}{4}$ in. thick.

2.28. Rework Problems 2.7 and 2.8, assuming in each case that there is also water, two feet deep, on the right hand side of the gate.

2.29. In Problem 2.7 find the resultant force vector of the water pressure on the gate. That is, what single force can replace the force of the water insofar as the reactions at the hinge and position-holding force F are determined by this resultant force vector?

2.30. A cart of water rolls down an inclined plane as shown in Fig. 2-36. Find the inclination of the free surface, neglecting friction of the wheels and wind resistance effects.

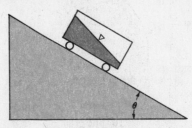

Fig. 2-36

2.31. A vertical cylinder with a tight lid and completely filled with water rotates at an angular velocity ω about its axis. What is the pressure distribution in the cylinder? A small hole in the center of the lid exposes the fluid to the atmosphere.

2.32. In Problem 2.31 the cylinder rotates in a zero gravity field. What is the pressure distribution in the water then?

2.33. In a zero gravity field, a rocket fuel tank containing liquid fuel of density ρ undergoes a constant acceleration **a**. What is the pressure distribution in the tank? Does the shape of the tank matter? Explain.

2.34. A cylinder filled with two immiscible liquids of density ρ_1 and ρ_2, with $\rho_2 > \rho_1$, as shown in Fig. 2-37 rotates with angular velocity ω about its axis. Find the pressure distribution, the shape of the interface between the liquids, and the shape of the free surface.

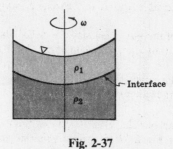

Fig. 2-37

2.35. Does the shape of the U-tube in Problem 2.11 matter? In terms of its performance, what parameters are important?

2.36. What happens in Problem 2.12 when the value of a_x increases? Can the tank be completely emptied by increasing a_x?

2.37. A balloon filled with helium (which is lighter than air) is tethered to the seat inside an automobile. If the windows are all closed, what is the air pressure distribution in the automobile if it accelerates forward at a constant value? Will the balloon tend to slant forward or backward relative to the car? If the balloon is filled with air and hung from the car ceiling, will the answer be the same? Explain.

2.38. One of the problems faced by the designers of the Buick Dynaflow torque-converter was the determination of the maximum tensile stress in the bolts which hold the pump cover to the pump housing as shown in Fig. 2-38.

The operation of the Dynaflow unit is as follows: The engine crankshaft is rigidly coupled to the pump housing and drives it at engine speed ω. The pump housing and cover constitute an oil-tight unit, a packing gland being provided around the output shaft which rotates at a speed less than or equal to that of the pump housing. This housing is kept full of oil at a constant "charging" pressure by means of an auxiliary

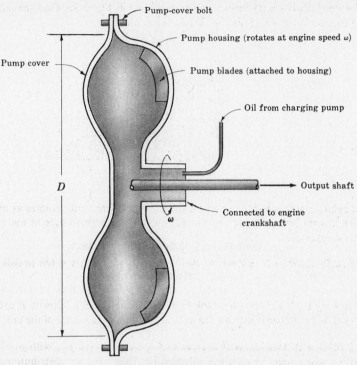

Fig. 2-38

charging pump; for all practical purposes this charging pump delivers oil into the pump housing at its center of rotation.

Attached to the inside of the pump housing are radial blades or vanes (the pump) which circulate the oil as required to the stator (not shown) and to the turbine (not shown) which is fastened to the output shaft.

What tensile stress can be expected in a pump cover bolt? The diameter of the pump housing is D feet, and its angular velocity is ω radians per second.

2.39. A rectangular prism (i.e., soap bar shape) of density ρ_0 floats at the interface of two immiscible liquids of density ρ_1 and ρ_2 as shown in Fig. 2-39. In terms of relevant parameters find

(a) The depth at which the prism floats in the lower liquid

(b) The criteria for the prism to float to the top of the upper liquid

(c) The criteria for the prism to sink to the bottom of the lower liquid

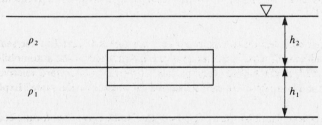

Fig. 2-39

2.40. A thin glass tube of radius a is inserted down into a free surface of a certain liquid (Fig. 2-40). It is observed that the liquid rises a distance H in the tube. The angle θ depends on the "wetting" of the glass by the liquid and is a constant parameter (available from tabulated data) that depends only on the material of the tube and the liquid. If θ is positive as shown, the liquid is said to wet the surface. If θ is negative such that the

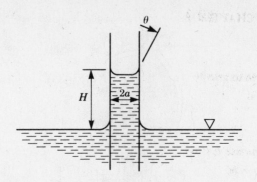

Fig. 2-40

miniscus is convex upward, the liquid is said not to wet the tube. For a nonwetting situation, will the liquid rise or fall in the tube? The free surface in the tube will be approximately spheroidal in shape. For a given positive value of θ, T of the air-liquid interface, and a, find an expression for H. The rise of liquid in a small tube is known as capillarity and explains "wicking" of liquid through porous media. *Hint:* What is the radius of the spheroidal surface?   *Ans.* $H = \dfrac{2T \cos \theta}{\rho g a}$

2.41. Try floating a thin sewing needle on the surface of a bowl of water. With care it can be done. Explain why the needle does not sink (surface tension effects). Make a sketch of the surface elevation of the water near the needle and explain the shape. Does the water wet the needle? Is θ positive or negative (referring to Fig. 2-40)?

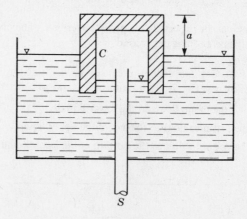

Fig. 2-41

NOMENCLATURE FOR CHAPTER 2

$\mathbf{a}$ = accelaration

$\mathbf{F}$ = force

g = acceleration due to gravity

h = height of a fluid

p = pressure

R = gas constant

T = temperature, surface tension

γ = specific weight = ρg

ρ = density

ω = angular velocity

Chapter 3

Mathematical Models of Fluid Motion

3.1 INTRODUCTION AND METHOD

In this chapter we will develop model representations of fluid motion in mathematical form. It must be kept in mind that these models are merely approximations of the actual flow situation, and an understanding of the limitations and proper use of the equations of this chapter is essential. We will carefully derive these mathematical models, emphasizing their physical meaning.

An actual fluid is made up of molecules with empty space between them. However, in general, when one establishes the mathematical models, it is convenient to assume that the fluid is a continuous medium—a continuum. This is the approach we will follow, and in doing so we have immediately begun to abstract about the actual makeup of the fluid and are developing a model representation of the fluid itself. It should be mentioned that while most fluid flow may be treated by the continuum approach, mathematical models taking into account the statistical behavior of the individual fluid molecules have been developed and are necessary for studies of rarefied gases. Examples of the particle approach to fluid motion may be found in books by Chapman and Cowling and by Curtiss and Bird; see references 4 and 5.

In this chapter we will first derive the basic fluid equations in integral form for a control volume, then obtain the differential equations of fluid flow by applying the integral equations to an elemental volume. As was mentioned in Chapter 1, there are five basic variables in fluid flow, three components of velocity and two thermodynamic properties. Hence there are five basic equations which describe the flow; three components of the momentum equation, the continuity equation, and the energy equation. In general, the energy equation becomes uncoupled in incompressible flow since the density is constant. In turbulent flow the situation is somewhat more complex and a closed set of equations cannot generally be developed. We will not discuss the basic equations of turbulent flow until Chapter 9. Certain constitutive equations, such as an equation of state, may also be used to allow the introduction of an additional thermodynamic property.

3.2 INTEGRAL EQUATIONS

In the study of fluid motion, we are concerned with four basic laws:

(a) Conservation of mass
(b) Newton's second law of motion
(c) Conservation of energy (the first law of thermodynamics)
(d) Second law of thermodynamics.

These laws apply to a fixed quantity of matter (system) which maintains its identity as it undergoes a change in conditions. In fact, one cannot meaningfully apply the basic laws until a definite system is identified.

It is not usually convenient to identify and follow fixed quantities of matter in the analysis of fluid motion. Rather, it is customary to adopt a field theoretic point of view and identify a definite fixed region or volume in space called a control volume. However, the four basic laws do not apply to fixed volumes but to fixed quantities of matter. Thus our immediate task is to derive equations which apply to control volumes from the known expressions for systems. We ask the question then: what are the properties of the fluid within the fixed control volume at any given time, regardless of the fact that the fluid in the control volume is constantly changing?

We will now proceed to develop the integral equations of fluid motion. The detailed derivations of these equations are presented here in order that one may better understand them in applying them to physical situations. The methods of deriving the control volume equations from the known basic laws follow a basic pattern with slight modifications because of the different physical quantities involved. After the equations are developed they will be applied to solving physical problems which occur in fluid motion.

Conservation of Mass: We will focus our attention on the flow field represented by the streamlines of Fig. 3-1. Let us consider a certain quantity of matter at some time t enclosed by the solid line. At some later time $t + \Delta t$, the boundary of the system has a new physical location as represented by the dotted line.

Considering the regions denoted by A, B and C, we have the system occupying the region A at time t and at time $t + \Delta t$ it occupies regions B and A-C. Letting m represent the mass contained in the different regions and at different times with the appropriate subscripts,

$$m_A(t) = m_A(t + \Delta t) - m_C(t + \Delta t) + m_B(t + \Delta t)$$

Rearranging and dividing by Δt,

$$\frac{m_A(t + \Delta t) - m_A(t)}{\Delta t} = \frac{m_C(t + \Delta t) - m_B(t + \Delta t)}{\Delta t}$$

Taking the limit as $\Delta t \to 0$, the left side becomes

$$\lim_{\Delta t \to 0} \frac{m_A(t + \Delta t) - m_A(t)}{\Delta t} = \frac{\partial}{\partial t} (m)_{\text{c.v.}} = \frac{\partial}{\partial t} \int_{\text{C.V.}} \rho \, d\mathcal{U}$$

where ρ is the mass density, $\mathcal{U}$ indicates volume, and C.V. designates the control volume fixed in space and bounded by the control surface (C.S.). The right side of the equation is

$$\lim_{\Delta t \to 0} \left\{ \frac{m_C(t + \Delta t)}{\Delta t} - \frac{m_B(t + \Delta t)}{\Delta t} \right\} = \dot{m}_{\text{in}} - \dot{m}_{\text{out}}$$

which becomes

$$\dot{m}_{\text{in}} - \dot{m}_{\text{out}} = \int_{A_{\text{in}}} \rho V \cos \alpha \, dA - \int_{A_{\text{out}}} \rho V \cos \alpha \, dA = - \int_{\text{C.S.}} \rho \mathbf{V} \cdot d\mathbf{A}$$

where $\dot{m}_{\text{in}}$ and $\dot{m}_{\text{out}}$ represent the mass rate of flow in and out of the control volume and $\mathbf{V}$ is the velocity vector. V is the magnitude of the velocity vector and α is the angle between the velocity vector and the

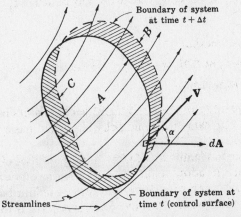

Fig. 3-1. System moving through a control volume.

outward normal. Then the continuity equation for the control volume becomes

$$\int_{\text{c.s.}} \rho \mathbf{V} \cdot d\mathbf{A} = -\frac{\partial}{\partial t} \int_{\text{c.v.}} \rho \, d\mathcal{V} \qquad (3.1)$$

Equation (3.1) is the integral form of the continuity equation and physically says that the net rate of mass flow out of the control surface is equal to the time rate of decrease of mass inside the control volume. We will examine equation (3.1) by first considering some general simplifications, then some specific examples.

Since the control volume is fixed, the right side of (3.1) is zero for steady flow ($\partial \rho / \partial t = 0$) giving

$$\int_{\text{c.s.}} \rho \mathbf{V} \cdot d\mathbf{A} = 0 \qquad (3.2)$$

For incompressible flow we have

$$\int_{\text{c.s.}} \mathbf{V} \cdot d\mathbf{A} = 0$$

Considering the steady flow of Fig. 3-2, where the fluid enters section 1 and leaves sections 2 and 3, we have

$$\int_{\text{c.s.}} \rho \mathbf{V} \cdot d\mathbf{A} = 0$$

$$\int_{A_2} \rho \mathbf{V} \cdot d\mathbf{A} + \int_{A_3} \rho \mathbf{V} \cdot d\mathbf{A} + \int_{A_1} \rho \mathbf{V} \cdot d\mathbf{A} = 0$$

Assuming that the velocity is normal to all surfaces where fluid crosses,

$$\int_{A_2} \rho_2 V_2 \, dA + \int_{A_3} \rho_3 V_3 \, dA - \int_{A_1} \rho_1 V_1 \, dA = 0$$

If the densities and velocities are uniform over their respective areas,

$$\rho_2 V_2 A_2 + \rho_3 A_3 V_3 - \rho_1 A_1 V_1 = 0 \qquad (3.3)$$

For a single pipe with no third exit the equation becomes

$$\rho_2 A_2 V_2 = \rho_1 A_1 V_1 \qquad (3.4)$$

The assumptions made in arriving at equation (3.4) are (a) steady flow, (b) velocities normal to the surfaces, (c) velocity and density constant over the respective areas, and (d) one exit and one inlet to the control volume.

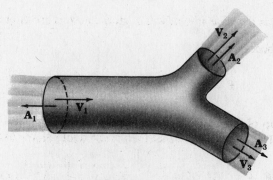

Fig. 3-2. Continuity in branching flow.

Momentum: We will now develop the momentum equation for the control volume. This equation is one of the most important mathematical relationships of fluid motion. It enables us to deal with problems involving forces of fluids on solid surfaces and other fluids, such as force on a pipe bend, thrust of a jet engine, lift and drag on an airplane wing, and many others.

The net force **F** acting on a particle or system of particles of fixed mass is given by Newton's second law

$$\mathbf{F} = \frac{d\mathbf{M}}{dt} \tag{3.5}$$

where **M** is the total linear momentum of the system. If we assume that the force is constant over a time Δt, we may write

$$\mathbf{F}\,\Delta t = \Delta \mathbf{M} \tag{3.6}$$

The right hand side of equation (3.6) is, referring to Fig. 3-1,

$$\Delta \mathbf{M} = \mathbf{M}_A(t + \Delta t) - \mathbf{M}_C(t + \Delta t) + \mathbf{M}_B(t + \Delta t) - \mathbf{M}_A(t)$$

Rearranging and dividing by Δt,

$$\frac{\Delta \mathbf{M}}{\Delta t} = \frac{\mathbf{M}_A(t + \Delta t) - \mathbf{M}_A(t)}{\Delta t} + \frac{\mathbf{M}_B(t + \Delta t) - \mathbf{M}_C(t + \Delta t)}{\Delta t} \tag{3.7}$$

By taking the limit of equation (3.7) as $\Delta t \to 0$, the first term on the right becomes

$$\lim_{\Delta t \to 0} \frac{\mathbf{M}_A(t + \Delta t) - \mathbf{M}_A(t)}{\Delta t} = \frac{\partial}{\partial t}\,(\mathbf{M})_{\text{c.v.}} = \frac{\partial}{\partial t}\int_{\text{c.v.}} \rho \mathbf{V}\, d\mathcal{V}$$

and the second term becomes

$$\lim_{\Delta t \to 0}\left[\frac{\mathbf{M}_B(t + \Delta t) - \mathbf{M}_C(t + \Delta t)}{\Delta t}\right] = \lim_{\Delta t \to 0}\left\{\frac{\left[\sum_B \Delta \mathbf{M}(t + \Delta t)\right]_B}{\Delta t} - \frac{\left[\sum_C \Delta \mathbf{M}(t + \Delta t)\right]_C}{\Delta t}\right\}$$

$$= \sum_B \Delta \dot{\mathbf{M}} - \sum_C \Delta \dot{\mathbf{M}} = \left[\sum \Delta \dot{m}\mathbf{V}\right]_{\text{out}} - \left[\sum \Delta \dot{m}\mathbf{V}\right]_{\text{in}}$$

$$= \int_{\text{c.s.}} \mathbf{V}\rho \mathbf{V} \cdot d\mathbf{A}$$

where $\sum_B \Delta \mathbf{M}(t + \Delta t)$ is the momentum associated with the mass that has crossed the boundary into region B in time Δt. $\sum_B \Delta \dot{\mathbf{M}}$ is the time rate at which momentum is crossing the surface into region B at time t. Thus equation (3.6) becomes

$$\mathbf{F} = \frac{\partial}{\partial t}\int_{\text{c.v.}} \mathbf{V}\rho\, d\mathcal{V} + \int_{\text{c.s.}} \mathbf{V}\rho \mathbf{V} \cdot d\mathbf{A} \tag{3.8}$$

The total force **F** is made up of the total surface force $\mathbf{F}_s$ (pressure and shear) and a body force **B** which is a force per unit volume. The momentum equation for a control volume becomes

$$\mathbf{F}_s + \int_{\text{c.v.}} \mathbf{B}\, d\mathcal{V} = \frac{\partial}{\partial t}\int_{\text{c.v.}} \mathbf{V}\rho\, d\mathcal{V} + \int_{\text{c.s.}} \mathbf{V}\rho \mathbf{V} \cdot d\mathbf{A} \tag{3.9}$$

It should be emphasized that this equation is valid only when referred to axes moving without acceleration, since the usual form of Newton's laws holds under these conditions.[1]

For steady flow and negligible body forces, equation (3.9) becomes

$$\mathbf{F}_s = \int_{\text{C.S.}} \mathbf{V} \rho \mathbf{V} \cdot d\mathbf{A} \qquad (3.10)$$

Further, if we assume that the density and velocity are uniform over the areas where the fluid is crossing the control surface, for one entrance 1 and one exit 2, we have

$$\sum F_x = \dot{m}(V_{x_2} - V_{x_1}), \qquad \sum F_y = \dot{m}(V_{y_2} - V_{y_1}), \qquad \sum F_z = \dot{m}(V_{z_2} - V_{z_1})$$

Angular Momentum: Rather than give a rigorous derivation of the angular momentum equation, we will merely write it down and discuss its physical meaning. For a derivation, see reference 11 or 13.

Let us again write the linear momentum equation (3.9)

$$\mathbf{F}_s + \int_{\text{C.V.}} \mathbf{B}\, d\mathcal{V} = \frac{\partial}{\partial t} \int_{\text{C.V.}} \mathbf{V} \rho\, d\mathcal{V} + \int_{\text{C.S.}} \mathbf{V} \rho \mathbf{V} \cdot d\mathbf{A}$$

Then write equation (3.9) with a position vector $\mathbf{r}$ taken as cross product with the vector of each term, giving

$$\int_{\text{C.S.}} \mathbf{r} \times d\mathbf{F}_s + \int_{\text{C.V.}} \mathbf{r} \times \mathbf{B}\, d\mathcal{V} = \frac{\partial}{\partial t} \int_{\text{C.V.}} \mathbf{r} \times \mathbf{V} \rho\, d\mathcal{V} + \int_{\text{C.S.}} \mathbf{r} \times \mathbf{V} \rho \mathbf{V} \cdot d\mathbf{A} \qquad (3.11)$$

which is the angular momentum equation. We will now look at the physical interpretation of each term. Refer to Fig. 3-3. It should be emphasized that $\mathbf{V}$ is the absolute velocity relative to the fixed coordinate system in which the control volume is stationary.

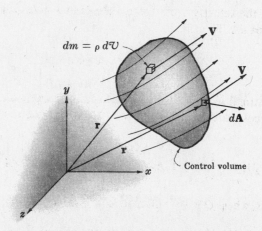

Fig. 3-3. Angular momentum in a control volume.

The integrand of the first term $\mathbf{r} \times d\mathbf{F}_s$ gives the moment around the origin attributed to the force $d\mathbf{F}_s$ at the control surface. The integrand of the second term of (3.11) is the moment about the origin due to the body force acting on the infinitesimal volume element $d\mathcal{V}$. The integrand of the third term is the angular momentum of the infinitesimal mass element $\rho\, d\mathcal{V}$. The integration gives the total angular momentum of the mass within the control volume. The last term is the rate of efflux of angular momentum through the control surface.

---

[1] For the case of an accelerating control volume without rotation, a term such as $-m\ddot{\mathbf{R}}$ is added to the left side of equation (3.9) where m is the total mass in the control volume and $\ddot{\mathbf{R}}$ is the acceleration relative to the inertial references. The derivation of the complete equation is given in reference 13.

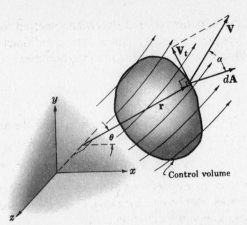

Fig. 3-4. Angular momentum about the z axis.

In applying equation (3.11) the scalar component form is used. For example, if we choose to write the equation for the z axis as shown in Fig. 3-4, we have (for steady flow and negligible body forces)

$$T_z = \int_{C.S.} (\mathbf{r} \times \mathbf{V})_z (\rho \mathbf{V} \cdot d\mathbf{A}) \qquad (3.12)$$

where T_z represents the net torque on the control volume about the z axis and

$$(\mathbf{r} \times \mathbf{V})_z = rV_t, \qquad \rho \mathbf{V} \cdot d\mathbf{A} = \rho V \cos \alpha \, dA$$

where V_t is the component of the velocity vector perpendicular to the z axis and α is the angle between the velocity vector and the area $d\mathbf{A}$. Then

$$T_z = \int_{C.S.} \rho r V_t V \cos \alpha \, dA \qquad (3.13)$$

Let us assume that the entire flow enters the control volume at an area A_1 and leaves at an area A_2, over each of which ρ, V and $\cos \alpha$ are uniform. Define $r_2 V_{t2}$ as the mean value of rV_t over A_2 and $r_1 V_{t1}$ as the mean value over A_1.

$$r_2 V_{t2} = \frac{1}{A_2} \int_{A_2} rV_t \, dA, \qquad r_1 V_{t1} = \frac{1}{A_1} \int_{A_1} rV_t \, dA$$

The continuity equation gives, where Q is the volumetric flow rate,

$$\rho_1 A_1 V_1 \cos \alpha_1 = \rho_2 A_2 V_2 \cos \alpha_2 = \rho_1 Q_1$$

Then equation (3.13) becomes

$$T_z = \rho_1 Q_1 (r_2 V_{t2} - r_1 V_{t1}) \qquad (3.14)$$

Equation (3.14) is applicable to the rotor of a turbomachine since the absolute flow in the interior of the rotor is cyclic and we may assume with a fair approximation that ρ, V and $\cos \alpha$ are uniform over the entrance section 1 and the exit section 2 (see Fig. 3-5 and 3-6). In the figures, $\mathbf{v}$ refers to the rotor (relative velocity), and $\mathbf{V}$ as usual denotes velocity relative to the earth (absolute velocity). For clarity, only one blade is shown.

If we neglect bearing friction, the drag of the fluid on the outside of the rotor, and fluid shearing stresses at 1 and 2, then the external torque T on the shaft equals the resultant external torque on either system. For the turbine, however, T is clockwise, while for the compressor it is counterclockwise. Hence

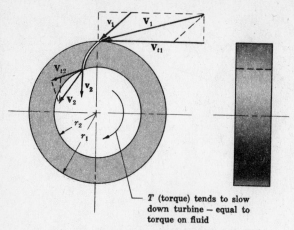

Fig. 3-5. Velocity diagrams for a radial flow turbine runner.

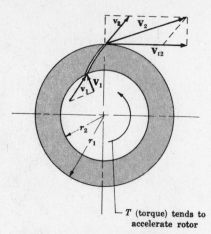

Fig. 3-6. Velocity diagrams for a radial flow impeller.

for the turbine,

$$T = \rho_1 Q_1 (r_1 V_{t1} - r_2 V_{t2}) \qquad (3.15)$$

and for the compressor or pump impeller (V_{t1} is usually zero here),

$$T = \rho_1 Q_1 (r_2 V_{t2} - r_1 V_{t1}) \qquad (3.16)$$

The turbomachines shown here are radial flow devices, i.e. the flow through the rotor is in a radial direction or normal to the axis of rotation. Turbomachines may also be axial flow (flow through the rotor parallel to the axis of rotation) or mixed flow (somewhere intermediate between axial and radial flow).

Energy: The mathematical statement of the first law of thermodynamics[1] is

$$Q - W = \Delta E \qquad (3.17)$$

where Q = heat added to the system
W = work done by the system
ΔE = change in energy of the system.

Again, as in the previous sections, it is strongly emphasized that this law applies to a system. Thus our task, as before, is to determine the mathematical expressions for a control volume.

Heat and work of equation (3.17) involve an interaction of the system with other systems. Energy, however, is energy associated with the mass of the system and is customarily separated into three parts,

$$E = U + \tfrac{1}{2}m^2 + mgz$$

where U = internal energy associated with molecular and atomic behavior

$\tfrac{1}{2}mV^2$ = kinetic energy

mgz = potential energy associated with the location in the earth's gravitational field.

We will write equation (3.17) on a unit mass basis,

$$q - w = \Delta e \qquad (3.18)$$

where $q = Q/m$, $w = W/m$, $e = E/m$.

---

[1] For definitions of heat and work see, for example, references 7 and 15.

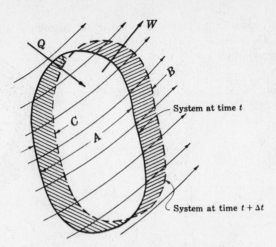

Fig. 3-7. Control volume for energy balance.

Consider the system at time t as shown in Fig. 3-7. At some later time $t + \Delta t$ the system has moved to another location. The energy equation for the system for this change is

$$Q - W = E_f - E_i$$

where E_f is the final energy of the system and E_i the initial energy. Dividing by Δt,

$$\frac{Q}{\Delta t} - \frac{W}{\Delta t} = \frac{E_f - E_i}{\Delta t} \tag{3.19}$$

Let us evaluate the right hand side of this equation:

$$\frac{E_f - E_i}{\Delta t} = \frac{E_A(t + \Delta t) - E_C(t + \Delta t) + E_B(t + \Delta t) - E_A(t)}{\Delta t}$$

$$= \frac{E_A(t + \Delta t) - E_A(t)}{\Delta t} + \frac{E_B(t + \Delta t) - E_C(t + \Delta t)}{\Delta t}$$

The first term on the right becomes, as $\Delta t \to 0$,

$$\lim_{\Delta t \to 0} \frac{E_A(t + \Delta t) - E_A(t)}{\Delta t} = \frac{\partial}{\partial t} (E)_{\text{c.v.}} = \frac{\partial}{\partial t} \int e \, dm = \frac{\partial}{\partial t} \int_{\text{c.v.}} e\rho \, d\mathcal{V}$$

The last term becomes

$$\frac{E_B(t + \Delta t) - E_C(t + \Delta t)}{\Delta t} = \frac{(\sum \Delta me)_B|_{t+\Delta t}}{\Delta t} - \frac{(\sum \Delta me)_C|_{t+\Delta t}}{\Delta t}$$

where the summation is for the mass crossing the surface, Δm is typical mass, and e is the stored energy associated with the mass Δm. In the limit as $\Delta t \to 0$ the last equation becomes

$$\lim_{\Delta t \to 0} \frac{E_B(t + \Delta t) + E_C(t + \Delta t)}{\Delta t} = \int_{\text{out}} e \, d\dot{m} - \int_{\text{in}} e \, d\dot{m} = \int_{\text{c.s.}} e\rho \mathbf{V} \cdot d\mathbf{A}$$

Thus we have

$$\lim_{\Delta t \to 0} \frac{E_f - E_i}{\Delta t} = \frac{\partial}{\partial t} \int_{\text{c.v.}} e\rho \, d\mathcal{V} + \int_{\text{c.s.}} e\rho \mathbf{V} \cdot d\mathbf{A}$$

Work may be done at the boundary of the system by normal and tangential stresses. We refer to the work done at the boundary of the system due to the normal stresses (hydrostatic pressure) as flow work.

This is the work done on element $(\Delta m)_B$, say, in moving it out of region A in time Δt and is equal to $p\,dA\,\Delta x$. However, $dA\,\Delta x$ is the volume of mass element Δm and may be written as $(\Delta m)_B/\rho$. Thus the flow work for both outflow and inflow is

$$\left(\frac{dW}{dt}\right)_{\text{flow work}} = \lim_{\Delta t \to 0}\left(\frac{\Delta W}{\Delta t}\right)_{\text{flow work}} = \lim_{\Delta t \to 0}\left[\frac{\sum (p/\rho)(\Delta m)_B|_{t+\Delta t}}{\Delta t} - \frac{\sum (p/\rho)(\Delta m)_C|_{t+\Delta t}}{\Delta t}\right]$$

$$= \int_{\text{C.S.}} (p/\rho)\rho\mathbf{V}\cdot d\mathbf{A}$$

The energy equation becomes

$$\frac{dQ}{dt} - \frac{dWs}{dt} = \frac{\partial}{\partial t}\int_{\text{C.V.}} e\rho\,d\mho + \int_{\text{C.S.}} (e + p/\rho)\rho\mathbf{V}\cdot d\mathbf{A} \qquad (3.20)$$

where

$$e = u + \tfrac{1}{2}V^2 + gz$$

and (dW_S/dt) is the time rate of work for all work except flow work. Equation (3.20) states that the time rate of heat added to the system minus the work done by the system (other than flow work) is equal to the time rate of change of the stored energy in the control volume plus the net rate of efflux of stored energy and flow work out of the control volume.

Let us consider one-dimensional steady flow for the device shown in Fig. 3-8. The shear work done over the cross section of the rotating shaft is called shaft work. The shear work done at all other parts of the boundary is zero, because the velocity is either zero or normal to the shear force. Thus from equation (3.20)

$$\frac{dQ}{dt} - \frac{dW_S}{dt} = \int_{A_1+A_2}\left(\frac{p}{\rho} + u + \frac{V^2}{2} + gz\right)\rho V\,dA$$

Since the flow is one-dimensional p, V, u, ρ are uniform over A_1 and A_2. If in addition to these conditions we neglect z variations over the areas, we have

$$\frac{dQ}{dt} - \frac{dW_S}{dt} = \left(\frac{p_2}{\rho_2} + u_2 + \frac{V_2^2}{2} + gz_2\right)\rho_2 A_2 V_2 - \left(\frac{p_1}{\rho_1} + u_1 + \frac{V_1^2}{2} + gz_1\right)\rho_1 A_1 V_1$$

From the continuity equation we have for one-dimensional steady flow,

$$\rho_1 A_1 V_1 = \rho_2 A_2 V_2 = \frac{dm}{dt} = \text{mass flow rate}$$

which upon substitution into the energy equation gives

$$\frac{dQ}{dt} - \frac{dW_S}{dt} = \left[\left(\frac{p_2}{\rho_2} - \frac{p_1}{\rho_1}\right) + (u_2 - u_1) + \frac{V_2^2 - V_1^2}{2} + g(z_2 + z_1)\right]\frac{dm}{dt} \qquad (3.21)$$

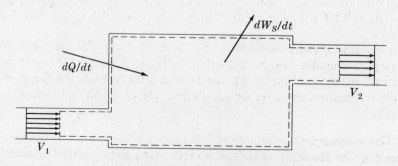

Fig. 3-8. A flow device with heat and work.

Writing equation (3.21) so that each term is referred to unit mass of the flowing fluid,

$$q - w_S = \left(\frac{p_2}{\rho_2} - \frac{p_1}{\rho_1}\right) + (u_2 - u_1) + \frac{V_2^2 - V_1^2}{2} + g(z_2 - z_1) \qquad (3.22a)$$

which is the usual one-dimensional steady flow energy balance of thermodynamics.

This equation may be written in terms of specific enthalpy h, where $h = p/\rho + u$.

$$q - w_S = (h_2 - h_1) + \frac{V_2^2 - V_1^2}{2} + g(z_2 - z_1)$$

By rearranging equation (3.22a) and assuming incompressible flow,

$$-w_S = \frac{p_2 - p_1}{\rho} + \frac{V_2^2 - V_1^2}{2} + g(z_2 - z_1) + (u_2 - u_1 - q)$$

In most actual flows, all of the quantities of the preceding equation may be measured by direct means except the internal energy and heat transfer. This is particularly true for flow of liquids in pipes, for example. Thus the usual procedure is to define

$$gH_L = u_2 - u_1 - q$$

so that $$-w_S = \frac{p_2 - p_1}{\rho} + \frac{V_2^2 - V_1^2}{2} + g(z_2 - z_1) + gH_L \qquad (3.22b)$$

where H_L is referred to as the "loss" or "head loss" and represents the conversion of mechanical energy into thermal energy. As the fluid flows through a pump or pipe it undergoes shearing deformation, since the fluid in contact with a solid surface does not slip. Shearing stresses are set up in the viscous fluid as a result of these deformations. The temperature rises above the value it would have in frictionless flow. The temperature rise tends to increase both $u_2 - u_1$ and the heat transferred to the surroundings.

For a frictionless flow of an incompressible fluid with zero shaft work, (3.22b) becomes

$$\frac{p_2 - p_1}{\rho} + \frac{V_2^2 - V_1^2}{2} + g(z_2 - z_1) = 0 \qquad (3.22c)$$

since H_L is then zero. We will want to compare this equation with Bernoulli's equation which will be derived later from the momentum equation. Actually, we can only prove that $(u_2 - u_1 - q)$ is a head loss term by comparing (3.22a) with the first integral of the equation of motion for an incompressible fluid, which is identical to (3.22b) and which is a generalized Bernoulli equation obtained directly from the momentum equation and discussed in Section 3.3. It is very important to realize that (3.22b), but not (3.22a), may be derived strictly from the momentum equation. Equation (3.22a), is a general energy equation for compressible flow, but once we say the flow is incompressible and lump $(u_2 - u_1 - q)$ as a frictionless loss term we have removed all true thermodynamic information and are left with only a mechanical energy balance redundant with the equation of motion. For frictionless incompressible flow, (3.22c) is identical with the ordinary Bernoulli equation, and the other part of (3.22b), $H_L = 0$, is the first law. That is, $q - \Delta u = 0$ for a unit mass of flowing fluid is the first law of thermodynamics and is independent of (3.22c).

Contained in the general energy equation (3.22a) is the first law of thermodynamics and a balance of mechanical energy. Remember that the balance of mechanical energy (the rate of increase of kinetic energy and potential energy is equal to the rate at which the translational forces do work) may be obtained from the momentum equation and is independent of the thermodynamic balance which is known as the first law.

Second Law of Thermodynamics: The second law for a control volume is determined by the same procedure as was done for the other basic laws. Rather than going through the entire derivation, only the results will be given. One may refer to references 8 and 10 for a detailed derivation.

The second law for a system is

$$dS - \frac{dQ}{T} \geqq 0$$

(where S is the entropy of the system), which says that the entropy change, minus the heat transferred to the system divided by the temperature, is equal to or greater than zero. By the same procedure as in the previous sections, the control volume form of the second law is obtained

$$\frac{\partial}{\partial t} \int_{\text{c.v.}} s\rho \, d\mathcal{V} + \int_{\text{c.s.}} s\rho \mathbf{V} \cdot d\mathbf{A} - \int_{\text{c.s.}} \frac{\mathbf{q}}{T} \cdot d\mathbf{A} \geqq 0 \qquad (3.23)$$

where $\mathbf{q}$ is the heat flux vector, which is the rate of heat transfer per unit area. s is the specific entropy, that is, the entropy per unit mass.

If it is assumed that a fluid flows steadily and adiabatically with one inlet and one outlet (e.g. Fig. 3-8), then equation (3.23) becomes

$$s_2 - s_1 \geqq 0$$

If it is further assumed that the process is reversible, then

$$s_2 - s_1 = 0$$

which says that the process is isentropic.

Entropy is a property and is related to other properties. The second law finds particular usefulness in the flow of gases.

3.3   DIFFERENTIAL EQUATIONS

In the previous section the basic laws which apply to a system were used to establish integral equations applicable to a control volume. Now we will use the basic laws to derive the differential equations of fluid dynamics. There are several ways to do this, although three different methods are commonly employed. First, one can formally, by use of the vector calculus, obtain differential expressions from integral expressions. No additional physical reasoning is necessary and the procedure is purely mathematical and rigorous. Second, we may apply the integral relationships to an elemental volume and obtain differential expressions by taking the limit as the volume becomes infinitesimal. The third method is to apply the basic system equations directly to an elemental volume, and in so doing, of course, obtain the basic integral expressions for an elemental volume along the way.

To illustrate the methods we will derive the differential continuity equation by the first method, the momentum equation by the second, and the energy equation by the third.

One important point to remember here is that the coordinates apply to points in space and time, and that coordinates do not denote the location of individual particles. That is, we ask: what are the properties and velocity $\mathbf{V}$ as a function of position and time? For example, $\mathbf{V} = \mathbf{V}(\mathbf{r}, t)$ or $\mathbf{V} = \mathbf{V}(x, y, z, t)$ in cartesian coordinates. Such field coordinates are known as Eulerian, as opposed to Lagrangian which are used in rigid body dynamics. In Lagrangian coordinates, acceleration is simply $\ddot{\mathbf{r}}$ and $\mathbf{r}$ denotes the position of a particle or fixed point on a rigid body. In fluid mechanics one may use Lagrangian coordinates, but such a description is not particularly useful and Eulerian coordinates are used almost exclusively. Clearly then, acceleration cannot be $\ddot{\mathbf{r}}$ in Eulerian coordinates since such a quantity is meaningless; $\mathbf{r}$ is a fixed point in space and its derivatives are meaningless. However, as the fluid flows past any point in question, it does have acceleration which can be written in terms of the fluid velocity. The appropriate expression will be derived presently.

Continuity (Conservation of Mass)

Beginning with equation (3.1),

$$\int_{\text{c.s.}} \rho \mathbf{V} \cdot d\mathbf{A} = - \frac{\partial}{\partial t} \int_{\text{c.v.}} \rho \, d\mho \tag{3.1}$$

we may formally apply Gauss' theorem to convert the left side to a volume integral. Hence (3.1) may be written

$$\int_{\text{c.v.}} \nabla \cdot (\rho \mathbf{V}) \, d\mho + \frac{\partial}{\partial t} \int_{\text{c.v.}} \rho \, d\mho = \int_{\text{c.v.}} [\nabla \cdot (\rho \mathbf{V}) + \partial \rho / \partial t] \, d\mho = 0 \tag{3.24}$$

Since the control volume is arbitrary the integrand must be zero, so that we obtain the differential form of the continuity equation as

$$\partial \rho / \partial t + \nabla \cdot (\rho \mathbf{V}) = 0 \tag{3.25}$$

which could also be obtained by applying (3.1) directly to an elemental volume.

For steady flow, $\partial \rho / \partial t = 0$ and we have

$$\nabla \cdot (\rho \mathbf{V}) = 0 \tag{3.26}$$

and for incompressible flow

$$\nabla \cdot \mathbf{V} = 0 \tag{3.27}$$

In cartesian coordinates the continuity equation is

$$\frac{\partial \rho}{\partial t} + \frac{\partial}{\partial x}(\rho u) + \frac{\partial}{\partial y}(\rho v) + \frac{\partial}{\partial z}(\rho w) = 0 \tag{3.28}$$

which for incompressible flow is

$$\frac{\partial u}{\partial x} + \frac{\partial v}{\partial y} + \frac{\partial w}{\partial z} = 0 \tag{3.29}$$

In cartesian tensor notation the continuity equation is

$$\frac{\partial \rho}{\partial t} + \frac{\partial}{\partial x_i}(\rho u_i) = 0 \tag{3.30}$$

A brief review of cartesian tensor notation is given in the Appendix.

Momentum Equation

We will derive the momentum equation by applying the integral form to an elemental volume. We will use a small cube so that the result will be valid only in cartesian coordinates. The general vector form is best derived by the formal mathematical method, but we will give only the result of that derivation here. Of course, equations for any coordinate system (cylindrical, spherical, etc.) could be derived just as we will for cartesian coordinates by choosing the appropriate elemental volume to begin with.

Before beginning the derivation, let us first review the concept of the stress tensor. We will want to account for the shear stresses and normal stresses (including pressure) in the momentum balance. (After our derivation we will relate these stresses to the velocity components to obtain the final form of the momentum equation.)

Consider a cube of fluid as shown in Fig. 3-9. Stresses are denoted as σ_{ij}. The first subscript denotes the face on which the stress acts, and the second subscript denotes the direction of the stress on the positive face. On the negative face the stresses are equal in magnitude but opposite in direction to those on the positive face. (The face is indicated as the plane perpendicular to the axis of the subscript. For instance, the 1 face is perpendicular to the x or x_1 axis.) The idea of positive and negative faces is indicated in Fig. 3-9. The stresses may vary through a fluid so that their gradients may exist. However,

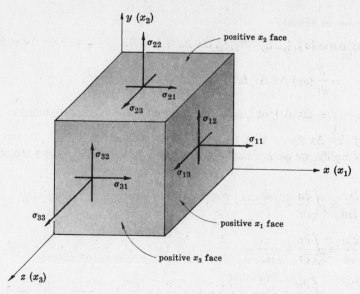

Fig. 3-9. Stresses at a point in space. The positive faces are shown; the opposite ones are the negative faces. The stresses on the negative faces are equal to but in the opposite direction to those on the positive faces.

the array of stresses σ_{ij} (there are nine of them) may be considered to exist at a point in space and be a function then of $\mathbf{r}$ and t. The stress tensor then may be written as

$$\sigma_{ij} = \begin{bmatrix} \sigma_{11} & \sigma_{12} & \sigma_{13} \\ \sigma_{21} & \sigma_{22} & \sigma_{23} \\ \sigma_{31} & \sigma_{32} & \sigma_{33} \end{bmatrix} \tag{3.31}$$

The stress tensor must be symmetric, that is, $\sigma_{ij} = \sigma_{ji}$, otherwise if the elemental volume were shrunk to infinitesimal size it would have to rotate with infinite angular velocity. Hence there are six independent components of the stress tensor.

We can now apply the integral momentum equation (3.9) to an elemental cube, where the stresses compose the external force $\mathbf{F}_s$. Referring to Fig. 3-10 we can write out the momentum balance for the x

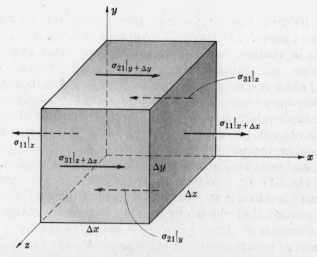

Fig. 3-10. Elemental cube for momentum equation derivation. Only the forces in the x direction are shown.

direction as

$$(\sigma_{11}|_{x+\Delta x} - \sigma_{11}|_x)\,\Delta y\,\Delta z + (\sigma_{21}|_{y+\Delta y} - \sigma_{21}|_y)\,\Delta x\,\Delta z + (\sigma_{31}|_{z+\Delta z} - \sigma_{31}|_z)\,\Delta x\,\Delta y + B_x\,\Delta x\,\Delta y\,\Delta z$$

$$= \frac{\partial}{\partial t}(u\rho)\,\Delta x\,\Delta y\,\Delta z + \Delta y\,\Delta z(\rho u^2|_{x+\Delta x} - \rho u^2|_x)$$

$$+ \Delta x\,\Delta z(\rho uv|_{y+\Delta y} - \rho uv|_y) + \Delta x\,\Delta y(\rho uw|_{z+\Delta z} - \rho uw|_z)$$

If we divide through by $\Delta x\,\Delta y\,\Delta z$ and take the limit as Δx, Δy and $\Delta z \to 0$, and combine with the continuity (3.28) to simplify, we obtain the following. (The y and z components are obtained in a similar manner.)

$$\rho\frac{Du}{Dt} = \rho\left(\frac{\partial u}{\partial t} + u\frac{\partial u}{\partial x} + v\frac{\partial u}{\partial y} + w\frac{\partial u}{\partial z}\right) = \frac{\partial\sigma_{11}}{\partial x} + \frac{\partial\sigma_{12}}{\partial y} + \frac{\partial\sigma_{13}}{\partial z} + B_x$$

$$\rho\frac{Dv}{Dt} = \rho\left(\frac{\partial v}{\partial t} + u\frac{\partial v}{\partial x} + v\frac{\partial v}{\partial y} + w\frac{\partial v}{\partial z}\right) = \frac{\partial\sigma_{21}}{\partial x} + \frac{\partial\sigma_{22}}{\partial y} + \frac{\partial\sigma_{23}}{\partial z} + B_y \qquad (3.32)$$

$$\rho\frac{Dw}{Dt} = \rho\left(\frac{\partial w}{\partial t} + u\frac{\partial w}{\partial x} + v\frac{\partial w}{\partial y} + w\frac{\partial w}{\partial z}\right) = \frac{\partial\sigma_{31}}{\partial x} + \frac{\partial\sigma_{32}}{\partial y} + \frac{\partial\sigma_{33}}{\partial z} + B_z$$

or in cartesian tensor notation,

$$\rho\frac{Du_i}{Dt} = \rho\left(\frac{\partial u_i}{\partial t} + u_j\frac{\partial u_i}{\partial x_j}\right) = \frac{\partial\sigma_{ij}}{\partial x_j} + B_i \qquad (3.33)$$

The terms on the left side are the acceleration. The $\partial/\partial t$ term is an unsteady term and indicates a change with time at a fixed point in space. The LHS terms other than $\partial/\partial t$ are known as convective acceleration terms and are present because we are working in Eulerian coordinates.

The D/Dt operator is called the substantial or material derivative. In general, D/Dt is a vector operator and consequently the components of D/Dt operating on a vector are *not* the same as D/Dt operating on the scalar components of a vector, *except in cartesian coordinates*. Equations (3.32) may be generalized to vector form valid in any coordinate system. Then the acceleration is $\rho\,D\mathbf{V}/Dt$ where

$$\frac{D\mathbf{V}}{Dt} = \frac{\partial\mathbf{V}}{\partial t} + (\mathbf{V}\cdot\nabla)\mathbf{V} = \frac{\partial\mathbf{V}}{\partial t} + \nabla(V^2/2) - \mathbf{V}\times(\nabla\times\mathbf{V})$$

The vector operator $D\mathbf{V}/Dt$ in various coordinate systems is listed in the Appendix. Care must be exercised in evaluating the convective term $(\mathbf{V}\cdot\nabla)\mathbf{V}$. Although this notation is frequently used, it may be expanded easily only in cartesian coordinates (and is really not a proper vector formulation). An equivalent but better and proper vector expression for the convective acceleration is $[\nabla(V^2/2) - \mathbf{V}\times(\nabla\times\mathbf{V})]$ which may be readily expanded into components in any coordinate system.

The material derivative operator may also be applied to a scalar quantity, such as temperature. The physical meaning of the result (which is then a scalar) is that it represents the rate of change of that quantity, in time, that an observer moving with the fluid would measure at any particular location in space and instant of time where the derivative is evaluated.

To illustrate the physical meaning of convective acceleration, consider a fluid flowing in a converging channel as shown in Fig. 3-11. The velocity at point B is greater than at point A because the area at B is smaller. Thus as a particle moves from A toward B it tends to undergo an acceleration due to area change (convective acceleration). If in addition we impose a time rate of change of mass flow rate (i.e., the mass flow rate at one instant of time is 3 lb/sec, say, and 5 seconds later it is 4 lb/sec), then each point in the flow will undergo a local increase in velocity or local acceleration. Thus local acceleration occurs because the flow is unsteady and convective acceleration occurs because the area changes. It is important to note that the acceleration is a vector, and whether we choose to express it in Eulerian or

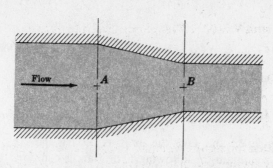

Fig. 3-11. Converging channel.

Lagrangian coordinates does not change its value. In other words, at a point in space at a given time, the fluid at that point has a certain velocity and acceleration, magnitude and direction. Hence $D\mathbf{V}/Dt = \ddot{\mathbf{r}}$ where $\mathbf{r}$ is the Lagrangian coordinate and D/Dt is expressed in terms of Eulerian coordinates.

Most frequently we shall consider steady flows. This means that for the most part the local acceleration will be zero and that particle acceleration will be due to convective effects alone.

The Momentum Equation for Frictionless Flow

In frictionless flow, there are no shear stresses and the normal stresses are just the pressure which is isotropic. Since we have defined the normal stresses (σ_{11}, σ_{22} and σ_{33}) as positive if they are tensions, we now set

$$\sigma_{11} = \sigma_{22} = \sigma_{33} = -p$$

so that the equations of motion in cartesian form are

$$\rho\left(\frac{\partial u_i}{\partial t} + u_j\frac{\partial u_i}{\partial x_j}\right) = -\frac{\partial p}{\partial x_i} + B_i \qquad (3.34)$$

or in component form

$$\rho\left(\frac{\partial u}{\partial t} + u\frac{\partial u}{\partial x} + v\frac{\partial u}{\partial y} + w\frac{\partial u}{\partial z}\right) = -\frac{\partial p}{\partial x} + B_x$$

$$\rho\left(\frac{\partial v}{\partial t} + u\frac{\partial v}{\partial x} + v\frac{\partial v}{\partial y} + w\frac{\partial v}{\partial z}\right) = -\frac{\partial p}{\partial y} + B_y \qquad (3.35)$$

$$\rho\left(\frac{\partial w}{\partial t} + u\frac{\partial w}{\partial x} + v\frac{\partial w}{\partial y} + w\frac{\partial w}{\partial z}\right) = -\frac{\partial p}{\partial z} + B_z$$

These equations are known as the Euler equations for frictionless flow. If the flow is incompressible, the density is constant and the three Euler equations plus the continuity equation constitute four equations in four unknowns, u, v, w, and p. The solution is often difficult, however, because the convective acceleration terms make the equations nonlinear.

In general vector form the Euler equations are

$$\frac{D\mathbf{V}}{Dt} = \frac{\partial \mathbf{V}}{\partial t} + (\mathbf{V} \cdot \nabla)\mathbf{V} = \rho\left[\frac{\partial \mathbf{V}}{\partial t} + \nabla\left(\frac{V^2}{2}\right) - \mathbf{V} \times (\nabla \times \mathbf{V})\right] = -\nabla p + \mathbf{B} \qquad (3.36)$$

An important first integral of the equation of motion can be obtained by integrating (3.36) between two points along a streamline. Let $d\mathbf{s}$ be an element of length along a streamline. Then

$$\rho\left[\frac{\partial \mathbf{V}}{\partial t} \cdot d\mathbf{s} + \nabla\left(\frac{V^2}{2}\right) \cdot d\mathbf{s} - \mathbf{V} \times (\nabla \times \mathbf{V}) \cdot d\mathbf{s}\right] = -(\nabla p + \rho\nabla\psi) \cdot d\mathbf{s} \qquad (3.37)$$

Since $\mathbf{V}$ is parallel to $d\mathbf{s}$, the term $\mathbf{V} \times (\nabla \times \mathbf{V})$ is perpendicular to $d\mathbf{s}$ and

$$\mathbf{V} \times (\nabla \times \mathbf{V}) \cdot d\mathbf{s} = 0$$

Hence we obtain for frictionless flow,

$$\int_1^2 \frac{\partial \mathbf{V}}{\partial t} \cdot d\mathbf{s} + \frac{V_2^2 - V_1^2}{2} + \int_1^2 \frac{dp}{\rho} + \psi_2 - \psi_1 = 0 \tag{3.38}$$

Usually this equation is written for steady state and the gravitational potential ψ is gz (where z is the elevation above an arbitrary datum), so that the force of gravity is $-\rho g \hat{z}$ where $\hat{z}$ is a unit vector pointing vertically). Then (3.38) becomes

$$\frac{V_2^2 - V_1^2}{2} + \int_1^2 \frac{dp}{\rho} + g(z_2 - z_1) = 0 \tag{3.39}$$

which is called the generalized Bernoulli or Euler equation. For incompressible flow,

$$\frac{V_2^2 - V_1^2}{2} + \frac{p_2 - p_1}{\rho} + g(z_2 - z_1) = 0 \tag{3.40}$$

which is the Bernoulli equation, a very important integral form of the equation of motion. Remember, (3.40) is based on steady, frictionless, incompressible flow with gravity being the only body force, and it holds along a streamline.

If we wish to integrate (3.37) between any two points in the flow field (between different streamlines), we can only get the Euler or Bernoulli equation if $\nabla \times \mathbf{V} = 0$. This term is known as the rotation of the fluid, and if $\nabla \times \mathbf{V} = 0$ the flow is said to be irrotational. The criterion for irrotationality will be discussed in Chapter 6.

It is interesting to note that (3.40) is identical with (3.22c) derived from energy considerations. By comparing (3.40) to (3.22b) and (3.22c) we can see that $(u_2 - u_1 - q)$ must be a loss term, and in fact (3.22b) and (3.22c) can be derived from the momentum equation. (3.22b) including w_S may be derived by integrating the generalized form of (3.36), taking into account the viscous terms (which gives rise to H_L) and assuming $\mathbf{B}$ to have a non-conservative part (which gives rise to w_S). This generalized form of the momentum equation will be given by equation (3.58) in Section 3.5 where we discuss the Navier-Stokes equations. We will not carry out the integration here, since it is much simpler to use the integral approach to obtain (3.22b) as we have already done.

Let us now turn to frictional flow, but first a word about stress and strain rates in fluids.

3.4 KINEMATICS AND STRESS-STRAIN RATE RELATIONSHIPS IN FLUIDS

If we wish to extend our study to viscous fluids, we must first discuss stress and strain rates in more detail. First let us consider the motion of a small system of fluid of arbitrary shape, say a cube. The motion of the cube may be divided into two types, a rigid body motion, and a deformation motion. The rigid body motion may be further divided into a translation and rotation. We usually describe this motion by keeping track of the location of the center of mass of the system and the rotation of the system. As the system undergoes rigid body motion, it can also deform. The deformation of the system of fluid may be completely specified by describing the dilatation rate (volumetric expansion or contraction rate) and the shear strain rate of the system. These are shown schematically in Fig. 3-12.

Now, quantitative expressions for all these quantities can be obtained in terms of the velocity vector $\mathbf{V}$ (or u_i) and its derivatives as follows:

Translation

The translation rate is simply given by the velocity vector $\mathbf{V}$ and corresponds to the velocity of the center of mass of the fluid element as it becomes infinitesimally small.

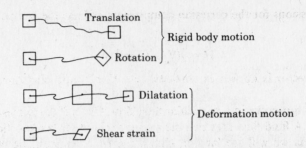

Fig. 3-12. Types of motion which, when superimposed, completely describe the kinematics of a small fluid elemental system. The cubic shape is arbitrary. Although shown in 2-D, the motion may occur in 3-D.

Rotation

The rotation rate (angular velocity) is a physical property which can be defined at a point. A finite system of fluid has an average angular velocity, and as the system becomes infinitesimal (in the limit) the angular velocity vector Ω becomes precisely defined. Ω is exactly equal to $\frac{1}{2}(\nabla \times \mathbf{V})$. This can be seen easily by examining a rotation in two dimensions. Consider a square shown in the xy plane in Fig. 3-13. We draw diagonals which remain perpendicular under rotation and average the angular velocity of the sides to find Ω_z for the element. The angular velocity of the side Δx is $(v|_{x+\Delta x} - v|_x)/\Delta x$ and the angular velocity of the side Δy is $-[u|_{y+\Delta y} - u|_y]/\Delta y$. We average these two expressions to find the angular velocity of the square and in the limit as Δx and $\Delta y \to 0$

$$\Omega_z = \frac{1}{2}\left(\frac{\partial v}{\partial x} - \frac{\partial u}{\partial y}\right) \tag{3.41}$$

and similarly

$$\Omega_x = \frac{1}{2}\left(\frac{\partial w}{\partial y} - \frac{\partial v}{\partial z}\right) \qquad \Omega_y = \frac{1}{2}\left(\frac{\partial u}{\partial z} - \frac{\partial w}{\partial x}\right)$$

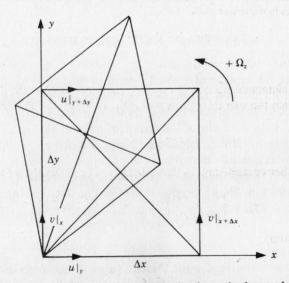

Fig. 3-13. A rotating and shearing cube shown in the xy plane.

We recognize these expressions for the cartesian components of Ω as the components of the curl of the velocity so that

$$\Omega = \tfrac{1}{2}(\nabla \times \mathbf{V}) = \tfrac{1}{2}\omega \qquad (3.42)$$

The curl of the velocity vector is known as vorticity, $\omega = \nabla \times \mathbf{V}$, and plays an important role in fluid mechanics.

Rotation may or may not occur in a particular fluid flow. Generally, viscous or frictional effects in a fluid give rise to vorticity. A fluid flow in which the angular velocity Ω (and vorticity ω) is zero is known as irrotational flow and as we will see later in Chapter 6 is a characteristic of aerodynamic flow. A simple example of irrotational flow is a whirlpool, which is known as a potential vortex in fluid mechanics.

A simple way to visualize vorticity is to float a small object on a free surface of water and note if it rotates as it moves with the water. A vortex (whirlpool) formed in a sink or bathtub gives rise to irrotational flow, and a floating object will not rotate but will remain parallel to itself as it circles around the drain. (The object translates in a circle without rotation.)

Vorticity meters are often made of a small rigid floating cross with an attached arrow which can be observed for rotation. The reason for a cross instead of a small rod is that the average angular velocity of two perpendicular lines must be measured to indicate angular velocity of a small element. For example, in a rotational viscous shear flow a rod would align itself with the flow and cease to rotate while a cross would continue to rotate (Fig. 3-14).

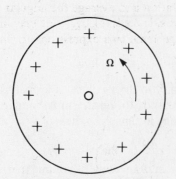
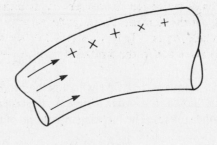

(a) Irrotational flow in a sink vortex.   (b) Rotational flow of a viscous fluid
     The object does not rotate as it   in a pipe.
     translates in a circular path.

Fig.   3-14

Shear Strain Deformation

Again referring to the elemental square of Fig. 3-13, we can see that the shear strain rate in the xy plane, γ_{xy}, is the rate at which the two sides close toward each other and is simply

$$\gamma_{xy} = \left(\frac{\partial u}{\partial y} + \frac{\partial v}{\partial x}\right) \qquad (3.43)$$

By similar reasoning the other components in three dimensions are readily written as

$$\gamma_{xz} = \left(\frac{\partial u}{\partial z} + \frac{\partial w}{\partial x}\right) \qquad \gamma_{yz} = \left(\frac{\partial v}{\partial z} + \frac{\partial w}{\partial y}\right)$$

or in cartesian tensor notation:

$$\gamma_{ij} = \left(\frac{\partial u_i}{\partial x_j} + \frac{\partial u_j}{\partial x_i}\right) = \gamma_{ji} \qquad (3.44)$$

which is, by inspection, symmetric because interchanging i and j does not change the equation. The shear strain rate cannot be easily written in vector form. The above expressions are valid only in cartesian coordinates, and one must be careful if working in other coordinate systems. The components in other coordinate systems are listed in the Appendix.

Dilatation

The dilatation which represents the rate of expansion or contraction of a small system of fluid can be expressed in terms of velocity by referring to Fig. 3-15.

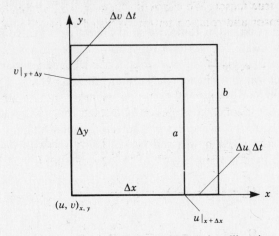

Fig. 3-15. An expanding cube illustrating dilatation.

We consider a square Δx by Δy which is the projection onto the xy plane of a small cube with edges Δx, Δy, and Δz, shown in a. After a time Δt the square expands to the size in b. Similar figures can be shown in the other planes. The increase in volume of the cube in time Δt is then

$$\Delta u \, \Delta t \, \Delta y \, \Delta z + \Delta v \, \Delta t \, \Delta x \, \Delta z + \Delta w \, \Delta t \, \Delta x \, \Delta y$$

Dividing by $\Delta x \, \Delta y \, \Delta z \, \Delta t$ and taking the limit we get the rate of increase of volume (per unit volume of fluid) as

$$\phi = \frac{\partial u}{\partial x} + \frac{\partial v}{\partial y} + \frac{\partial w}{\partial z} = \frac{\partial u_i}{\partial x_i} \tag{3.45}$$

which in vector form may be written

$$\phi = \nabla \cdot \mathbf{V} \tag{3.46}$$

The numerical value of the dilatation is independent of the coordinate system used.[1]

Stress-Strain Rate Relationships in Fluids

In a fluid the velocity gradient tensor is written as $\partial u_i / \partial x_j$. The symmetrical part of this tensor is the strain rate tensor, and the antisymmetrical part is the rotation tensor, which is related to $\nabla \times \mathbf{V}$. We denote the strain rate tensor as e_{ij} and the rotation tensor as Ω_{ij}. The velocity gradient tensor then may be written

$$\frac{\partial u_i}{\partial x_j} = e_{ij} + \Omega_{ij} \tag{3.47}$$

The terms of the rotation tensor are related to the angular velocity Ω_i of an infinitesimal fluid particle by

$$\Omega_1 = \Omega_{32} = -\Omega_{23}, \qquad \Omega_2 = \Omega_{13} = -\Omega_{31}, \qquad \Omega_3 = \Omega_{21} = -\Omega_{12} \tag{3.48}$$

---

[1] Mathematically ϕ is the first invariant of the strain rate tensor.

where Ω_1, Ω_2, and Ω_3 are the components of the angular velocity vector Ω (an axial vector). Ω_{ij} are the components of the corresponding antisymmetric rotation tensor ($\Omega_{ij} = -\Omega_{ji}$). It can be shown by purely mathematical means that to every axial vector there corresponds an antisymmetric tensor.

In vector form, the following important relationship may be written

$$\Omega = \tfrac{1}{2}\nabla \times V = \tfrac{1}{2}\omega \tag{3.49}$$

The physical angular velocity of the fluid is equal to one-half of the curl of the velocity vector.

The diagonal (normal) components of the strain rate tensor may be identified directly with the true normal strain rates. However, the off-diagonal terms (shear rate components) e_{ij}, $i \neq j$, are equal to one-half the true rate of shear strain components, which we denote as γ_{ij}. The one-half factor is necessary in order to make e_{ij} a true tensor. We can write them $e_{ii} = \gamma_{ii}$ and $e_{ij} = \tfrac{1}{2}\gamma_{ij}$ ($i \neq j$). The cartesian components of the strain tensor and rotation tensor are listed below. (For a more complete listing see reference 6.)

$$e_{xx} = \frac{\partial u}{\partial x}; \qquad e_{xy} = e_{yx} = \frac{1}{2}\left(\frac{\partial u}{\partial y} + \frac{\partial v}{\partial x}\right)$$

$$e_{yy} = \frac{\partial v}{\partial y}; \qquad e_{yz} = e_{zy} = \frac{1}{2}\left(\frac{\partial v}{\partial z} + \frac{\partial w}{\partial y}\right) \tag{3.50}$$

$$e_{zz} = \frac{\partial w}{\partial z}; \qquad e_{xz} = e_{zx} = \frac{1}{2}\left(\frac{\partial w}{\partial x} + \frac{\partial u}{\partial z}\right)$$

$$\Omega_x = \Omega_{zy} = -\Omega_{yz} = \frac{1}{2}\left(\frac{\partial w}{\partial y} - \frac{\partial v}{\partial z}\right) = \frac{1}{2}\omega_x$$

$$\Omega_y = \Omega_{xz} = -\Omega_{zx} = \frac{1}{2}\left(\frac{\partial u}{\partial z} - \frac{\partial w}{\partial x}\right) = \frac{1}{2}\omega_y$$

$$\Omega_z = \Omega_{yx} = -\Omega_{xy} = \frac{1}{2}\left(\frac{\partial v}{\partial x} - \frac{\partial u}{\partial y}\right) = \frac{1}{2}\omega_z$$

Written out in full in cartesian tensor notation, where the 1, 2, 3 correspond to x, y, z, respectively, $\partial u_i / \partial x_j$ may be expressed as the sum $e_{ij} + \Omega_{ij}$. The sum of the diagonal components is ϕ which may be recognized as the trace of $\partial u_i / \partial x_j$.

$$
\begin{bmatrix}
\dfrac{\partial u_1}{\partial x_1} & \dfrac{\partial u_1}{\partial x_2} & \dfrac{\partial u_1}{\partial x_3} \\[2mm]
\dfrac{\partial u_2}{\partial x_1} & \dfrac{\partial u_2}{\partial x_2} & \dfrac{\partial u_2}{\partial x_3} \\[2mm]
\dfrac{\partial u_3}{\partial x_1} & \dfrac{\partial u_3}{\partial x_2} & \dfrac{\partial u_3}{\partial x_3}
\end{bmatrix}
=
\begin{bmatrix}
\dfrac{\partial u_1}{\partial x_1} & 0 & 0 \\[2mm]
0 & \dfrac{\partial u_2}{\partial x_2} & 0 \\[2mm]
0 & 0 & \dfrac{\partial u_3}{\partial x_3}
\end{bmatrix}
$$

$$
+
\begin{bmatrix}
0 & \dfrac{1}{2}\left(\dfrac{\partial u_1}{\partial x_2} + \dfrac{\partial u_2}{\partial x_1}\right) & \dfrac{1}{2}\left(\dfrac{\partial u_1}{\partial x_3} + \dfrac{\partial u_3}{\partial x_1}\right) \\[3mm]
\dfrac{1}{2}\left(\dfrac{\partial u_2}{\partial x_1} + \dfrac{\partial u_1}{\partial x_2}\right) & 0 & \dfrac{1}{2}\left(\dfrac{\partial u_2}{\partial x_3} + \dfrac{\partial u_3}{\partial x_2}\right) \\[3mm]
\dfrac{1}{2}\left(\dfrac{\partial u_3}{\partial x_1} + \dfrac{\partial u_1}{\partial x_3}\right) & \dfrac{1}{2}\left(\dfrac{\partial u_3}{\partial x_2} + \dfrac{\partial u_2}{\partial x_3}\right) & 0
\end{bmatrix}
$$

$$
+
\begin{bmatrix}
0 & \dfrac{1}{2}\left(\dfrac{\partial u_1}{\partial x_2} - \dfrac{\partial u_2}{\partial x_1}\right) & \dfrac{1}{2}\left(\dfrac{\partial u_1}{\partial x_3} - \dfrac{\partial u_3}{\partial x_1}\right) \\[3mm]
-\dfrac{1}{2}\left(\dfrac{\partial u_1}{\partial x_2} - \dfrac{\partial u_2}{\partial x_1}\right) & 0 & \dfrac{1}{2}\left(\dfrac{\partial u_2}{\partial x_3} - \dfrac{\partial u_3}{\partial x_2}\right) \\[3mm]
-\dfrac{1}{2}\left(\dfrac{\partial u_1}{\partial x_3} - \dfrac{\partial u_3}{\partial x_1}\right) & -\dfrac{1}{2}\left(\dfrac{\partial u_2}{\partial x_3} - \dfrac{\partial u_3}{\partial x_2}\right) & 0
\end{bmatrix}
\tag{3.51}
$$

Now, for a Newtonian fluid, by definition the stress tensor and strain rate tensor are linearly related. The most general relationship can be shown to be of the following form (in cartesian tensor notation)

$$\sigma_{ij} = -p\delta_{ij} + \sigma'_{ij} = -p\delta_{ij} + 2\mu e_{ij} + \delta_{ij}\lambda\phi \qquad (3.52)$$

σ'_{ij} is known as the deviatoric stress tensor and is the difference between the total stress tensor and the isotropic pressure p. The mechanical pressure p is defined as $p = -\frac{1}{3}(\sigma_{11} + \sigma_{22} + \sigma_{33}) = -\frac{1}{3}\sigma_{ii}$ and is an invariant of the stress tensor. The normal components of the deviatoric stress tensor, σ'_{11}, σ'_{22}, and σ'_{33}, are the differences between the actual normal stresses σ_{11}, σ_{22}, σ_{33}, and their average. These normal components of σ'_{ij}, which are a measure of the anisotropy of the stress tensor, are present only in a moving fluid and are usually negligibly small. It should be noted that the mechanical pressure p defined here cannot always be identified with the thermodynamic pressure, but the difference is usually of little consequence from an engineering point of view. ϕ is the fluid dilatation defined as $\nabla \cdot \mathbf{V}$ (or $\partial u_k/\partial x_k$) which is zero for an incompressible fluid. δ_{ij} is the Kronecker delta. λ is a second coefficient of viscosity. Another second coefficient of viscosity is defined as $\zeta = \lambda + \frac{2}{3}\mu$ and is zero for a monatomic gas. Equation (3.52) may be written out as

$$\sigma_{ij} = -p\delta_{ij} + \mu\left(\frac{\partial u_i}{\partial x_i} + \frac{\partial u_j}{\partial x_i} - \frac{2}{3}\delta_{ij}\frac{\partial u_k}{\partial x_k}\right) + \zeta\delta_{ij}\frac{\partial u_k}{\partial x_k} \qquad (3.53)$$

The reader is referred to the references for a complete derivation of this expression.

Now this expression for σ_{ij} may be sustituted into the momentum equation to yield the full equation of motion for a viscous fluid. These are known as the Navier-Stokes equations of motion.

3.5 THE NAVIER-STOKES EQUATIONS

The Navier-Stokes equations are the complete equations of motion for a viscous Newtonian fluid. Using (3.53), (3.33) becomes

$$\rho\left(\frac{\partial u_i}{\partial t} + u_j\frac{\partial u_i}{\partial x_j}\right) = -\frac{\partial p}{\partial x_i} + B_i + \frac{\partial}{\partial x_j}\left[\mu\left(\frac{\partial u_i}{\partial x_j} + \frac{\partial u_j}{\partial x_i} - \frac{2}{3}\delta_{ij}\frac{\partial u_k}{\partial x_k}\right)\right] + \frac{\partial}{\partial x_i}\left(\zeta\frac{\partial u_k}{\partial x_k}\right) \qquad (3.54)$$

Usually, the viscosity may be removed from the derivatives with negligible error. For an incompressible fluid ($\nabla \cdot \mathbf{V} = 0$) these equations reduce to

$$\rho\frac{D\mathbf{V}}{Dt} = -\nabla p + \mathbf{B} + \mu\nabla^2\mathbf{V} \qquad (3.55)$$

which is a very important equation and should become quite familiar to the fluid dynamicist. For a gravitational body force, $\mathbf{B}$ may be replaced by $-\rho\nabla\psi$. It is *most important* to remember that D/Dt and ∇^2 are vector operators and may *not* be applied to the velocity components except in cartesian coordinates. Rather, the operations must first be performed, then components taken. The components of this equation in various coordinates are listed in the Appendix and in reference 6. The $\nabla^2\mathbf{V}$ term may be evaluated explicitly if it is written as $\nabla^2\mathbf{V} = \nabla(\nabla \cdot \mathbf{V}) - \nabla \times (\nabla \times \mathbf{V})$.

Although equation (3.55) appears simple, its solution is complicated by the nonlinear terms on the left hand side and in fact represents one of the major challenges in fluid mechanics.

Now we can compare the complete Navier-Stokes equations with the Euler equations for frictionless flow. The differences are the additional terms involving the viscosity. If the fluid is incompressible these are all of the form $\partial u_i/\partial x_j$ ($i \neq j$). Thus for an incompressible fluid, even if μ is not zero, the viscous terms are negligible if derivatives of the form $\partial u_i/\partial x_j$ are small, and the equations reduce to the Euler equations.

Therefore the Navier-Stokes equations reduce to Euler's equations if the viscosity is small or the derivatives of velocities with respect to directions other than the velocity direction are small. The latter approximation is very important in fluid mechanics. It is assumed that for many flows there are two regions: one close to a solid surface where the viscous terms are important (where Navier-Stokes equations must be used), and a region away from the surface where Euler's equations are a good approximation. These two distinct regions are the subjects of two later chapters: "Boundary Layer Flow" (Chapter 5) and "Potential Flow" (Chapter 6). The manner of treating these two subjects is entirely different as a result of the different forms of the equations.

Other forms of the equation of motion are sometimes used, particularly for incompressible flow where ρ is assumed to be constant. For incompressible flow, the vector equation of motion along with the continuity equation constitutes four equations in four unknowns, the three velocity components and pressure.

As an alternative to solving for the so-called primitive variables, $\mathbf{V}$ and p, other variables may be introduced which describe the flow field and from which $\mathbf{V}$ and p may be found.

Of particular interest is the equation of motion in terms of vorticity ω. If we take the curl of all terms in the vector equation (3.55) (with $\mathbf{B} = -\rho\nabla\psi$) and make use of the fact that the curl of a gradient of a scalar is zero (so that the pressure and body force terms go away), we find

$$\frac{\partial\omega}{\partial t} - \nabla \times (\mathbf{V} \times \omega) = \nu\nabla^2\omega \qquad (3.56)$$

where ν is the kinematic viscosity, μ/ρ. Expanding and using the continuity equation $\nabla \cdot \mathbf{V} = 0$ and the fact that the $\nabla \cdot \omega = \nabla \cdot (\nabla \times \mathbf{V}) = 0$ (since the divergence of the curl of any vector must be zero), we have

$$\frac{\partial\omega}{\partial t} - (\omega \cdot \mathbf{V})\mathbf{V} + (\mathbf{V} \cdot \nabla)\omega = \nu\nabla^2\omega \qquad (3.57)$$

and by definition of the D/Dt operation, written as $\partial/\partial t + (\mathbf{V} \cdot \nabla)$, we obtain

$$\frac{D\omega}{Dt} = (\omega \cdot \nabla)\mathbf{V} + \nu\nabla^2\omega \qquad (3.58)$$

Along with the continuity equation this equation may be used to describe the velocity field of an incompressible viscous flow (the pressure may be determined from the original equation of motion).

Physically this equation tells us how vorticity is transported in a moving fluid. The first term on the right hand side is a transport term and the second term is a diffusion term which depends on ν, the kinematic viscosity, which is also known as the viscous diffusivity. If ν is assumed zero (which it is in an inviscid flow where the viscous effects are negligible), this equation tells us that vorticity does not diffuse through the fluid and indeed the vorticity field is "frozen" into the fluid. Exactly as a magnetic or electric field can be mapped by field lines, so can a vector vorticity field. The field lines represent direction of the vorticity vector at each point, and the spacing of the field lines represents the strength of the vorticity. If the vorticity is "frozen in," the lines are affixed to the same fluid molecules as the fluid moves. This concept may be expressed by stating that fluid lines coincide with vorticity lines. A "fluid line" is a line made of a chain of molecules of fluid and maintains its identity even though it moves and twists about. Many theorems may be derived from this physical concept, and the reader is referred to Chapter 6 for more details.

In two-dimensional flow $(\omega \cdot \nabla)\mathbf{V} = 0$ identically (since ω and $\mathbf{V}$ are perpendicular here) and the equation reduces to

$$\frac{D\omega}{Dt} = \nu\nabla^2\omega \qquad (3.59)$$

In two-dimensional flow of an incompressible viscous fluid, a stream function may be defined. The physical meaning of the stream function is discussed in Chapter 6, but its use in solving the viscous equations of motion will be introduced here. The stream function ψ is a scalar function defined such that

$$u = -\frac{\partial\psi}{\partial y}; \qquad v = \frac{\partial\psi}{\partial x}$$

(ψ is often used as a symbol for both stream function and gravitational potential.)

This definition inherently satisfies the continuity equation $\partial u/\partial x + \partial v/\partial y = 0$, and the specification of the stream function is one way of describing two-dimensional incompressible flow. By substituting the above definition into the vorticity equation for two-dimensional flow we obtain

$$\frac{\partial}{\partial t}(\nabla^2 \psi) - \frac{\partial \psi}{\partial y}\frac{\partial}{\partial x}(\nabla^2 \psi) + \frac{\partial \psi}{\partial x}\frac{\partial}{\partial y}(\nabla^2 \psi) + \nu \nabla^4 \psi = 0 \qquad (3.60)$$

which has the advantage of being a single scalar equation that describes the flow (rather than a vector equation), but the boundary conditions are often difficult to formulate.

We will not pursue these equations here but will remark that the vorticity and stream-function formulations are often convenient for numerical analysis of fluid flow.

3.6 THE ENERGY EQUATION

The energy equation will be derived by considering the balance on a control volume, and then formally obtaining the differential form in a purely mathematical manner. We begin in a manner similar to the integral derivations earlier in this chapter, but we may now consider a control volume immediately without looking at a moving system. Consider the control shown in Fig. 3-16.

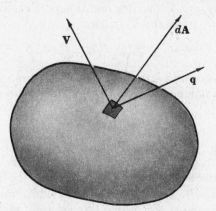

Fig. 3-16.   The control volume for the energy balance.

We assume the velocity vector $\mathbf{V}$ and rate of heat flux vector $\mathbf{q}$ as shown at a point on the surface. $\mathbf{q}$ is a vector and represents the rate of heat flux by conduction and radiation. The total rate of heat flow into the control volume dQ/dt would be $-\int_{\text{C.S.}} \mathbf{q} \cdot d\mathbf{A} = -\int_{\text{C.S.}} q_i \, d\mathbf{A}_i$.

Work done by the fluid in the volume may be again divided into two parts, reversible work done by the pressure at the control surface and irreversible work done by the shear stresses on the surface. In terms of the stress tensor, the rate of total work done by the fluid in the control volume is (in cartesian tensor form)

$$\frac{dW}{dt} = -\int_{\text{C.S.}} u_i \sigma_{ji} \, dA_j$$

Now we can say that the net rate of increase of total energy (including kinetic, internal and potential) in the control volume is equal to the rate at which heat flows into the control volume, plus the rate of internal heat generation, minus the rate at which the fluid does work on its surroundings. Hence we have

$$\frac{\partial}{\partial t}\int_{\text{C.V.}} \rho e \, d\mathcal{V} + \int_{\text{C.S.}} \rho e u_i \, dA_i = -\int_{\text{C.S.}} q_i \, dA_i + \int_{\text{C.S.}} u_i \sigma_{ij} \, dA_j + \int_{\text{C.V.}} q''' \, d\mathcal{V}$$

Remember that e, the total energy per unit mass, is given by $e = V^2/2 + u + \psi$ where ψ is the gravitational potential denoted as gz in the integral development earlier. q''' is the internal heat generation rate per unit volume. The stress tensor may be divided into a pressure part and a shear part as $\sigma_{ij} = -p\delta_{ij} + \sigma'_{ij}$. By using Gauss' theorem the energy equation then becomes

$$\int_{\text{c.v.}} \left[\frac{\partial}{\partial t}(\rho e) + \frac{\partial}{\partial x_i}(\rho e u_i) + \frac{\partial}{\partial x_i} q_i + q''' + \frac{\partial}{\partial x_i}(p u_i) - \frac{\partial}{\partial x_i}(u_j \sigma'_{ij}) \right] d\mathcal{U} = 0$$

Since the volume is arbitrary, the integrand itself must be zero. By expanding the terms and combining with the continuity equation, we get

$$\rho \frac{De}{Dt} = -\frac{\partial}{\partial x_i} q_i - p \frac{\partial u_i}{\partial x_i} - u_i \frac{\partial p}{\partial x_i} + u_i \frac{\partial \sigma'_{ji}}{\partial x_j} + \Phi + q''' \tag{3.61}$$

where $\Phi = \sigma'_{ij} \partial u_j / \partial x_i$ is the dissipation function[1] and is the rate at which the deviatoric stresses do irreversible work on the fluid. It is given in terms of u_i in the Appendix. The material derivative here operates on a scalar and we need not be concerned about its vector properties as we were in the equation of motion. The forms of D/Dt in various coordinate systems are listed in the Appendix. Physically the derivative is the rate of change of a property or parameter that one would see if one were riding along with the fluid. If instead of considering a control volume we were to consider a particular fixed particle of fluid, we would use $\partial/\partial t$ instead. The form of D/Dt comes about because we are really writing a balance, instantaneously on a system of fluid (in the control volume) but in Eulerian coordinates.

This equation is useful, but it has hidden in it the momentum equation which may be subtracted out to simplify it. We notice that on the left side are the rates of increase of kinetic and potential energy, and on the right side the third and fourth terms are the rate at which the stresses (including pressure) do work. By taking the vector dot product of $\mathbf{V}$ with the equation of motion (3.33), assuming the body force B_i is the conservative gravitational force $-\rho \, \partial\psi/\partial x_i$, we obtain

$$\rho u_i \frac{Du_i}{Dt} = \rho \frac{D}{Dt}\left(\frac{V^2}{2}\right) = -u_i \frac{\partial p}{\partial x_i} + u_i \frac{\partial \sigma'_{ij}}{\partial x_i} - \rho u_i \frac{\partial \psi}{\partial x_i} \tag{3.62}$$

which is the mechanical energy equation. Since ψ is constant in time we may add $\rho \, \partial\psi/\partial t$ to the left side (since it is zero) and obtain

$$\rho \frac{D}{Dt}(V^2/2 + \psi) = -u_i \frac{\partial p}{\partial x_i} + u_i \frac{\partial \sigma'_{ij}}{\partial x_j}$$

which may be subtracted out of the energy equation (3.61) to yield

$$\rho \frac{Du}{Dt} = -p \frac{\partial u_i}{\partial x_i} - \frac{\partial q_i}{\partial x_i} + \Phi + q''' \tag{3.63}$$

which is the final form of the energy equation. u here is the internal energy per unit mass.

Various simplifications of (3.63) may be effected. Let us introduce Fourier's law of heat conduction, $\mathbf{q} = -\kappa\nabla T$, and assume constant thermal conductivity and a perfect gas ($du = c_v \, dT$). Also, $\partial u_i / \partial x_i = \nabla \cdot \mathbf{V}$, and $\nabla \cdot \nabla T = \nabla^2 T$, so that we can write finally

$$\rho c_v \frac{DT}{Dt} = -p\nabla \cdot \mathbf{V} + \kappa\nabla^2 T - \nabla \cdot \mathbf{q}_r + q''' + \Phi \tag{3.64}$$

where $\mathbf{q}_r$ is the radiation heat flux vector. It should be noted that c_v may be a function of temperature and even then occurs outside (*not inside*) the derivative.

Equation (3.64) may be put into other forms. If we introduce the specific enthalpy $h = u + p/\rho$, the equation may be written

$$\rho \frac{Dh}{Dt} = \frac{Dp}{Dt} + \kappa\nabla^2 T - \nabla \cdot \mathbf{q}_r + q''' + \Phi \tag{3.65}$$

---

[1] Sometimes the dissipation term is written as $\mu\Phi$, and Φ then is different from ours by a factor μ.

and for a perfect gas it becomes

$$\rho c_p \frac{DT}{Dt} = \frac{Dp}{Dt} + \kappa \nabla^2 T - \nabla \cdot \mathbf{q}_r + q''' + \Phi \qquad (3.66)$$

It should be noted here that c_v and c_p are outside the derivative (even if c_v and c_p are functions of T). This follows from the definitions of c_v and c_p.

An incompressible fluid, we must remember, is not a perfect gas and both h and u depend on T and p. It is best then to start with (3.63) and set $\nabla \cdot \mathbf{V} = 0$. c_v and c_p are nearly equal for most liquids (liquid metals being an exception) and to a degree of accuracy, which is often adequate, u is only a weak function of p in liquids so that $du \approx c_v \, dT \approx c_p \, dT$. Then we obtain for liquids the approximate equation

$$\rho c_v \frac{DT}{Dt} = \kappa \nabla^2 T - \nabla \cdot \mathbf{q}_r + q''' + \Phi \qquad (3.67)$$

If higher accuracy is desired, thermodynamics property tables must be used in order to evaluate u or h in equation (3.63) or (3.65). Either of these equations is valid, the choice of u or h depending on the situation.

Sometimes it is useful to use the total equation (3.61), particularly for frictionless fluids. For frictionless fluids where $\tau'_{ij} = 0$ (3.61) becomes (setting $\psi = gz$, using Fourier's law, and introducing the enthalpy):

$$\rho \frac{D}{Dt} (V^2/2 + gz + h) = \frac{\partial p}{\partial t} + \kappa \nabla^2 T - \nabla \cdot \mathbf{q}_r + q'''$$

$$\rho \frac{D}{Dt} (V^2/2 + gz + u) = -p\nabla \cdot \mathbf{V} - \mathbf{V} \cdot \nabla p + \kappa \nabla^2 T - \nabla \cdot \mathbf{q}_r + q''' \qquad (3.68)$$

Usually $\mathbf{q}_r$ is negligible and q''' is zero except for cases involving such things as resistive heating. It is seldom in fluid dynamics that $\mathbf{q}_r$ and q''' must be considered, but we have included them for completeness.

It is interesting to refer back now to the comparison of equations (3.22b), (3.22c) and (3.40). Equation (3.22b) is a true energy equation. But now we know that since the total energy equation [(3.61) or (3.68)] has momentum hidden in it, equation (3.22c) is actually the momentum equation and is identical in meaning to equation (3.40), the Bernoulli equation.

The Second Law of Thermodynamics and Entropy Production

We can make equation (3.23) into an equality by introducing the concept of entropy production rate per unit volume, θ. We find, introducing the heat flux vector $\mathbf{q}$,

$$\int_{c.v.} \theta \, d\mathcal{U} - \int_{c.s.} \frac{\mathbf{q}}{T} \cdot d\mathbf{A} = \frac{\partial}{\partial t} \int_{c.v.} \rho s \, d\mathcal{U} + \int_{c.s.} \rho s \mathbf{V} \cdot d\mathbf{A} \qquad (3.69)$$

Using Gauss' law and the continuity equation, we obtain

$$\theta - \nabla \cdot (\mathbf{q}/T) = \rho \frac{Ds}{Dt} \qquad (3.70)$$

By introducing the energy equation and using the basic thermodynamic relationship

$$\rho \frac{Du}{Dt} = \rho T \frac{Ds}{Dt} - p\nabla \cdot \mathbf{V} \qquad (3.71)$$

which is the Eulerian coordinate form of

$$du = T \, ds - p \, d(1/\rho)$$

Equation (3.70) can be put into the form (assuming Fourier's law holds, and neglecting radiation)

$$\theta = \frac{\Phi}{T} + \frac{\mathbf{q} \cdot \mathbf{q}}{\kappa T^2} = \frac{\Phi}{T} + \kappa \frac{(\nabla T)^2}{T^2} \tag{3.72}$$

which shows the entropy production rate in terms of the coupled irreversibilities of heat currents and viscous dissipation.

3.7 RELATIONSHIPS AMONG MOMENTUM, ENERGY, THERMODYNAMICS, AND THE BERNOULLI EQUATION

We mentioned in Section 3.2 when we derived the general energy equation that it contained two independent energy balances: a thermodynamic first law balance among work, heat, and internal energy, and a separate purely mechanical energy balance.

We can now derive this mechanical energy relationship directly from the equation of motion. This balance states that the rate of increase of kinetic and potential energy in the fluid is equal to the rate at which external forces do work on the fluid. No thermodynamic information is involved. This mechanical energy balance in general form is known as the generalized Bernoulli equation. It can be obtained by integrating the equation of motion for general unsteady motion (Navier-Stokes equation) along a streamline. For simplicity, let us assume that the viscosity is constant. Further, let the body force due to gravity be expressed as $-\rho \nabla \psi$ per unit volume. In vector form:

$$\rho \left[\frac{\partial \mathbf{V}}{\partial t} + \nabla \left(\frac{V^2}{2} \right) - \mathbf{V} \times (\nabla \times \mathbf{V}) \right] = -\nabla p + \mu \nabla^2 \mathbf{V} - \rho \nabla \psi$$

Consider a streamline and an elemental length ds along it. Division of the equation by ρ puts it on a unit mass basis. By forming the vector dot product of ds with the equation of motion and integrating between two points 1 and 2 in the direction of flow along a streamline we find

$$\int_1^2 \frac{\partial \mathbf{V}}{\partial t} \cdot d\mathbf{s} + \int_1^2 \nabla \left(\frac{V^2}{2} \right) \cdot d\mathbf{s} - \int_1^2 [\mathbf{V} \times (\nabla \times \mathbf{V}) \cdot d\mathbf{s}] = -\int_1^2 \left(\frac{\nabla p}{\rho} \right) \cdot d\mathbf{s} - \int_1^2 \nabla \psi \cdot d\mathbf{s} + \int_1^2 \frac{\mu}{\rho} \nabla^2 \mathbf{V} \cdot d\mathbf{s}$$

$\mathbf{V} \times (\nabla \times \mathbf{V})$ is perpendicular to the streamline so that its dot product with $d\mathbf{s}$ is zero. Although it is beyond the scope of our presentation to prove it, the viscous term represents the work done on the unit mass of flowing fluid by friction forces that retard the motion as the fluid flows from points 1 to 2. Hence the equation then can be written on a unit mass basis as

$$\int_1^2 \frac{\partial \mathbf{V}}{\partial t} \cdot d\mathbf{s} + \frac{V_2^2 - V_1^2}{2} + \int_1^2 \frac{dp}{\rho} + (\psi_2 - \psi_1) + w_f = 0 \tag{3.73}$$

where w_f is work done by retarding friction forces as the fluid travels from point 1 to 2.

The equation may be extended to a more general form if we consider flow in a pipe or conduit which has machinery within the pipe. This machinery may be connected by a shaft (through the pipe) to an external power source or sink as, for example, a turbine, pump, or paddle wheel. Or, we could imagine a portion of the pipe wall that can move, as a continuous belt moving over a hole in the pipe either driving the fluid or being driven by the fluid. This work transmitted to or from the fluid across the pipe wall is denoted as w_s per unit mass of flowing fluid and is known as "shaft" work. By convention, w_s is positive if work is done by the flowing fluid and is negative if work is done on the fluid. The shaft work w_s should not be confused with w_f. w_s represents work crossing the boundary of the control volume (pipe) while w_f is work done by frictional forces inside the pipe.

Equation (3.73) may be written for each streamline through a pipe, conduit, or machine housing and averaged by mass flow rate weighting, or if the properties are essentially uniform over the cross section (as is the case in turbulent flow), equation (3.73) is valid for the bulk pipe flow, as shown in Fig. 3-17. By integrating the equation of motion through the machinery in the pipe the forces (shear and pressure) exerted by the machine elements appear as local nonconservative body forces in the fluid and

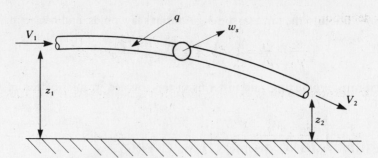

Fig. 3-17. Flow through a pipe. w_s and q are the shaft work out and heat into a unit mass of flowing fluid, respectively. z is the elevation above an arbitrary datum line.

may be added to the equation of motion. The explicit integration is again not easy since streamlines may lose their integrity through the machinery. Taking account of work done by these forces, and also shaft work which is transmitted through the pipe wall, there results the following equation known as the extended Bernoulli equation. We restrict the equation to steady flow and to flow where the velocity and properties are uniform over the pipe cross section. Further we express the gravity potential ψ as $-gz$, where z is the elevation above an arbitrary datum.

$$\frac{V_2^2 - V_1^2}{z} + \int_1^2 \frac{dp}{\rho} + g(z_2 - z_1) + w_s + w_f = 0 \tag{3.74}$$

The power associated with w_s is simply $\dot{m} w_s$ where $\dot{m}$ is the mass flow rate. w_f is always a positive number, but w_s may be positive or negative—positive for a turbine and negative for a pump.

It is customary to express w_f as a "head loss" in terms of "equivalent head" of the flowing fluid: $w_f = gH_L$, where H_L is the head loss (in dimensions of length). As a specific example, consider a pipe of uniform cross sectional area and with a flowing fluid of constant density (continuity says that $V_1 = V_2$) and constant elevation $z_2 = z_1$, and $w_s = 0$. Then H_L represents the pressure loss due to the friction effects. A differential manometer filled with the flowing fluid would show a height at point 2 lower than at point 1 by amount H_L.

Equation (3.74) can then be written

$$+ w_s = -\int_1^2 \frac{dp}{\rho} + \frac{V_1^2 - V_2^2}{z} + g(z_1 - z_2) - gH_L \tag{3.75}$$

For an incompressible fluid we have

$$+ w_s = \frac{p_1 - p_2}{\rho} + \frac{V_1^2 - V_2^2}{2} + g(z_1 - z_2) - gH_L \tag{3.76}$$

For turbine action $(p_1/\rho + V_1^2/2 + gz_1) > (p_2/\rho + V_2^2/2 + gz_2)$ and w_s is a positive number; hence we see that the head loss reduces the power output for given conditions at the inlet and outlet conditions. For pump action $(p_1/\rho + V_1^2/2 + gz_1) < (p_2/\rho + V_2^2/2 + gz_2)$ and w_s is negative; hence the presence of head loss requires the pump input power to be larger for given inlet and outlet conditions.

The total energy equation (3.22) appears to be very similar to equation (3.75) and was derived under the same assumptions. While either equation gives an expression for w_s, one or the other is more convenient to use under different circumstances. Use of the total energy equation requires knowledge about q and Δu while the extended Bernoulli equation requires specification of H_L. Further, the energy equation has a term $p_2/\rho_2 - p_1/\rho_1$ (for either incompressible or compressible flow) and the Bernoulli equation $\int_1^2 dp/\rho$. These are not generally the same. The relationship between the two equations may be seen as follows. By definition

$$\frac{p_2}{\rho_2} - \frac{p_1}{\rho_1} = \int_1^2 d(p/\rho) = \int_1^2 \frac{dp}{\rho} + \int_1^2 p\,d(1/\rho)$$

We may substitute this expression for $(p_2/\rho_2 - p_1/\rho_1)$ into the energy equation (3.22a) to obtain

$$-w_s = \int_1^2 \frac{dp}{\rho} + \frac{V_2^2 - V_1^2}{z} + g(z_2 - z_1) + \int_1^2 p\,d(1/\rho) + (u_2 - u_1) - q \qquad (3.77)$$

This equation is written on a basis of unit mass (i.e., a system of flowing fluid). Relative to an observer "moving with the fluid," the first law of thermodynamics for a "system" between states 1 and 2 may be stated:

$$q - w = u_2 - u_1 \qquad (3.78)$$

where w is the total work done by the system on its surroundings and includes the reversible work $\int_1^2 p\,d(1/\rho)$ and irreversible work which is the same as $(-w_f)$. w_f is exactly the same as we defined in the Bernoulli equation and must be a positive number by virtue of the second law of thermodynamics. That is, $+w_f$ is work done "on" the system by frictional forces which in essence play the same role as heat added to the system and is, thermodynamically, dissipation. Putting this together, equation (3.78) becomes

$$w_f = (u_2 - u_1) + \int_1^2 p\,d(1/\rho) - q$$

Now if this equation is subtracted from the above energy equation (3.77), we obtain the extended Bernoulli equation (3.74). Also since $w_f = gH_L$, we recognize that gH_L must be equal to $(u_2 - u_1)$ $+ \int_1^2 p(1/\rho) - q$. For an incompressible fluid (with constant ρ), this reduces to

$$gH_L = (u_2 - u_1) - q$$

We see that the extended Bernoulli equation and the general energy equation are equivalent. The energy equation contains thermodynamic information that is expressed in terms of equivalent irreversible work in the extended Bernoulli equation. In other words the energy equation contains the sum of the two independent energy balances, one mechanical and the other thermodynamical, but the Bernoulli equation is a purely mechanical energy balance. Either equation may be more convenient depending on the circumstances. The Bernoulli equation is useful for incompressible liquid flow, and the total energy equation is often more useful for flow of compressible fluids in heat engines where the enthalpy change is an important consideration.

Later in Chapters 4 and 5 we will see how the head loss may be calculated in practical situations.

3.8 SUMMARY, APPLICATIONS AND PROBLEMS

The derivations of the equations of this chapter have been long and tedious. These equations cannot be treated as "handy formulae". The shotgun approach of picking several equations, plugging in numbers and praying that one will yield the desired result may and probably will be disastrous. Understanding of the meaning of the equations and the limitations or restrictions of the special forms of the equations is of the utmost importance.

The procedure of this chapter was to start with the basic laws, write the corresponding system equations, derive the control volume integral equations, derive the differential equations, and then present some simplified or working forms of these equations. This is summarized in the following:

Basic Laws	→ System Equations →	Integral (Control Volume) Equations	or	Differential Equations
1. Continuity	1. $\dfrac{d}{dt}(m) = 0$	1. Eq. (3.1)		1. Eq. (3.25)
2. Momentum (linear)	2. $\dfrac{d}{dt}(m\mathbf{V}) = \mathbf{F}$	2. Eq. (3.9)		2. Eq. (3.32), (3.34), (3.54)
3. Momentum (angular)	3. $\dfrac{d}{dt}(m\mathbf{r} \times \mathbf{V}) = \mathbf{r} \times \mathbf{F}$	3. Eq. (3.11)		3. —
4. Energy	4. $\dfrac{d}{dt}(E) = \dfrac{dQ}{dt} - \dfrac{dW}{dt}$	4. Eq. (3.20)		4. Eq. (3.64), (3.65)
5. Second Law of Thermodynamics	5. $dS - \dfrac{dQ}{T} \geqq 0$	5. Eq. (3.23)		5. Eq. (3.69), (3.72)

Integral Equations

All four integral equations for the control volume have the form

$$\text{External effects} = \frac{\partial}{\partial t}\int_{\text{C.V.}} \xi\rho \; d\mathcal{V} + \int_{\text{C.S.}} \xi\rho\mathbf{V}\cdot d\mathbf{A} \tag{3.79}$$

where

Basic Law	External Effects	ξ
Continuity	0	1
Linear Momentum	$\mathbf{F}_S + \displaystyle\int_{\text{C.V.}} \mathbf{B} \; d\mathcal{V}$	$\mathbf{V}$
Angular Momentum	$\displaystyle\int_{\text{C.S.}} \mathbf{r} \times d\mathbf{F}_S + \int_{\text{C.V.}} \mathbf{r} \times \mathbf{B} \; d\mathcal{V}$	$\mathbf{r} \times \mathbf{V}$
Energy	$\dfrac{dQ}{dt} - \dfrac{dW}{dt}$	e

We can readily see the similarities in these equations by summarizing them in this manner. Also, we can look back and observe that the methods of deriving the control volume equations were essentially the same.

Special Forms of Integral Equations

1. Continuity—one-dimensional steady flow:

$$\rho_1 A_1 V_1 = \rho_2 A_2 V_2 \tag{3.4}$$

2. Momentum—one-dimensional steady flow with the density and velocity uniform over entrance and exit areas:

$$F_x = \dot{m}(V_{2x} - V_{1x}), \qquad F_y = \dot{m}(V_{2y} - V_{1y}), \qquad F_z = \dot{m}(V_{2z} - V_{1z})$$

3. Angular momentum—velocity uniform over A_1 and A_2, steady flow through a rotor:

$$T_z = \rho_1 Q_1 (r_2 V_{t2} - r_1 V_{t1}) \tag{3.14}$$

4. Energy—one-dimensional, steady flow:

$$q - w_S = (p_2/\rho_2 - p_1/\rho_1) + (u_2 - u_1) + (V_2^2 - V_1^2)/2 + g(z_2 - z_1) \tag{3.22a}$$

5. Extended Bernoulli equation:

$$w_s = -\int_1^2 \frac{dp}{\rho} + \frac{V_1^2 - V_2^2}{2} + g(z_1 - z_2) - gH_L \tag{3.75}$$

Differential Equations

1. Continuity:

$$\frac{\partial \rho}{\partial t} + \nabla \cdot (\rho \mathbf{V}) = 0 \tag{3.25}$$

which for incompressible flow is

$$\nabla \cdot \mathbf{V} = 0 \tag{3.27}$$

2. Momentum:

$$\rho \left(\frac{\partial u_i}{\partial t} + u_j \frac{\partial u_i}{\partial x_j} \right) = -\frac{\partial p}{\partial x_i} + B_i + \frac{\partial}{\partial x_j} \left[\mu \left(\frac{\partial u_i}{\partial x_j} + \frac{\partial u_j}{\partial x_i} - \frac{2}{3} \delta_{ij} \frac{\partial u_k}{\partial x_k} \right) \right] + \frac{\partial}{\partial x_i} \left(\zeta \frac{\partial u_k}{\partial x_k} \right) \tag{3.54}$$

For constant density and viscosity the momentum equation is

$$\rho \frac{D\mathbf{V}}{Dt} = -\nabla p + \mathbf{B} + \mu \nabla^2 \mathbf{V} \tag{3.55}$$

3. Energy equation:

$$\rho c_v \frac{DT}{Dt} = -p\nabla \cdot \mathbf{V} + \kappa \nabla^2 T - \nabla \cdot \mathbf{q}_r + \Phi + q''' \tag{3.64}$$

For more detailed equations in other coordinate systems see the Appendix and reference 6.

Approach to Solving Problems

In this chapter we have developed some of the general and important equations of fluid mechanics. After one has attained an understanding of these equations, the next step is to be able to apply them in solving physical problems. While there is no substitute for experience in working problems, a few general guidelines at this point should be helpful.

Let us consider the possible approaches in dealing with real physical problems. In general, there are three:

1. Integral equations—gross effects
2. Differential equations—distributions
3. Start with basic laws and develop equations appropriate to the particular situation.

The integral equations are usually appropriate when one desires to determine the gross effects, e.g., the total force of a flowing fluid on a pipe wall or a rotor vane. The differential equations are usually appropriate when the distributive conditions are desired, e.g., the velocity and pressure distribution (field) around an aerodynamic body. The possibility of developing the equations from basic laws for any particular problem should never be precluded (although the equations could always be obtained from the basic integral or differential equations under the appropriate assumptions). We will find this method particularly appropriate in the development of the boundary layer equations.

The equations of this chapter are the tools which we will use to solve real physical problems (or perhaps to better understand nature). The starting point is the physical problem. And the solution is never (or rarely) an exact description of the real physical situation, but rather an approximation whose accuracy depends on the accuracy of the mathematical model. For example, the procedure for solution may be outlined as follows:

1. Specific physical problem.

2. Physical model representation—we represent the real physical problem with a new but simplified problem. For example, we may assume one-dimensional flow, steady flow, fluid obeys ideal gas equation of state, etc. Thus the assumptions make up the model.

3. Mathematical model—we set down the equations corresponding to the physical model.

4. Solution—although there are simple exact solutions to many real practical problems in fluid mechanics, most problems in fluid mechanics are very difficult from a mathematical point of view. Solutions are usually effected by making simplifying assumptions which reduce the equations to a solvable set or by recourse to numerical analysis. The remainder of this book will be devoted to various classifications of fluid flow which are modeled by different sets of assumptions.

5. Experiment—the final step is to determine the correctness of the model by experiment. This step is very important in fluid mechanics. Experiment will lead to improvements in the model.

References

1. Batchelor, G. K., *An Introduction to Fluid Dynamics*, Cambridge, 1967.

2. Bird, R. B., Stewart, W. E., and Lightfoot, E. N., *Transport Phenomena*, John Wiley, 1960.

3. Brenkert, K. Jr., *Elementary Theoretical Fluid Mechanics*, John Wiley, 1960.

4. Chapman, S., and Cowling, T. G., *The Mathematical Theory of Non-Uniform Gases*, Cambridge University Press, 1961.

5. Curtiss, C. F., and Bird, R. B., *Molecular Theory of Gases and Liquids*, John Wiley, 1954.

6. Hughes, W. F., and Gaylord, E. W., *Basic Equations of Engineering Science*, Schaum's Outline Series, McGraw-Hill, 1964.

7. Keenan, J. H., *Thermodynamics*, John Wiley, 1941.

8. Landau, L. D., and Lifshitz, E. M., *Fluid Mechanics*, Addison-Wesley, 1959.

9. Pai, S. I., *Viscous Flow Theory*, Vols. I and II, Van Nostrand, 1956.

10. Reynolds, W. C., *Thermodynamics*, McGraw-Hill, 1965.

11. Sabersky, R. H., Acosta, A. J., and Hauptmann, E. G., *Fluid Flow*, 3rd ed., Macmillan Co., 1971.

12. Schlichting, H., *Boundary Layer Theory*, 7th ed., McGraw-Hill, 1986.

13. Shames, I. H., *Mechanics of Fluids*, McGraw-Hill, 1962.

14. Sommerfeld, A., *Dynamics of Deformable Media*, Academic Press, 1950.

15. Van Wylen, G. J., and Sonntag, R. E., *Fundamentals of Classical Thermodynamics*, 3rd ed., John Wiley, 1985.

16. White, F. M., *Viscous Fluid Flow*, McGraw-Hill, 1974.

Solved Problems

3.1. Consider the steady flow of a compressible fluid through a pipe bend as shown in Fig. 3-18. Determine the force of the fluid on the pipe between sections 1 and 2.

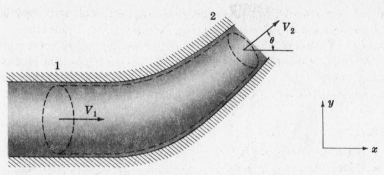

Fig. 3-18

First, choose a control volume for the fluid as shown by the dotted line. Equation (*3.9*) for steady flow is

$$\mathbf{F}_s + \int_{\text{C.V.}} \boldsymbol{B} \, d\mathcal{V} = \int_{\text{C.S.}} \mathbf{V}\rho\mathbf{V} \cdot d\mathbf{A}$$

It will be assumed that the pressures and velocities are uniform over the areas A_1 and A_2. The surface forces for the x and y directions respectively are

$$F_{sx} = p_1 A_1 - p_2 A_2 \cos\theta + F_{px}$$

$$F_{sy} = -p_2 A_2 \sin\theta + F_{py}$$

where p is the pressure and F_{px} and F_{py} are the unknown forces of the pipe wall on the fluid. The only body force is due to gravity and is equal to the weight of the fluid between sections 1 and 2. For many problems the weight is negligible compared to other forces, particularly for gases; it will be neglected here. The momentum flux terms become

$$\int_{\text{C.S.}} V_x \rho\mathbf{V} \cdot d\mathbf{A} = \rho_2 A_2 V_2^2 \cos\theta - \rho_1 A_1 V_1^2$$

$$\int_{\text{C.S.}} V_y \rho\mathbf{V} \cdot d\mathbf{A} = \rho_2 A_2 V_2^2 \sin\theta$$

Then the equations become

$$F_{px} = p_2 A_2 \cos\theta - p_1 A_1 + \rho_2 A_2 V_2^2 \cos\theta - \rho_1 A_1 V_1^2$$

$$F_{py} = p_2 A_2 \sin\theta + \rho_2 A_2 V_2^2 \sin\theta$$

The components of the force of the fluid on the pipe, say, R_x and R_y, would be opposite F_{px} and F_{py} and therefore we have

$$R_x = p_1 A_1 - p_2 A_2 \cos\theta + \rho_1 A_1 V_1 (V_1 - V_2 \cos\theta)$$

$$R_y = -p_2 A_2 \sin\theta - \rho_2 A_2 V_2^2 \sin\theta$$

3.2. Consider the steady flow of a water jet impinging on a stationary vane as shown in Fig. 3-19. Determine the force required to hold the vane in place. Assuming the jet follows the surface of the vane in a smooth manner and has the same shape and area at the exit, then the momentum equation can be written for the control volume shown in Fig. 3-19.

The jet will be exposed to atmospheric pressure over the entire surface plus an additional force where the jet makes contact with the vane. This additional force is precisely the force required to turn the fluid jet. The atmospheric pressure on the jet side will be canceled by the atmospheric pressure on the opposite side.

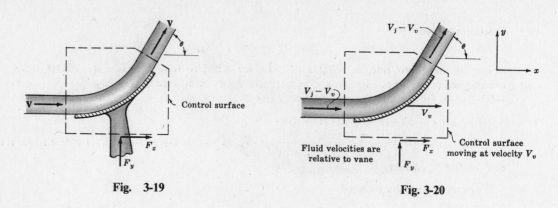

Fig. 3-19   Fig. 3-20

Thus we have

$$F_x = \rho A V^2 \cos \theta - \rho A V^2, \qquad F_y = \rho A V^2 \sin \theta$$

which is also equal to the force required to hold the vane in place.

3.3. In Problem 3.2 consider that the vane moves in the positive x direction at some velocity V_v which is less than the jet velocity V_j. Fig. 3-20 shows the moving vane. Find the force necessary to keep the vane moving at constant velocity, that is, to prevent it from accelerating.

We must write the momentum equation in a frame of reference at rest with respect to the vane. Thus all velocities must be the velocities relative to the vane. The velocity of the fluid relative to the vane entering the control volume is $V_j - V_v$, and if we neglect friction of the vane Bernoulli's equation tells us that the velocity leaving relative to the vane is the same, $V_j - V_v$ at the angle θ. Since the velocity is the same, the cross sectional area of flow A must be the same. The momentum equation gives

$$F_x = \rho A (V_j - V_v)^2 \cos \theta - \rho A (V_j - V_v)^2, \qquad F_y = \rho A (V_j - V_v)^2 \sin \theta$$

which gives the net force required to keep the vane from accelerating. Since the vane is in equilibrium, the force of the fluid on the vane must be $-F_x$ and $-F_y$ and hence the reaction force of the vane on the fluid is simply F_x and F_y.

Alternatively, we could have taken the control surface just on the inner surface of the vane so that it included only fluid. Then the analysis is exactly the same but we get immediately that the force on the fluid (exerted by the vane) is F_x and F_y.

3.4. A vane shown in Fig. 3-21 moves at a constant velocity $u = 30$ fps and receives a jet of water which leaves a nozzle at a velocity $V = 100$ fps. The nozzle has an exit area of 0.04 ft². Find the total force on the moving vane.

Take the control volume around the vane as shown in Fig. 3-22. The velocities relative to the control volume are $V - u = 100 - 30 = 70$ fps in the x direction entering and $V - u = 70$ fps leaving in the direction shown. Thus the force on the fluid is determined from the steady flow momentum equation. For the x

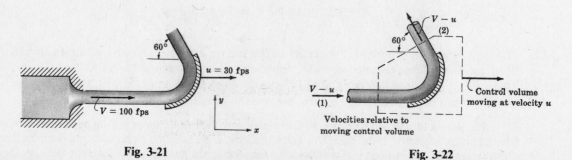

Fig. 3-21   Fig. 3-22

direction we have

$$F_x = \dot{m}(V_{2x} - V_{1x})$$

where the mass flow rate, $\dot{m}$, relative to the control surface is $\dot{m} = \rho A(V - u) = \rho A(100 - 30)$. The x components of velocities entering and leaving relative to the control volume are $V_{1x} = V - u$ and $V_{2x} = -(V - u) \cos 60°$. Substituting into the momentum equation,

$$F_x = \rho A(V - u)\{-(V - u) \cos 60° - (V - u)\}$$

$$= -\rho A(V - u)^2\{\cos 60° + 1\} = -(62.4/32.2)(0.04)(100 - 30)^2(\cos 60° + 1) = -569 \text{ lb}_f \text{ on fluid}$$

or $F_x = 569 \text{ lb}_f$ on vane.

For the y direction we have $$F_y = \dot{m}(V_{2y} - V_{1y})$$

The velocities entering and leaving in the y direction are $V_{1y} = 0$ and $V_{2y} = (V - u) \sin 60°$. Substituting into the momentum equation,

$$F_y = \rho A(V - u)\{(V - u) \sin 60° - 0\}$$

$$= \rho A(V - u)^2 \sin 60° = (62.4/32.2)(100 - 30)^2 \sin 60° = 329 \text{ lb}_f \text{ on fluid}$$

or $F_y = -329 \text{ lb}_f$ on vane.

The magnitude of the force on the vane is

$$|\mathbf{F}| = \{F_x^2 + F_y^2\}^{1/2} = [(569)^2 + (-329)^2]^{1/2} = 658 \text{ lb}_f$$

The direction of the force is $\tan \theta = -329/569 = -0.578$ or $\theta = -30°$, the negative sign indicating that the force acts downward. Finally, the forces on the vane are as shown in Fig. 3-23.

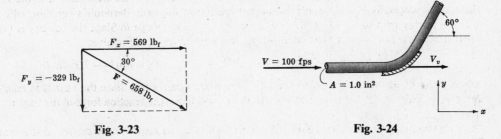

Fig. 3-23   **Fig. 3-24**

3.5.   A water jet having a velocity of 100 fps strikes a vane moving at a velocity of 40 fps as shown in Fig. 3-24. Determine (a) the power transmitted to the vane and (b) the absolute velocity of the jet leaving the vane.

(a)   The power transmitted to the vane is equal to the product of force in the direction of motion of the vane and the velocity of the vane. Thus,

$$\text{Power} = F_x V_v$$

The force of the fluid on the vane may be determined from the momentum equation for a control volume that moves at a velocity equal to that of the vane as shown in Fig. 3-25.

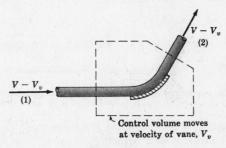

Fig. 3-25

The momentum equation for the control volume is

$$F_x = \dot{m}(V_{2x} - V_{1x}) \text{ on fluid}$$

F_x here is also the force exerted on the vane by the external mounting as was explained in Problem 3.3. The mass flow rate relative to the control surface is $\dot{m} = \rho A(V - V_v)$. The velocities for the x direction entering and leaving the control volume are $V_{1x} = V - V_v$ and $V_{2x} = (V - V_v) \cos 60°$. Substituting,

$$F_x = \rho A(V - V_v)\{(V - V_v) \cos 60° - (V - V_v)\} = \rho A(V - V_v)^2 \{\cos 60° - 1\}$$

$$= (62.4/32.2)(100/12 - 40/12)^2(\cos 60° - 1) = -6.48 \text{ lb}_f \text{ on fluid}$$

or $F_x = 6.48$ lb$_f$ on vane. Thus the power $= 6.48(40) = 259$ ft-lb$_f$/sec $= 0.472$ horsepower.

(b)   The velocity leaving the vane may be found by adding the velocity of the fluid relative to the vane to the velocity of the vane. Thus the absolute velocity of the fluid leaving the vane in the x direction is

$$V_{2x} = V_{x\,\text{relative}} + V_v = (100 - 40) \cos 60° + 40 = 70 \text{ fps}$$

The velocity of the fluid leaving the vane in the y direction is

$$V_{2y} = (100 - 40) \sin 60° = 51.9 \text{ fps}$$

The absolute velocity of the fluid leaving the vane is

$$|\mathbf{V}_2| = (V_{2x}^2 + V_{2y}^2)^{1/2}$$

$$= [(70)^2 + (51.9)^2]^{1/2} = 87.2 \text{ fps}$$

The direction of fluid leaving the vane is given by $\tan \theta = 51.9/70 = 0.741$ or $\theta = 36.6°$. The velocity vector diagram for the fluid leaving the vane is shown in Fig. 3-26.

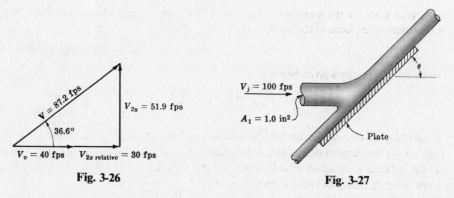

Fig. 3-26   Fig. 3-27

3.6.   Determine the force of the water jet on the inclined plate as a function of the angle θ as shown in Fig. 3-27.

Consider the control volume as shown in Fig. 3-28 where the fluid is entering at section 1 and leaving at sections 2 and 3. The mass flow rate of the fluid entering the control volume is equal to the mass flow rate of the fluid leaving the control volume, giving

$$\rho A_1 V_j = \rho A_2 V_2 + \rho A_3 V_3 \qquad \text{or} \qquad A_1 V_j = A_2 V_2 + A_3 V_3$$

Next we write Bernoulli's equation along a streamline between sections 1 and 2 and between sections 1 and 3, yielding

$$p_1/\rho + \tfrac{1}{2}V_j^2 = p_2/\rho + \tfrac{1}{2}V_2^2$$

and

$$p_1/\rho + \tfrac{1}{2}V_j^2 = p_3/\rho + \tfrac{1}{2}V_3^2$$

where all elevation changes have been neglected. If it is assumed that the pressures at sections 1, 2 and 3 are all equal to atmospheric pressure, then Bernoulli's equation gives $V_j = V_2 = V_3$. Using the result that the velocities are equal, we have from the continuity equation $A_1 = A_2 + A_3$.

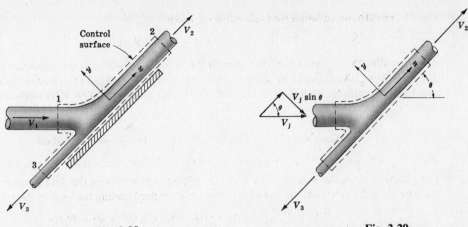

Fig. 3-28   Fig. 3-29

Now we write the steady flow momentum equation in the x direction along the plate neglecting any shear stress where the fluid comes in contact with the plate. Referring to Fig. 3-29 we have for F_x, the force of the plate on the fluid,

$$F_x = 0 = \rho A_2 V_2^2 - \rho A_3 V_3^2 - \rho A_1 V_j^2 \cos \theta$$

which simplifies to $A_1 \cos \theta = A_2 - A_3$. Now adding this equation to the expression obtained from continuity, we obtain

$$A_2 = \tfrac{1}{2}(1 + \cos \theta)A_1 \qquad \text{and} \qquad A_3 = \tfrac{1}{2}(1 - \cos \theta)A_1$$

Next, we write the momentum equation in the y direction normal to the plate. We have for F_y, the y component of the force of the plate on the fluid,

$$F_y = \rho A_1 V_j^2 \sin \theta$$

which for the numbers given becomes

$$F_y = (62.4/32.2)(1.0/144)(100)^2 \sin \theta = 135 \sin \theta \ \text{lb}_\text{f}$$

3.7. A converging two-dimensional channel shown in Fig. 3-30 has a linear area variation. The flow rate of an incompressible fluid is constant at $Q = 10 \ \text{ft}^3$ per sec per unit width of the channel. What is the acceleration as a function of the distance x? What is the acceleration at a point 1 ft from the beginning of the converging section?

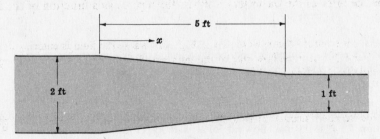

Fig. 3-30

We assume the flow to be quasi-one-dimensional and neglect variations across the channel. Then, from continuity, the flow rate Q is given by

$$Q = uA = u(2 - x/5)$$

where u, the velocity in the channel, is a function of x given by the above equation since Q is a constant. Then the acceleration is given by

$$a_x = \frac{\partial u}{\partial t} + u\,\frac{\partial u}{\partial x}$$

but since the flow is steady, $\partial u/\partial t = 0$, and by differentiating the expression for u we can find $\partial u/\partial x$ which is simply du/dx since u depends only on x. We obtain

$$a_x = u\,\frac{du}{dx} = \left(\frac{5Q}{10-x}\right)\!\left[\frac{5Q}{(10-x)^2}\right] = \frac{(5\times 10)^2}{(10-x)^3} = \frac{2500}{(10-x)^3}\ \text{ft/sec}^2$$

where x is measured in feet. At $x = 1$ ft, $a_x = 3.43$ ft/sec^2.

3.8. In the preceding problem, suppose the flow is unsteady and increasing at a rate of 2 ft^3 per sec^2 (per unit width of the channel). What is the acceleration now at the instant when $Q = 10$ ft^3 per sec per ft of channel width?

The acceleration is given by $a_x = \partial u/\partial t + u\,\partial u/\partial x$, and $u\,\partial u/\partial x$ is the same as in the preceding problem. But now $\partial u/\partial t$ is not zero. We find it as

$$\frac{\partial Q}{\partial t} = \frac{\partial u}{\partial t}\,(2 - x/5) = 2\ \text{ft}^3/\text{sec}^2/\text{ft}$$

Hence $\partial u/\partial t = 10/(10-x)$ ft/sec^2, which is 1.11 ft/sec^2 at $x = 1$ ft. Then the value of a_x is found by adding $\partial u/\partial t$ and $u\,\partial u/\partial x$ to give $1.11 + 3.43 = 4.54$ ft/sec^2 as the acceleration at the station $x = 1$ ft.

3.9. A jet engine shown in Fig. 3-31 is being tested statically on a test stand. The inlet velocity is 500 fps and the exhaust gases leave at a velocity of 3500 fps. The air at the inlet and the exhaust gases at the exit are at atmospheric pressure. The fuel-to-air ratio is 1/50 and the inlet and exhaust areas are both 2 ft^2. The density of air entering is 0.0024 slugs/ft^3. Determine the force T_x required to hold the jet engine stationary.

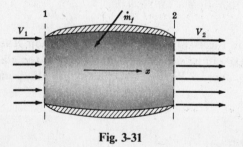

Fig. 3-31

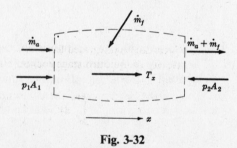

Fig. 3-32

Consider a control volume around the fluid inside the engine between sections 1 and 2 as indicated in Fig. 3-32. The steady flow momentum equation is

$$F_x = (\dot m_a + \dot m_f)V_{2x} - \dot m_a V_{1x}$$

where F_x is the total external force acting on the engine and includes the pressure forces and the mounting stand reaction T_x. $\dot m^a$ is the mass flow rate of air and $\dot m_f$ is the mass flow rate of fuel. By continuity, $\dot m_a$ must be the same at the inlet and outlet. We will assume that the fuel has negligible momentum in the x direction when it enters the engine. Then we can write

$$p_1 A_1 - p_2 A_2 + T_x = (\dot m_a + \dot m_f)V_{2x} - \dot m_a V_{1x}$$

But since the inlet and outlet areas are the same and we assume p_1 and p_2 are both atmospheric, we have

$$T_x = (\dot m_f + \dot m_a)V_{2x} - \dot m_a V_{1x} = (\dot m_f/\dot m_a + 1)\dot m_a V_{2x} - \dot m_a V_{1x}$$

Assuming the inlet air to be at standard conditions with $\rho_1 = 0.0024$ slugs/ft^3, the air mass flow rate is $\dot{m}_a = \rho_1 A_1 V_1 = (0.0024)(2)(500) = 2.40$ slugs/sec. Then we have for T_x,

$$T_x = (1/50 + 1)(2.40)(3500) - (2.40)(500) = 7300 \text{ lb}_f$$

The static thrust on the engine would be equal and opposite in direction to T_x.

3.10. Consider a three-stage rocket for space exploration. The first two stages will carry the third stage rocket to such a distance from the earth for launching that gravitational attraction will be negligible during its travel. Frictional resistance will be negligible also. This third stage rocket will be launched from the second stage rocket just as the second stage rocket attains its maximum height, at which time, of course, its velocity with respect to the earth will be zero. The engine of the final rocket is so designed that the velocity of the exhaust jet with respect to the rocket is maintained at a constant value of V_0 ft/sec. What is the final velocity of the third stage rocket in terms of relevant parameters?

In Fig. 3-33 consider a control volume in the shape of a cylinder which extends from the earth out far beyond the third stage rocket. The control volume will contain the third stage rocket, its fuel, and all the material exhausted from the third stage rocket. Then the momentum of the rocket plus fuel plus all exhausted material remains constant. The control volume has no external force acting on it. Hence, if M = mass of third stage rocket, m = instantaneous mass of fuel in the rocket, and the mass of fuel in the rocket is changing at a rate dm/dt which is a negative number since m is decreasing. If V is the velocity of the rocket, then the z component of momentum of the rocket plus fuel in the rocket at time t is $(M + m)V$ and the z component of momentum of the exhausted fuel is $\int_0^t -\dfrac{dm}{dt}(V - V_0)\, dt$, so that the momentum balance becomes

$$0 = \frac{d}{dt}\left[(M + m)V\right] - \frac{dm}{dt}(V - V_0) = (M + m)\frac{dV}{dt} + V_0\frac{dm}{dt}$$

where V is the absolute velocity of the rocket in the z direction. But $m = m_0 + (dm/dt)t$ where m_0 is the original mass of fuel at time $t = 0$ when the third stage is fired, and dm/dt is assumed a constant which we write simply as $\dot{m}$. Hence we have

$$0 = (M + m_0 + \dot{m}t)\frac{dV}{dt} + V_0\dot{m}$$

which can be integrated from $V = 0$ at $t = 0$ to $V = V_f$ when the fuel is exhausted and $m = 0$. V_f is the final velocity of the third stage rocket. The fuel is finally consumed at time $t = m_0/(-\dot{m})$. Remember $\dot{m}$ is negative.

$$\int_0^{V_f} dV = -\int_0^{-m_0/\dot{m}} \frac{V_0\dot{m}}{M + m_0 + \dot{m}t}\, dt \qquad \text{and} \qquad V_f = V_0 \ln\frac{M + m_0}{M}$$

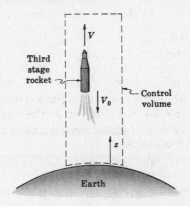

Fig. 3-33

3.11. A simplified two-dimensional boundary layer of water over one side of a flat plate is shown in Fig. 3-34. At the leading edge the velocity is uniform and equal to U. Downstream at the trailing edge the velocity profile is as shown. Find the shear force of the fluid on the plate by using an overall control volume approach.

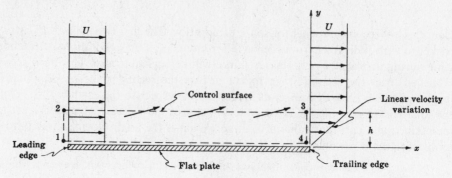

Fig. 3-34

We choose a rectangular control volume as shown. We write

$$\mathbf{F} = \int_{C.S.} (\rho \mathbf{V}) \mathbf{V} \cdot d\mathbf{A}$$

where $\mathbf{F}$ is the total force on the fluid. But we assume the pressure to be uniform throughout the flow and if we write the momentum equation in a direction parallel to the plate (x direction), F_x is the shear force exerted by the plate on the fluid. Across the surface 2-3 fluid flows with a small velocity normal to the plate and with a component parallel to the plate of nearly U. Hence F_x per unit width of the plate is

$$F_x = \underbrace{-\rho U^2 h}_{1\text{-}2} + \underbrace{\int_0^h (\rho U^2 y^2/h^2)\, dy}_{3\text{-}4} + \underbrace{\dot{m}U}_{2\text{-}3}$$

where $\dot{m}$ is the mass flow rate per unit width through surface 2-3 and is given by $\dot{m} = \rho U h - \rho U h/2 = \rho U h/2$. Hence

$$F_x = -\rho U^2 h + \rho U^2 h/3 + \rho U^2 h/2 = -\rho U^2 h/6$$

The fact that F_x is negative indicates that it is a retarding force on the fluid (acting in the negative x direction). The force of the fluid on the plate is equal and opposite to F_x.

3.12. A centrifugal pump shown in Fig. 3-35 pumps water at a rate of 1.0 ft³/sec. The water enters the impeller in an axial direction. The diameter of the impeller is 10 in. and the vanes are 1.0 in. high and radial at the outside diameter. Determine the power input to the rotor which turns at 1000 rpm.

The power input is the product of the torque T of the shaft on the rotor and the angular velocity ω of the rotor. Power $= T\omega$. The angular velocity is given and the torque can be determined from the angular momentum equation for steady flow. We have for the torque about the axis of rotation,

$$T_z = \int_{C.S.} (\mathbf{r} \times \mathbf{V})_z \rho \mathbf{V} \cdot d\mathbf{A}$$

where the control surface is taken around the impeller as shown in Fig. 3-36 so that the torque appearing in the equation is the torque exerted on the rotor by the drive shaft. Assuming the velocities are uniform over the inlet and outlet surfaces, we have

$$T_z = \rho_1 Q_1 (r_2 V_{t2} - r_1 V_{t1})$$

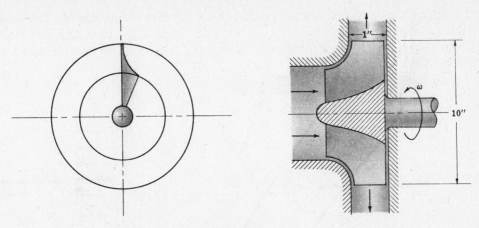

Fig. 3-35

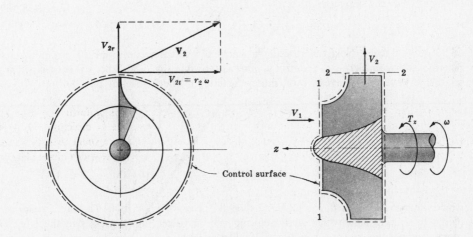

Fig. 3-36

where $\rho_1 Q_1 = \dot{m}$ is the mass rate of flow. Since the flow at the impeller inlet is in the axial direction, the tangential component of velocity there, V_{t1}, is zero. We need only, then, to determine the tangential component of velocity at the exit. Assuming there is no slippage at the impeller exit, that is, the fluid follows exactly the contour of the vanes, then the tangential velocity at the exit is $V_{t2} = r_2 \omega = (10/12)(1000 \times 2\pi/60) = 87.3$ ft/sec and the torque is $T_z = (62.4/32.2)(1.0)(10/12)(87.3) = 141$ lb$_f$-ft. The power required $= 141(1000 \times 2\pi/60) = 14{,}800$ ft-lb$_f$/sec or 27 horsepower.

3.13. Water flows horizontally out of a pipe through a 1/8 in. slit as shown in Fig. 3-37. The total flow rate is 1 ft^3/sec and the velocity varies linearly from a maximum at one end of the slit to zero at the other end. Determine the moment about the axis of the connecting vertical pipe resulting from the flow out of the slit.

First determine an expression for the velocity distribution. Let u_{max} be the maximum velocity at the inlet as shown in Fig. 3-37. Then we have

$$u = u_{max}(1 - x/6)$$

along the pipe. The value of u_{max} can be determined from the continuity equation. The total flow rate Q is given by

$$Q = \int_A u\, dA = \int_0^6 u_{max}(1 - x/6)(\tfrac{1}{8}/12)\, dx = 1$$

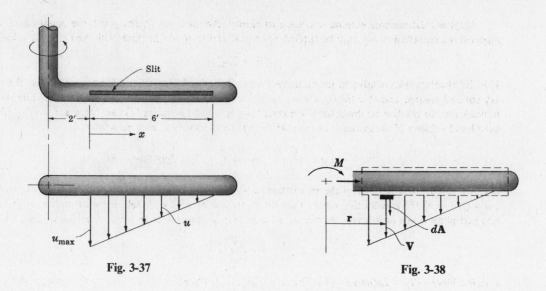

Fig. 3-37   Fig. 3-38

and $u_{max} = 32$ ft/sec. Next the reaction moment of the connecting pipe on the horizontal pipe can be determined by considering the control volume as shown in Fig. 3-38. Since $\mathbf{r}$ is perpendicular to $\mathbf{V}$ and $d\mathbf{A}$ is parallel to $\mathbf{V}$ we have, where M is the moment exerted by the connecting pipe,

$$M = \int_{c.s.} (\mathbf{r} \times \mathbf{V})\rho \mathbf{V} \cdot d\mathbf{A} = \int_{c.s.} (ru)\rho u \, dA$$

and since $r = 2 + x$ and $u = 32(1 - x/6)$, we have

$$M = (32)^2 \rho \int_0^6 (2 + x)(1 - x/6)^2 (\tfrac{1}{8}/12) \, dx$$

which, when evaluated with $\rho = 62.4/32.2$ slugs/ft^3, gives $M = 145$ lb$_f$-ft. This is the moment about the axis of the connecting pipe, which would be important in determining stresses in the pipe. The sign of the moment is positive as shown in Fig. 3-38. The moment exerted on the vertical connecting pipe by the horizontal pipe is equal and opposite in sign to M.

3.14.   The top view of a lawn sprinkler is shown in Fig. 3-39. It consists of two arms with right angle nozzles which rotate about a pivot in a horizontal plane. The flow rate is Q through each nozzle which has an exit area A. Find the absolute velocity of the water leaving the nozzle if (a) there is no friction to impede rotation and (b) there is a constant retarding frictional torque T_f acting on the sprinkler when it rotates.

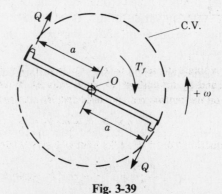

Fig. 3-39

A cylindrical control volume is chosen as shown. About point O, the pivot, the equation of angular momentum, equation (3.14), may be applied. (The flow enters under the pivot with zero angular velocity.)

$$T = 2\rho Q a V_t$$

V_t is the absolute (i.e., relative to the stationary control volume) tangential velocity as it crosses the stationary control volume and Q is the flow rate crossing the control volume surface. Q is the same relative to the moving arm or relative to the control volume. The factor of 2 appears because there are two nozzles. The peripheral velocity of the nozzle is $a\omega$ so that the absolute velocity V_t may be written

$$V_t = \left(a\omega - \frac{Q}{A}\right) \quad \text{so that} \quad T = 2\rho Q a(a\omega - V)$$

where V is Q/A, the velocity of the jet relative to the nozzle. For case (a), $T = 0$, so that $V_t = 0$, and hence $a\omega = Q/A = V_t$. Physically this means that the water leaves with zero velocity with respect to the ground and just plops down as it leaves the nozzle, wetting only a circle under the nozzles, and ω is given by

$$\omega = \frac{V}{a} = \frac{Q}{aA}$$

case (b). Here $-T_f = 2\rho Q a(a\omega - V)$. The minus sign appears since T_f is a retarding torque and hence ω is given by

$$\omega = \left(\frac{V}{a} - \frac{T_f}{2\rho Q a^2}\right)$$

The absolute velocity of the jet is $-(T_f/2\rho Q a)$. If T_f is increased to arrest the sprinker, so that $\omega = 0$, $(T_f = 2\rho Q^2 a/A)$, then the absolute velocity leaving the nozzle is a maximum and of course simply Q/A. We see that in order to sprinkle a circular area of ground, frictional torque must be applied to the rotor.

3.15. A hydraulic jump is a striking phenomenon which often occurs in nature. It is a sudden discontinuity in the depth of flowing liquid. In tidal rivers a jump may sometimes be observed standing or moving upstream. Jumps are easy to generate in the laboratory. A dinner plate held horizontally under a water faucet may be used to demonstrate a hydraulic jump. Allow the water to hit the center of the plate, then observe that it flows radially outward in a fast thin layer and suddenly increases in thickness before flowing over the edge of the plate.

Find the relationship between the upstream and downstream thicknesses in terms of relevant parameters as shown in Fig. 3-40.

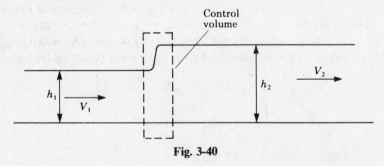

Fig. 3-40

Consider a control volume of depth w with the paper. The jump is stationary with respect to it. Assume the velocities V_1 and V_2 are uniform over the channel. By assuming the control volume to be very thin, the frictional forces on the bottom may be neglected. By continuity

$$w\rho h_1 V_1 = w\rho h_2 V_2 = Q$$

and by the momentum equation (considering the hydrostatic pressure force over each face):

$$\frac{\rho g h_1^2}{2} - \frac{\rho g h_2^2}{2} = \rho V_2^2 h_2 - \rho V_1^2 h_1$$

By combining these equations we obtain

$$(h_1 - h_2)\left(h_1 + h_2 - \frac{2h_1 V_1^2}{gh_2}\right) = 0$$

The solution $h_1 = h_2$ corresponds to uniform flow but by setting the second bracket to zero, we find h_2 in terms of h_1 and the flow rate Q as

$$\frac{h_2}{h_1} = -\frac{1}{2} + \sqrt{\frac{1}{4} + \frac{2Q^2}{w^2 gh_1^3}}$$

It can be shown from energy considerations and the second law of thermodynamics that $V_2 < V_1$, and $h_2 > h_1$ since energy must be lost by friction through the jump.

It is interesting to examine the result as h_1 approaches h_2 and the jump becomes a small surface wave. In that limit it can be seen from the above equations that $V_1 = V_2 = \sqrt{gh}$, where $h \approx h_1 \approx h_2$.

3.16. The water level in a reservoir is located a distance h above a turbine (Fig. 3.41). The water flows down through a pipe of diameter D and discharges through the turbine to a river. The discharge pipe is at the same elevation as the turbine. Find the maximum power that can be extracted from the turbine, neglecting all losses in the system.

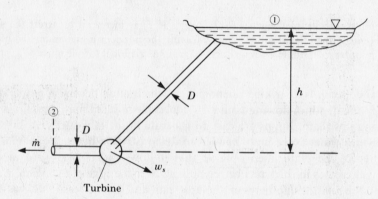

Fig. 3-41

We write the extended Bernoulli equation for an incompressible fluid between points 1 and 2, assuming the water level is constant and the velocity at point 1 is negligible, and neglecting losses.

$$w_s = \frac{p_1 - p_2}{\rho} + \frac{V_1^2 - V_2^2}{2} + g(z_1 - z_2)$$

Let $z_2 = 0$, $z_1 = h$, $V_1 = 0$, $p_1 = p_2 = 0$ (gage) so that

$$w_s = -\frac{V_2^2}{2} + gh$$

and the power output is $\dot{m}w_s = \rho V_2 A w_s = \rho V_2 \pi D^2 w_s/4 = T\omega$ where T is the torque output of the turbine and ω its angular speed. Then:

$$\text{Power output} = \dot{m}w_s = \frac{\rho V_2 \pi D^2 w_s}{4} = \frac{\rho V_2 \pi D^2}{4}\left(gh - \frac{V_2^2}{2}\right)$$

There are two values of V_2 where the power output is zero: $V_2 = 0$ and $V_2 = \sqrt{2gh}$. At $V_2 = 0$, there is a torque on the turbine but no flow, which corresponds to the turbine shaft being held stationary and not allowed to move ($\omega = 0$). At $V_2 = \sqrt{2gh}$, the turbine is allowed to rotate freely but produces no torque. We find the maximum value of power by finding the value of V_2 which maximizes the above expression, then substituting it back in to find the power.

$$\frac{d}{dV_2}(\text{power}) = \frac{\rho \pi D^2}{4}\left(gh - \frac{3V_2^2}{2}\right) = 0 \qquad V_2\Big|_{\text{max power}} = \sqrt{2gh/3}$$

and the maximum power is

$$\frac{\rho \pi D^2}{4}\left(\frac{2gh}{3}\right)^{3/2}$$

Supplementary Problems

3.17. One method of producing a vacuum is to discharge water through a venturi, and connect the system to be evacuated with the throat of the venturi as shown in Fig. 3-42. How much water must discharge through the venturi tube to produce a vacuum of 20 inches of mercury?

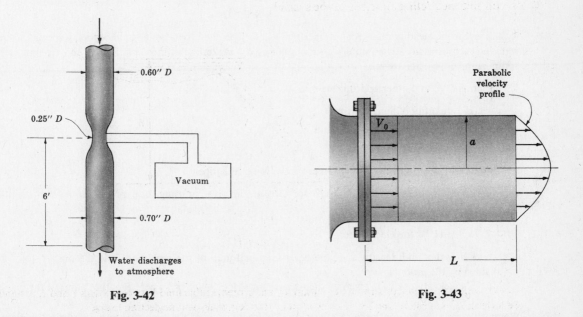

Fig. 3-42   Fig. 3-43

3.18. A circular pipe of radius a ft and length L ft is attached to a smoothly rounded outlet of a liquid reservoir by means of flanges and bolts as shown in Fig. 3-43. At the flange section the velocity is uniform over the cross section with magnitude V_0. At the outlet, which discharges into the atmosphere, the velocity profile is parabolic because of the friction in the pipe. What force must be supplied by the bolts to hold the pipe in place?

3.19. In the previous problem find the upstream pressure at the flange if the shear stress (due to viscous friction) on the wall is τ_0 lb/in^2.

3.20. Determine the force of the water jet on the inclined plate of Problem 3.6 if the plate is moving at a velocity V_p (which is less than the jet velocity) in the same direction as the water jet.

3.21. A ship is propelled at a constant speed of 25 mph by hydraulic jet propulsion. The total drag force is 3000 lb. The diameter of the jet exit section of the exhaust nozzle is 2 ft. Calculate the absolute velocity of the jet.

3.22. The water jet pump of Fig. 3-44 has a jet area $A_j = 0.05$ ft² and a jet velocity $V_j = 90$ ft/sec which entrains a secondary stream of water having a velocity $V_s = 10$ ft/sec in a constant area pipe of total area $A = 0.6$ ft². At section 2 the water is thoroughly mixed in such a way that the velocity is uniform. Neglect wall shear. Determine the pressure at section 2. The pressure at the nozzle is 10 psi.

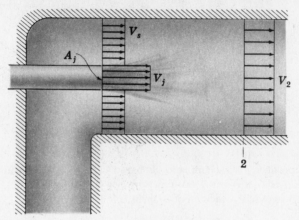

Fig. 3-44

3.23. A fluid is flowing steadily in a circular pipe. The velocity profile is parabolic in shape, being zero at the pipe wall and increasing to a maximum velocity V_c along the centerline of the pipe. Radial and circumferential components of velocity are zero.

1. In terms of V_c and other appropriate quantities, find expressions for

    (a)   quantity of fluid flowing

    (b)   axial momentum of the fluid

    (c)   kinetic energy of the fluid

    as the flow passes through any given cross-section of the pipe.

2. For a constant density fluid, find the error (percent) involved in using the expressions

    (a) $\rho A V^2$ for momentum

    (b) $\rho A V^3/2$ for kinetic energy

    where ρ is the fluid density, A is the cross-sectional area of the pipe, and V is the mass rate of flow divided by the product ρA.

    The velocity profile is given by the equation $u = V_c[1 - (r/R)^2]$ where R is the radius of the pipe.

3.24. Water enters a pipe from a large reservoir and on issuing out of the pipe strikes a 90° deflector plate as shown in Fig. 3-45. If a horizontal thrust of 200 lb is developed on the deflector, what is the horsepower developed by the turbine?

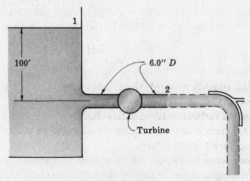

Fig. 3-45

3.25. Water flows steadily up the vertical pipe and enters the annular region between the circular plates as shown in Fig. 3-46. It then moves out radially, issuing out as a free sheet of water. (*a*) If we neglect friction entirely, what is the flow of water through the pipe if the pressure at *A* is 24.7 psi? (*b*) Determine the force in the pipe wall at section *A*. Neglect the weight of the pipe.

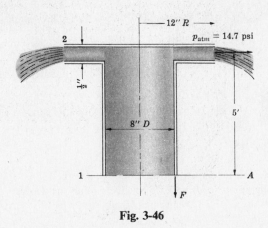

Fig. 3-46

3.26. A vehicle is moving at a velocity $V = 100$ ft/sec along level ground under the action of a constant force $F = 1000$ lb. At time $t = 0$, mass begins leaving the vehicle through a hole in the bottom. Assuming the mass leaves the vehicle vertically at a rate of 10 lb/sec and the vehicle continues to move under the action of the constant force F, determine the velocity of the vehicle after 20 sec. Initial mass of the vehicle is 2000 lb.

3.27. A container for water in the form of a wagon (Fig. 3-47) is being driven along a frictionless, level tract by a jet of water. The jet of water has a horizontal velocity of 30 ft/sec and the cross-sectional area of the jet is 2 in^2. The jet of water overtakes the wagon, hits the far end of it, and is deflected down into the body of the wagon. No water is spilled; all of it falls down into the wagon. At time zero, the wagon has a velocity of 10 ft/sec and the mass of the wagon and the water which is in it at that instant is 100 lb. Calculate the time required for the wagon to accelerate from a velocity of 10 ft/sec to a velocity of 20 ft/sec. Assume that the jet of water follows the wagon and falls into it during all this time.

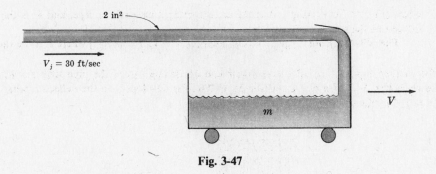

Fig. 3-47

3.28. When water is flowing steadily through the turbine sketched in Fig. 3-48, the turbine delivers 75 horsepower. The water flow rate is 1200 lb/sec.

The inlet pipe to the turbine is 12 in. inside diameter and the discharge pipe is 16 in. inside diameter. The rest of the piping system is not shown and the inlet and outlet pressures are not known. The change in elevation of the piping shown is not known.

A manometer connected across the inlet and outlet, as shown in the sketch, contains mercury. Calculate the difference in height, *h*, between the mercury columns under steady flow conditions and state whether the right leg or left leg of mercury is higher.

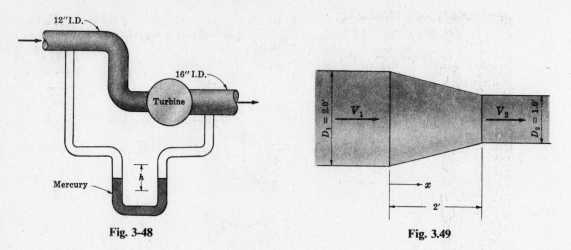

Fig. 3-48   Fig. 3.49

3.29. A pipe has a contraction as shown in Fig. 3-49. At a certain time the velocity of water entering, V_1, is 30 ft/sec and is changing at a rate of 2 ft/sec^2. Determine the pressure gradient (dp/dx) at $x = 1$ ft at the time when $V_1 = 30$ ft/sec, assuming the velocity is uniform over any cross-section.

3.30. Derive the momentum equation (in terms of stresses) in cylindrical coordinates by applying the integral form to the appropriate elemental volume.

3.31. Repeat Problem 3.30 for spherical coordinates.

3.32. Go through the details of the derivation of the forms of the energy equation.

3.33. In Problem 3.8 find the acceleration if the flow rate is decreasing instead of increasing (at the rate given).

3.34. A cylinder with a closed top has a tube, exposed to atmospheric pressure, extending some distance into the incompressible fluid within the tank of Fig. 3-50. The fluid discharges from a nozzle of area A_2 a distance h_1 from the bottom of the tube. If the cross-sectional area of the cylinder is A_1, find the flow rate Q as a function of time.

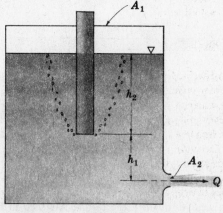

Fig. 3-50

3.35. Rework Problem 3.11 by choosing a control volume such that the upper surface lies on a streamline.

3.36. A cylinder is placed in a uniform flow of incompressible fluid of density ρ (Fig. 3-51). Velocity profiles are measured upstream and downstream as shown. Find the drag force on the cylinder per unit length. *Hint.* Take a large control volume about the cylinder.

Downstream a distance of
10 diameters, the velocity
data is:

y/D	u/U
0	0.5
0.5	0.6
1.0	0.8
1.5	0.9
2.0	1.0

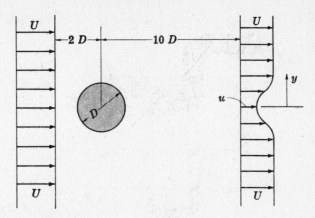

Fig. 3-51

3.37. A connecting rod bearing is to be lubricated by feeding oil of specific gravity 0.8 into the center of the crankshaft at atmospheric pressure. From there, the oil flows through a small drilled passage to the connecting rod bearing. At what minimum rpm of the crankshaft will there be positive oil pressure at the connecting rod bearing when the crank arm is in a vertical position as shown in Fig. 3-52?

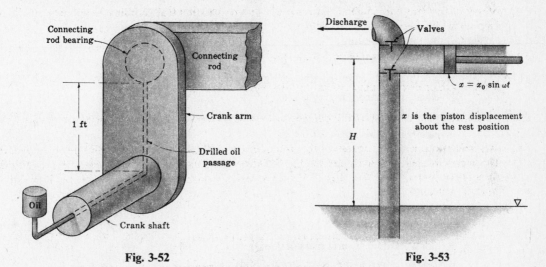

Fig. 3-52   Fig. 3-53

3.38. A water pump shown in Fig. 3-53 pumps water from a reservoir up to a height H. The motion of the pump piston is sinusoidal. At very high piston velocities it is found in practice that the water cavitates in front of the piston, allowing the water to drop back to the reservoir. This is undesirable and requires the pump to be primed in order to get it going again.

   Find the maximum frequency of the piston such that the cavitation does not occur when the piston is at bottom dead center. Cavitation is defined as the condition when the pressure in the fluid drops to zero absolute, or at least tends to this value.

   This maximum frequency you find then will be a limiting value, and if it is surpassed the pressure on the face of the piston tries to drop below zero absolute; the result is cavitation and separation of the fluid.

NOMENCLATURE FOR CHAPTER 3

A = area

a   = acceleration vector

B   = body force vector per unit volume

D/Dt = substantial or material derivative

e = total energy per unit mass ($e = u + V^2/2 + gz$)

e_{ij} = strain rate tensor

E = total energy

g = local acceleration of gravity

H_L = head loss

m = mass

$\dot{m}$ = mass flux

$\mathbf{M}$ = momentum vector

p = pressure

q = heat transfer per unit mass of flowing fluid

$\mathbf{q}$ = heat flux vector, (directed rate of heat transfer per unit area)

$\mathbf{q}_r$ = radiation heat flux vector

q''' = internal heat generation rate per unit volume

Q = total heat transferred to system; volumetric flow rate

r = radial coordinate

s = entropy per unit mass

S = entropy of system

t = time

T = absolute temperature

T_z = torque about z axis

u = velocity in x direction, internal energy per unit mass

u_i = velocity in x_i direction ($i = 1, 2, 3$)

U = internal energy for system, velocity of a free stream

v = velocity in y direction

$\mathbf{V}$ = velocity vector

V_t = tangential velocity

$\mathcal{V}$ = volume

w = velocity in z direction

W = work done by system

w_s = shaft work done by unit mass of flowing fluid

w = work done by a unit mass of flowing fluid

x_i = coordinates (x_1, x_2, x_3)

x, y, z = coordinates—correspond to (x_1, x_2, x_3)

α = angle between normal to surface and velocity vector

γ_{ij} = shear strain rate tensor

δ_{ij} = Kronecker delta, $\delta_{ij} = \begin{cases} 1; i = j \\ 0; i \neq j \end{cases}$

θ = entropy production rate per unit volume

κ = thermal conductivity

μ = viscosity

ν = kinematic viscosity

ρ = density

σ_{ij} = stress tensor, σ'_{ij} = deviatoric stress tensor

Φ = dissipation function

ψ = gravitational potential; stream function

ω = vorticity vector

ω_{ij} = vorticity tensor

Ω = angular velocity vector

Ω_{ij} = rotation tensor

Chapter 4

Dimensional Analysis and Similitude

4.1 SIMILITUDE IN FLUID DYNAMICS

In fluid mechanics a great many important results may be obtained from dimensional arguments. The relevant parameters in any physical situation may be combined into independent dimensionless groups which characterize the flow. These dimensionless parameters are often precisely defined and given names in fluid dynamics.

These dimensionless parameters, or Π's as they are called, may be obtained by the method of dimensional analysis or directly from the governing differential equations as a consequence of their being put into dimensionless form.

Consider any fluid flow problem; for example, the flow over a solid object. The flow pattern and properties are determined by the geometrical shape of the object and the relevant fluid properties. We say that two flows are similar if they are geometrically similar and if all the relevant dimensionless parameters are the same for both flows. Consider a model (as in a wind tunnel) and a prototype. How do we relate measurements made on the model to the prototype? The answer is: by the methods of similitude, that is, by making the geometry similar and the Π's the same for both.

The meaning of similar flow and model-prototype correlation can be understood by considering the dimensionless form of the governing equations. Clearly, if all the relevant differential equations are made dimensionless, the size of the object is irrelevant if the shape is geometrically similar. However, the dimensionless parameters (which appear as dimensionless coefficients in the dimensionless differential equations) must be made the same for both flows. These parameters depend on the fluid properties and a characteristic physical dimension (used as a reference base) of the object. Hence in similar flows the describing differential equations are identical for the model and prototype. Measurements might then be made of any dimensionless variable, say, dimensionless pressure, on the model. The appropriate dimensionless pressure for the model would have the same value as for the prototype and by converting back to dimensional form, data taken on the model can be directly related to the prototype.

In summary, two flows are similar if the dimensionless parameters and variables are all the same regardless of the size of the flow pattern, if geometrical similarity is maintained.

In practice, it is not always possible to make all the dimensionless parameters the same, at the same time, for model and prototype. In such cases some compromise must be made; however, in most important flows sufficient similarity may be achieved to make the method useful. Often, only certain of the dimensionless parameters are important for a given flow, and the fact that all the parameters cannot be made the same does not matter.

We will, in the next section, examine the basic differential equations of fluid mechanics and see what the dimensionless differential equations look like. But first, there is another method used to find the dimensionless parameters for any given problem: dimensional analysis.

Dimensional analysis, or the Buckingham Pi Theorem as it is called, is a method of finding the relevant dimensionless parameters without knowing the relevant differential equations. This method requires that one know all the relevant variables in any given problem. Then these variables are combined together in as many independent dimensionless groups or Π's as possible. A systematic method of obtaining these Π's is given in Problem 4.8. The advantage of this system is that one need not know the governing equations or laws for complex problems, but then one must know all the variables and only those relevant variables must be introduced into the problem. Any extraneous variables or the omission of one relevant one will destroy the analysis. These restrictions are rather severe and it is usually safer to begin with the governing equations. At any rate, once these parameters are found, experimental data can be obtained and correlated in dimensionless form.

The number of Π's for a given problem is fixed and is usually, but not always, equal to the total number of variables minus the number of fundamental dimensions. In mechanics there are three fundamental dimensions which are independent. The actual choice is arbitrary, but the three most commonly chosen ones are mass, length and time, and in compressible flow the temperature enters so that there are four fundamental dimensions.

A basic set of Π's must be chosen so that they are independent. Actually, many Π's may be formed by multiplying or dividing other Π's, but the number of independent Π's for a given problem is fixed. Hence there is not a unique set of Π's for a problem; but in fluid mechanics there are certain dimensionless parameters which are customarily studied and have physical significance.

After a set of Π's is found they can be expressed in functional form, that is, any one Π can be written as a function of all the other Π's. This functional form cannot be found by dimensional analysis, but only by either solving the governing equations or by experimentation. In experimental work it is very convenient to express the results in dimensionless form as plots of Π's. Hence correlations among Π's may be determined. Aside from removing the scale of the experiment as a parameter, the dimensionless form reduces the number of variables by at least three. Dimensional analysis is a vital tool in all engineering and scientific experimentation.

4.2 PARAMETERS OF INCOMPRESSIBLE FLOW

In incompressible flow only the equations of motion and continuity are necessary to describe the flow. There are four dependent variables, the three velocity components and the pressure, which appear in four equations. In addition, there are the fluid properties, viscosity and density, and the gravitational potential which enter into the Π formulation.

If we introduce the following dimensionless variables, independent and dependent, (denoted by asterisks), the equations may be written in dimensionless or normalized form. Let

$$p^* = p/\rho V_0^2, \qquad \mathbf{V}^* = \mathbf{V}/V_0, \qquad t^* = t/t_0 = tV_0/L, \qquad \mathbf{r}^* = \mathbf{r}/L, \qquad \psi^* = \psi/gL \qquad (4.1)$$

where the subscript zero indicates a characteristic value and L is a characteristic dimension. For example, if we were to consider the external flow over a cylinder, L could be the diamter and V_0 the free stream velocity. ψ is the gravitational potential. In terms of these variables the equation of motion in vector form is

$$\frac{\partial \mathbf{V}^*}{\partial t^*} + \nabla^*(V^{*2}/2) - \mathbf{V}^* \times (\nabla \times \mathbf{V}^*) = -\nabla^* p^* + \frac{1}{Re} \nabla^{*2}\mathbf{V}^* - \frac{1}{F_r} \nabla^*\psi^* \qquad (4.2)$$

and continuity is

$$\nabla^* \cdot \mathbf{V}^* = 0 \qquad (4.3)$$

Here Re is the Reynolds number $\rho LV_0/\mu$ and F_r is the Froude number V_0^2/gL, where ρ is the mass density and μ is the absolute or dynamic viscosity. Both numbers are dimensionless and represent the two parameters which must be set equal in order to achieve similarity. The total number of relevant Π's for a given geometry would be the two numbers Re and F_r, the dimensionless independent variables r^* and t^* (if the problem is time independent) and one dependent dimensionless variable such as p^* or $\mathbf{V}^*$. The total number of variables and parameters is ten (including the components of $\mathbf{r}$ and $\mathbf{V}$ and the gravitational potential and the viscosity) so that the total number of independent Π's would be seven. Such is the case. Re and F_r together with t^*, the three components of r^*, and one other dimensionless dependent variable such as p^* or one component of $\mathbf{V}^*$ make seven. Once all these seven Π's are specified the others follow from the equations since the other Π's are not independent.

Now equations (4.2) and (4.3) are entirely dimensionless so that flow over objects of different sizes which are geometrically similar should have the same solutions in terms of these dimensionless variables if the two characteristic parameters of the fluid, the Reynolds number Re and the Froude number F_r, are the same.

These numbers have physical significance. The Reynolds number is a measure of the ratio of the inertia to viscous force. When the Reynolds number is small, $\text{Re} \ll 1$, the viscous forces dominate and when $\text{Re} \gg 1$ the inertia forces dominate. The Froude number F_r is a measure of the ratio of inertia to the gravity force. By examining equation (4.2) we see that when both Re and F_r are $\gg 1$ the inertia must be balanced by the pressure forces.

In a large class of flows the Froude number is rather large and gravity is unimportant and only the Reynolds number becomes of interest in modeling. One exception is in marine modeling of ships where gravity waves are very important. Unfortunately, in modeling of incompressible flow it is usually possible to make only either the Reynolds or Froude number the same for model and prototype.

In modeling we can set

$$\text{Re}_m = \left(\frac{\rho L V_0}{\mu}\right)_m = \left(\frac{\rho L V_0}{\mu}\right)_p = \text{Re}_p \qquad (4.4)$$

where the m and p represent model and prototype respectively, or we can set

$$F_{r_m} = \left(\frac{V_0^2}{gL}\right)_m = \left(\frac{V_0^2}{gL}\right)_p = F_{r_p} \qquad (4.5)$$

but these two conditions cannot be satisfied simultaneously. In aerodynamics the Reynolds number is important and condition (4.4) can be satisfied without regard to (4.5) (since gravity is unimportant and F_r is very large). However, in marine ship modeling the situation is more complex and two experiments must be run, one with the Re numbers equal (to determine viscous drag) and one with the F_r numbers equal (to determine the gravity wave drag).

In terms of the Buckingham Pi Theorem then, we would have Re and F_r and we could write the dependent dimensionless variables as functions of Re and F_r and the dimensionless independent variables. For example,

$$p^* = f(\text{Re}, F_r, \mathbf{r}^*, t^*)$$

or any steady function of p^* such as lift could be computed as a function of only Re and F_r, the variable $\mathbf{r}^*$ being integrated out. As we will see, the dimensionless lift and drag on an object will be functions only of the Reynolds number (for a given geometry, of course).

There is one other important point to understand now in connection with showing the equivalence of the buckingham Pi Theorem method and the equation normalization method. In the latter we have a set of characteristic parameters, such as Re and F_r, which appear as coefficients in the governing equations. For a given geometry we say that any dimensionless dependent variable depends only on these characteristic parameters and the dimensionless independent variables, as we have pointed out above in the case of pressure. Now, on the other hand, if we had started out from scratch without any equations and asked: for a given flow what variables are important and relevant in determining some aspect of the flow behavior over the body?, we would have had to pick all the relevant variables and properties of the flow (such as V_0, L, μ, p, $\mathbf{r}$, t, etc.) and combine them into an appropriate set of dimensionless quantities. Even if we are able to pick all the relevant variables and properties, we are still faced with the task of combining them in the most convenient manner and then giving them physical interpretation. For example, suppose we ask again: what is the pressure distribution in steady flow over a body and what does it depend on? We would get for a given shape,

$$p = f(L, V_0, \mathbf{r}, \mu, \rho, g) \qquad (4.6)$$

There are nine variables and hence they must reduce to six dimensionless ones. These are p^*, Re, F_r, $\mathbf{r}^*$. $\mathbf{r}^*$ is a vector and has three scalar components, each of which is an independent variable. F_r may of course be unimportant in the problem of interest. p^* is the pressure above the pressure in the free stream of the flow. If we need to know the absolute pressure, we would have to introduce an additional parameter, the pressure p_0 in the free stream and we would have as a result an additional dimensionless parameter, $p_0/\rho V_0^2$. This is not usually necessary since we are mainly interested in the pressure excess in an aerodynamic problem.

4.3 PARAMETERS OF COMPRESSIBLE FLOW

In compressible flow additional parameters enter. In addition to motion and continuity we need the energy equation and an equation of state. We introduce the following dimensionless variables

$$\rho^* = \rho/\rho_0, \qquad T^* = T/T_0, \qquad M_0 = V_0/a_0, \qquad a_0 = \sqrt{kRT_0}, \qquad \Phi^* = L\Phi/V_0^2 v\rho_0 \qquad (4.7)$$

in addition to those of the previous section. Here a_0 is the ordinary sonic speed, Φ is the viscous dissipation function, k is the ratio of specific heats c_p/c_v, R is the gas constant, T is the absolute temperature, and v is the kinematic viscosity μ/ρ.

The dimensionless equations then become

Continuity:
$$\frac{\partial \rho^*}{\partial t^*} + \nabla^* \cdot (\rho^* \mathbf{V}^*) = 0 \qquad (4.8)$$

Motion:

$$\rho^*\left[\frac{\partial V^*}{\partial t^*} + (\mathbf{V}^* \cdot \nabla)\mathbf{V}^*\right]$$

$$= -\nabla^* p^* + \frac{1}{\text{Re}'}\nabla^{*2}\mathbf{V}^* + \left(\frac{1}{\text{Re}'} + \frac{1}{3\text{Re}}\right)\nabla^*(\nabla^* \cdot \mathbf{V}^*) - \frac{\rho^*}{F_r}\nabla^*\psi^* \qquad (4.9)$$

Energy:
$$\rho^* \frac{DT^*}{Dt^*} = \frac{k(k-1)M_0^2 \Phi^*}{\text{Re}} - p^*\nabla^* \cdot \mathbf{V}^*[k(k-1)M_0^2] + \frac{k}{\text{Re}P_r}\nabla^{*2}T^* \qquad (4.10)$$

State:
$$p^* = \frac{1}{kM_0^2}\rho^* T^* \qquad (4.11)$$

In addition to the dimensionless coefficients Re and F_r which have appeared already in the incompressible analysis, three new dimensionless groups appear. These new numbers are the Mach number M_0 which is the ratio of the characteristic speed V_0 to the characteristic sonic speed a_0, the Prandtl number P_r which is the ratio of the kinematic viscosity v to the thermal diffusivity α, and k. Thus

$$M_0 = V_0/a_0 \qquad \text{and} \qquad P_r = v/\alpha = \mu c_p/\kappa$$

where μ is the viscosity, c_p the specific heat at constant pressure, κ the thermal conductivity, and α is the thermal diffusivity defined as $\kappa/\rho c_p$.

The Prandtl number plays an important role in convective heat transfer since, physically, the product $\text{Re}P_r$ is a measure of the ratio of heat transfer by convection to heat transfer by thermal conduction. For most gases P_r is of order unity and for liquids P_r may vary between 0.1 and 0.001. The value of Re must be small if heat conduction is to be important in a fluid. Usually in nature Re is quite large, much greater than unity, so that convection dominates. However, in some situations of engineering interest Re may be less than unity and both conduction and convection become important modes of heat transfer. Such is the case in a thin film of lubricating oil in a bearing, for example.

Many other combinations of the variables are possible to form other dimensionless groups, but they are not independent of the ones already mentioned. In certain engineering problems it is convenient to introduce other dimensionless groups. Sometimes the term Φ^* is written as $\kappa B_r/\text{Re}P_r \Phi^*$, where B_r is the Brinkman number defined as $B_r = \mu V_0^2/\kappa T_0$. Further, the ratio of B_r/P_r is sometimes called the Eckert number E_c. Clearly then the form of the Π's is not independent and we can choose only one of M_0, B_r or E_c in addition to Re, P_r and k (the ratio of specific heats c_p/c_v) in the energy equation. There is a second Reynolds number, Re', based on the second coefficient of viscosity but it is not important except for certain problems such as sound attenuation and dispersion.

To summarize, we model as follows for incompressible and compressible flow.

Re the same: viscous flow and subsonic aerodynamics.

M_0 and k the same: high speed compressible and supersonic flow.

Re and M_0 the same: compressible boundary layer.

P_r the same: heat conduction.

Re, M_0, and P_r the same: compressible boundary layer with heat conduction.

Re and F_r the same: marine ship modeling.

Usually only one number can be held the same for model and prototpye at the same time, so that several experiments may be necessary to model various effects of interest and such modeling may become rather complex and will not be discussed further here. Often, however, two parameters can be simultaneously made the same, as for example k and M_0. If air is used for the model and prototype, k is the same and by appropriate scaling M_0 can be made the same.

4.4   ADDITIONAL PARAMETERS INVOLVED IN FREE CONVECTIVE HEAT TRANSFER IN FLUIDS

The equation of motion may be complicated by one other factor, the buoyancy effects. These effects give rise to free convection if thermal gradients exist which cause density gradients. We assume here that the fluid is essentially incompressible, with a temperature coefficient of volume expansion β.

The equation of motion is

$$\rho[\partial \mathbf{V}/\partial t + (\mathbf{V} \cdot \nabla)\mathbf{V}] = -\rho\beta g(T - T_0) + \mu\nabla^2\mathbf{V} \qquad (4.12)$$

where the compressibility effects have been neglected. In such problems there is usually no reference velocity and we define a new dimensionless velocity, time and temperature as

$$\mathbf{V}^\dagger = \mathbf{V}L\rho/\mu, \qquad t^\dagger = t\mu/\rho L^2, \qquad \Theta = (T - T_0)/(T_1 - T_0) \qquad (4.13)$$

where $T_1 - T_0$ is a characteristic temperature difference of the system, so that (4.12) becomes

$$\partial\mathbf{V}^\dagger/\partial t^\dagger + (\mathbf{V}^\dagger \cdot \nabla^*)\mathbf{V}^\dagger = -\nabla^{*2}\mathbf{V}^\dagger + \Theta G_r \qquad (4.14)$$

G_r is the Grashof number, $g\rho_0^2\beta(T_1 - T_0)L^3/\mu^2$.

In free convection the Reynolds number is unimportant and does not enter. In the energy equation, dissipation is usually neglected and compressibility effects are unimportant so that in dimensionless form

$$\frac{D\Theta}{Dt^\dagger} = \frac{1}{P_r}\nabla^{*2}\Theta \qquad (4.15)$$

Hence in free convection, only the Grashof and Prandtl numbers are important.

References

1. Birkhoff, G., *Hydrodynamics*, Dover Publications, 1955.

2. Bridgman, P. W., *Dimensional Analysis*, Yale University Press, 1931.

3. Buckingham, E., On Physically Similar Systems, *Physical Review*, Vol. 4, Ser. 2, p. 345, 1914.

4. Kline, S. J., *Similitude and Approximation Theory*, McGraw-Hill, 1965.

5. Langhoar, H. L., *Dimensional Analysis and the Theory of Modeling*, John Wiley, 1951.

6. Taylor, E. S., *Dimensional Analysis for Engineers*, Oxford, 1974.

Solved Problems

4.1. In incompressible time dependent flow oscillating with period τ due to a body immersed in a fluid, what are the parameters which characterize the flow?

For incompressible flow, only the Reynolds number appears as a dimensionless coefficient. If we solve for a dimensionless period τ^* defined as $\tau V_0/L$, we can write

$$\tau^* = \tau V_0/L = f(\text{Re})$$

The dimensionless number $\tau V_0/L$ is known as the Strouhal number. Such oscillations occur in the shedding of Karman vortices in the wake and will be discussed in Chapter 5.

4.2. In problems involving surface tension T_s, what new parameters enter?

The equations of motion are the same as for flow without surface tension effects but the boundary conditions are different. At a liquid-air interface the pressure force must be balanced by the surface tension, and is not necessarily zero as it must be at a free surface without surface tension. A finite pressure difference, remember, can exist across an interface (if it has curvature) and be balanced by surface tension. Consider a two dimensional case as shown in Fig. 4-1. From equilibrium considerations,

$$2T_s(d\theta/2) = pR\,d\theta$$

where R is the radius of curvature of the surface and $\sin\theta \simeq \theta$ for small θ. Then we can write

$$T_s = p^*R^*(\rho_0 V_0^2 L)$$

having introduced the dimensionless pressure and radius of curvature as $p^* = p/\rho_0 V_0^2$ and $R^* = R/L$. We define a dimensionless surface tension T_s^* as

$$T_s^* = T_s/\rho_0 V_0^2 L$$

so that
$$T_s^* = p^*R^*$$

The dimensionless parameter $\rho_0 V_0^2 L/T_s$ is called the Weber number.

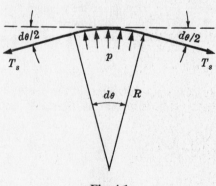

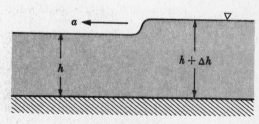

Fig. 4-1   Fig. 4-2

4.3. In a shallow liquid, small surface disturbances propagate as shown in Fig. 4-2. On what parameters does the infinitesimal surface wave speed depend? What dimensionless groups are relevant?

Consider a wave as shown. The speed a can depend on the depth h and the value of gravity g. The only dimensionless group that can be formed is

$$\Pi = a^2/gh$$

Here, since the number of variables is three and the number of fundamental units (length and time) are two, we have one Π. Since there is only one Π in the problem, we must say that it is a constant:

$$\Pi = a^2/gh = \text{constant}$$

It happens here that the constant is unity and $a = \sqrt{gh}$, but this result could not be determined from dimensional analysis.

Another point to remember here is that the density of the fluid does not enter. However, one might not know this in the beginning and its introduction could lead to irrelevant Π's and possible incorrect conclusions. It is always a problem to know just what variables are relevant if we do not begin with the basic governing equations.

It is interesting to compare the results here with Problem 3.15, where the solution was obtained analytically.

4.4. When a valve is closed suddenly in a pipe with flowing water, a water hammer wave is set up. Such waves can generate enormous pressures causing damage to the pipe. Using dimensional analysis, find the maximum pressure generated by this phenomenon.

We take as the relevant parameters: the maximum pressure $p_{\max}$, the density ρ, the initial flow velocity U_0, and the bulk modulus β (since the wave must be a compression wave of some sort). There are two possible independent Π's here. We may pick them as

$$p_{\max}/\beta \quad \text{and} \quad U_0^2 \rho/\beta$$

Hence we can write
$$p_{\max}/\beta = f(U_0^2 \rho/\beta)$$

This is as far as we can go with dimensional analysis. Actually it turns out that $p_{\max}/\beta = (U_0^2 \rho/\beta)^{1/2}$ so that $p_{\max} = \sqrt{\rho\beta}U_0$, but this fact cannot be obtained except by solution of the relevant differential equations.

As in the previous problem, we can say something about the speed of propagation of this water hammer wave. The wave speed a depends on the density ρ and bulk modulus β (if one is fortunate enough to guess this correctly), so that the only Π that can be formed is $a^2\rho/\beta$ and it must be a constant. Again, the actual speed is $\sqrt{\beta/\rho}$.

4.5. In subsonic aerodynamics, how can we model lift on an airfoil?

The lift on a body may be found by integrating the appropriate component of the normal pressure force on the body over its surface. The pressure is found from the equation of motion. In steady flow the dimensionless pressure depends on the location of the body, $\mathbf{r}^*$, and on the Reynolds number; however, usually the Reynolds number dependence is slight, and if the flow is streamlined, as in a true airfoil, the Reynolds number dependence may be neglected. For a given geometrical shape the pressure may be integrated over the body to eliminate the $\mathbf{r}^*$ dependence as we find the lift L. However, even for a given geometrical shape, the angle of attack θ (the relative angle at which the free stream flow approaches the body) affects the lift.

Referring to Fig. 4-3, it follows that

$$L = -\int_A p\hat{\mathbf{y}} \cdot d\mathbf{A}$$

where $\hat{\mathbf{y}}$ is a unit vector in the y direction, and hence

$$L^* = \frac{L}{\rho V_0^2 A_0} = -\int p^*\hat{\mathbf{y}} \cdot d\mathbf{A}^* = f(\theta)$$

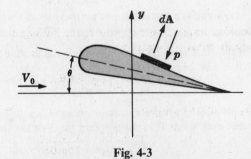

Fig. 4-3

where A_0 is a characteristic area of the airfoil and $A^* = A/A_0$. Therefore the lift may be written

$$L = C_L(\theta)\rho V_0^2 A_0/2$$

where C_L is known as the lift coefficient. The factor 2 is introduced by convention so that $\rho V_0^2/2$ may be interpreted as dynamic pressure. In experiments the lift coefficient is usually determined as a function of θ for a given shape airfoil.

4.6.   In subsonic aerodynamics, how do we model drag?

The drag is caused mainly by skin friction, although if the body is not streamlined a wake will occur and the low pressure in the wake will dominate the skin friction (viscous drag due to the boundary layer). In the flow over an airfoil there is also an induced drag which is brought about by the effective angle of attack changing due to the downwash. The lift, which must be at right angles to the local effective free stream flow, then has a component in the main free stream direction (that is, opposite to the direction of motion of the body through the fluid).

The induced drag D may be formulated in terms of a pressure integral, and as for lift we can write

$$D = C_D \rho V_0^2 A_0/2$$

when $C_D(\theta)$ is the drag coefficient.

If we consider bodies which are not streamlined and in which the major portion of drag is due to viscous drag and the low pressure wake, the flow is highly sensitive to the Reynolds number (unlike lift and induced drag) and we can write that the integrated skin friction and wake drag must be a function of Reynolds number (for a given geometrical shape). Of course we could still have an angle of attack dependence, but we assume now that the angle of attack is held constant as the Reynolds number is varied. Actually the angle of attack is not a useful variable then, since the body is not streamlined and the angle of attack is not very meaningful. A different angle of attack is considered a different geometry. Then the drag can be written in dimensionless form as

$$D^* = \frac{D}{\rho V_0^2 A_0/2} = f(\text{Re}) = C_D(\text{Re})$$

so that we retain the concept of drag coefficient but say that it may not be a constant for a given geometry but depends on Re. In the general case we can say for any airfoil with separating boundary layer or not,

$$C_D = f(\text{Re}, \theta)$$

4.7.   In a pipe the pressure drop depends on the wall friction which in turn depends on whether the flow is laminar or turbulent. If the flow is laminar the wall friction depends on the viscosity. If the flow is turbulent the wall friction depends on the Reynolds number and the roughness of the wall. How can this pressure drop be modeled?

The calculation of pressure drop, Δp, in a pipe of length L must be based on the equation of motion. Hence the dimensionless pressure drop per unit length along the pipe must depend on the Reynolds number and pipe roughness, which can be represented as the ratio of the mean height of surface asperities to the diameter, ϵ/D. Hence

$$\frac{\Delta p^*}{L^*} = h(\text{Re}, \epsilon/D)$$

if we define the dimensionless length L^* as L/D, where D is the pipe diameter and h is a function of Re and the roughness of the pipe. In terms of pressure p,

$$\frac{\Delta p^*}{L^*} = \frac{\Delta p}{\rho V^2}\frac{D}{L} = h(\text{Re}, \epsilon/D)$$

Now the head loss is H_L, so that we can write

$$H_L = \frac{\Delta p}{\rho g} = \frac{\Delta p^*}{L^*}\frac{V^2 L}{Dg}$$

Physically, H_L is the equivalent height differential of flowing fluid that produces the pressure drop Δp. That is, a column of fluid of height H_L produces a hydrostatic pressure equal to $\rho g H_L = \Delta p$. The dimensionless head loss H_L is

$$H_L^* = H_L \frac{Dg}{V^2 L} = h(\text{Re}, \epsilon/D) = f/2$$

The function (Re, roughness) is written as $f/2$ where f is the friction factor, a dimensionless number which depends on the Reynolds number and also on roughness if the flow is turbulent. f may be defined in terms of τ_0, the wall shear stress, as $f = 8\tau_0/\rho V^2$. Finally,

$$H_L = \frac{fL}{D} \frac{V^2}{2g}$$

Experimental data for friction factors are of vital importance in pipe flow problems.

4.8. Solve Problem 4.7 using the detailed Buckingham Pi method. Consider the flow of a fluid in a pipe. Determine the important Π parameters for such a flow.

First we have to guess the important physical quantities. We will assume that they are the following. M, L and T indicate dimensions of mass, length and time, respectively.

Physical Quantity	V	D	ρ	μ	ϵ	dp/dx
Dimension	LT^{-1}	L	ML^{-3}	$ML^{-1}T^{-1}$	L	$ML^{-2}T^{-2}$

where ϵ is some average height of the wall roughness and dp/dx is the pressure change per unit length in the flow direction.

We have 6 physical quantities involving 3 primary dimensions and therefore there will be a maximum of three Π terms. We will now determine the Π terms taking 4 physical quantities at a time.

$$V^{x_1} D^{y_1} \rho^{z_1} \mu = (LT^{-1})^{x_1} L^{y_1} (ML^{-3})^{z_1} ML^{-1} T^{-1} = M^0 L^0 T^0$$

Then $\qquad L \to x_1 + y_1 - 3z_1 - 1 = 0, \qquad M \to z_1 + 1 = 0, \qquad T \to -x_1 - 1 = 0$

from which $z_1 = -1, x_1 = -1, y_1 = -1$. Thus

$$\Pi_1 = \mu/\rho V D \qquad \text{or} \qquad \Pi_1 = \rho V D/\mu$$

By the same procedure we can take

$$D^{x_2} \rho^{y_2} \mu^{z_2} \epsilon = M^0 L^0 T^0 \qquad \text{and} \qquad V^{x_3} \rho^{y_3} D^{z_3} \, dp/dx = M^0 L^0 T^0$$

and obtain $\qquad\qquad\qquad \Pi_2 = \epsilon/D, \qquad \Pi_3 = \dfrac{(dp/dx)D}{\rho V^2}$

Then we have $\Pi_3 = h(\Pi_1, \Pi_2)$ and

$$\frac{(dp/dx)D}{\rho V^2} = h(\rho V D/\mu, \epsilon/D) = h(\text{Re}, \epsilon/D)$$

where h indicates a functional relationship.

The functional relation among these quantities is determined experimentally. These results are shown graphically in Chapter 5 and are called the Moody or Stanton diagram.

The usefulness of dimensionless quantities is readily apparent in this example. An enormous amount of experimental work would have been required if we were to vary each of the five physical quantities independently, whereas by reducing the number to three dimensionless quantities a complete experimental program can be reasonably performed.

It is customary to write the head loss in a pipe of length L as $H_L = \Delta p / \rho g$ where Δp is the pressure drop in the pipe. Then $\Delta p / L = dp/dx$ and we can write, using the results above,

$$H_L^* = H_L \frac{Dg}{V^2 L} = \frac{(dp/dx)D}{\rho V^2} = \text{a function of } (\text{Re}, \, \epsilon/D) = f/2$$

where f is the friction factor, defined by the above equation. In terms of f then, we can express the head loss as $H_L = (fL/D)(V^2/2g)$.

4.9. An airfoil of surface area 1 ft^2 is tested for lift L in a wind tunnel. At an angle of attack of $5°$ with standard air of density $0.0024 \text{ slugs/ft}^3$ at a speed of 100 ft/sec, the lift is measured as 7.0 pounds. What is the lift coefficient C_L? For a prototype wing of area 100 ft^2, what is the lift L_p at an air speed of 100 mph at the same angle of attack, $5°$?

C_L may be computed from the data taken on the model.

$$C_L = \left(\frac{L}{A \rho V_0^2 / 2} \right)_m = \frac{7.0}{1(0.0024)(100)^2/2} = 0.58$$

Then for the prototype C_L is the same and we find the lift L_p as

$$L_p = (C_L A \rho V_0^2 / 2)_p = 0.58(100)(0.0024)(100 \times 88/60)^2/2 = 1500 \text{ lb}$$

4.10. Discuss the modeling of a pump or fan for an incompressible fluid.

The modeling is essentially the same for a centrifugal or axial flow machine. The important parameters which describe the performance of a machine are the power input P, the head H produced across the machine, and the efficiency η. For a given geometrical design the performance is characterized by the following related variables: ρ, fluid density; ω, angular velocity of the rotor; D, mean diameter of the rotor; μ, fluid viscosity; Q, fluid flow rate.

P, gH and η are not independent but are functions of the variables listed above. It is convenient to introduce the head as gH instead of H since the product gH represents the shaft work per unit mass of fluid and is independent of g. Hence we can write

$$P = f_1(\rho, \, \omega, \, D, \, Q, \, \mu)$$

$$\eta = f_2(\rho, \, \omega, \, D, \, Q, \, \mu)$$

$$gH = f_3(\rho, \, \omega, \, D, \, Q, \, \mu)$$

We apply the Buckingham Π Theorem to each equation and obtain a convenient set of Π's as

$$P/\rho \omega^3 D^5 = f_4(Q/\omega D^3, \, \rho \omega D^2/\mu)$$

$$\eta = f_5(Q/\omega D^3, \, \rho \omega D^2/\mu)$$

$$gH/\omega^2 D^2 = f_6(Q/\omega D^3, \, \rho \omega D^2/\mu)$$

It is important to note here that the Π's $(P/\rho \omega^3 D^5, \, gH/\omega^2 D^2$ and $\eta)$ are not independent and are determined once the other Π's are known.

From experimental data it is known that viscosity and hence the Π $\rho \omega D^2/\mu$ is rather unimportant in determining the performance of a pump or fan. Hence it can be neglected and we obtain finally for a given design configuration,

$$gH/\omega^2 D^2 = f_7(Q/\omega D^3), \qquad P/\rho \omega^3 D^5 = f_8(Q/\omega D^3), \qquad \eta = f_9(Q/\omega D^3)$$

Experimental data are usually expressed as a plot of the Π's $gH/\omega^2 D^2$, $P/\rho \omega^3 D^5$ and η versus $Q/\omega D^3$ for a given machine.

Supplementary Problems

4.11. A model of a subsonic wing is tested in a laboratory wind tunnel and the following data taken.

Lift (pounds)	0	10	12	14	12	8
Angle of attack, θ	0°	5°	10°	15°	20°	25°

The model is tested with standard air at a speed of 100 ft/sec. The total area of the model wing is 1 ft². Plot the lift coefficient C_L vs. θ. For a prototype wing of area 100 ft², what would be the lift for an air speed of 100 mi/hr and an angle of attack of 5°?

4.12. A new type of interior finish has been developed for oil pipelines. This finish is applied to a sample 12″ I.D. pipe for testing. A 100 ft length of the sample pipe is tested with water (at room temperature) with varying flow rates. The following data were taken:

Pressure drop across pipe (ft of water)	0.011	0.078	0.28	15.3
Flow rate (gal/min)	172	475	950	7860

Plot the friction factor f (as defined in Problem 4.7) vs. the Reynolds number. For a real pipe of 12″ I.D., what pressure drop in psi would occur per mile of pipe for a flow rate of 5000 gal/min of crude oil? Convert the pressure drop into feet of oil and into feet of water.

4.13. A model of a hemispherically shaped dome cover is to be tested in a wind tunnel for lift and the lift coefficient C_L determined. The real dome is subjected to 100 mph gales in the Arctic at temperatures of −40°F. If the model is to be made 1 ft in diameter and the real dome is 50 ft in diameter, what should be the wind tunnel speed if standard air is used? Does the speed matter? Could you achieve Reynolds number modeling here?

4.14. A nonretractable streamlined landing gear for a small plane is to be modeled $\frac{1}{3}$ scale for drag in a small wind tunnel. The wind tunnel uses standard air. The frontal projected area of the model is 0.5 ft². If we wish to determine the drag on the prototype operating at 100 mph, the cruising speed, at what speed should the wind tunnel be operated? Operating at the proper speed, the wind tunnel gives a drag of 20 pounds. What would be the drag on the actual gear at the cruising speed? What is the drag coefficient C_D?

4.15. In Fig. 4-4 is shown a weir which is a notch in a dam or barrier in an open channel and which is used for measuring the flow rate. The notch may be of varying shape and the height H of water in the notch is a measure of the flow rate. Show that for a triangular notch as in the figure, if the water is fairly deep and the effects of surface tension and viscosity are neglected the flow rate Q can be approximated by $Q = C\sqrt{g}H^{5/2}$ where C is a constant. From experiment it has been found that C can be represented by $C = 0.44 \tan \frac{1}{2}\theta$.

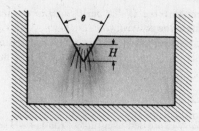

Fig. 4-4

4.16. The flow of a liquid through a pipe may be measured by a flat plate orifice inserted in the pipe as shown in Fig. 4-5. A pressure gage on each side of the plate gives the pressure drop across the orifice, and from this data the flow rate may be calculated for a given fluid. If we assume that the flow rate depends on Δp across the orifice, the fluid density ρ, the pipe diameter D, and the orifice diameter D_0, show that the flow rate may be expressed as $Q = CD_0\sqrt{2\Delta p/\rho}$. Is C a constant? Explain.

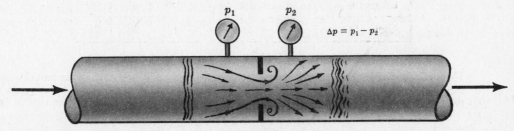

Fig. 4-5

4.17. In astronomical and cosmological calculations where the distances and times involved are so large compared to our ordinary units, it is sometimes convenient to employ a special system of units.

For one such system of units the following universal constants are taken as unity for simplicity in making calculations: velocity of light, $c = 1$; mass of the sun, $M = 1$; Newtonian gravitational constant, $K = 1$. (K is defined as follows: Two masses attract each other according to the gravitational law $F = KM_1M_2/r^2$. F is usually given in dynes, the masses in grams, and r in centimeters.)

It is desired to determine the fundamental units of M, L and T in this new system where the given universal constants are unity, and to relate these new units to our ordinary system of units of grams, cm, sec, etc. That is, how many cm make up the new astronomical unit, and how many sec make the new astronomical unit of time? Do the same for mass.

Data: Mass of sun, 1.98×10^{33} grams
Velocity of light, 10^{10} cm/sec
Gravitational constant (K), 6.67×10^{-8} gym^{-1} cm^3 sec^{-2}

Note: What about the force; will it still be dynes even in the new system of units?

4.18. Discuss the modeling of a propeller. Assume the thrust depends on the diameter D, the fluid density ρ and viscosity μ, the velocity V of the propeller through the fluid, and the angular velocity ω of the propeller.

4.19. The propeller of a ship 800 ft long is 10 ft in diameter and rotates at 100 rpm. If a model ship 8 ft long is to be tested in a towing tank, discuss the conditions of the propeller test in order to achieve meaningful modeling.

4.20. A small ship 100 ft long is designed to move at 20 mph through fresh water. Find the kinematic viscosity of a liquid suitable for similitude tests with a model 5 ft long.

4.21. Analyze a turbine using an analysis similar to that of Problem 4.10. The results are the same except that the terms must be interpreted in terms of the performance of a turbine instead of a pump.

4.22. A pump tested at 1000 rpm delivers 5 ft^3/sec of water against a head of 200 ft. The power necessary to run the pump is 200 hp. Calculate the efficiency of the pump. A geometrically similar pump of three times the diameter is made to run at 500 rpm. Find the flow rate, head and power for the same efficiency.

4.23. A rather well-known cookbook gives the following table for turkey baking.

Weight of turkey, lb	6-10	10-16	18-25
Time required per pound, min/lb	20-25	18-20	15-18

From dimensional analysis considerations you should be able to obtain this information from just one bit of experimental information. That is, from only the information that a 6-10 pound turkey takes 20-25 min/lb to bake, you should be able to construct the table and extrapolate to heavier turkeys or fill in a more complete table.

Hint. Consider two geometrically similar turkeys, both at an initial temperature T_0 and to be baked to a temperature T_c in an oven at temperature T_s. A dimensionless temperature may be defined as $\theta = (T - T_0)/(T_s - T_0)$, and a dimensionless time may be defined as $\tau = \alpha t/L^2$ where α is the thermal diffusivity of turkeys, L is the characteristic dimension, say the length, and t is time. It can be shown (do it) that the following formula is valid:

$$(\text{baking time}) \times (\text{weight of turkey})^{-2/3} = \text{constant}$$

4.24. In experiments on heat leakage through the walls of home freezers, a test was made on a specially insulated box about $3 \times 3 \times 4$ ft in size. With the box initially at $70°F$ an incandescent light bulb was used to maintain a constant flow of heat to the interior of the box. The outside surface of the box was maintained at $70°F$. Data on the temperature rise of the inside surface over the outside surface are given below.

(a) Plot the data on temperature rise versus time in terms of dimensionless groups so that it can be applied to other problems of a similar nature. One group should contain temperature but not time and the other group should contain time but not temperature.

(b) Determine the temperature difference that should be expected at the end of 15 minutes under conditions exactly the same as those of the test except with a wall thickness only one-half as great.

Data:

Time, hours	0	0.5	1.0	2.0	3.0	5.0	10
Temperature difference, °F	0	15.9	22.5	30.0	35.5	42.0	48.3

Rate of heat flow per unit area = 6.09 BTU/hr-ft^2
Thickness of insulation = 2 in.
Conductivity of insulation = 0.0209 BTU/ft-hr-°F
Thermal heat capacity of insulation = 4.85 BTU/ft^3-°F

NOMENCLATURE FOR CHAPTER 4

a = sonic speed
B_r = Brinkman number = $\mu V_0^2/kT_0$
C_D = drag coefficient
C_L = lift coefficient
c_p = specific heat at constant pressure
c_v = specific heat at constant volume
D = drag force, diameter
E_c = Eckert number = B_r/P_r
f = friction factor
F_r = Froude number = V_0^2/gL
G_r = Grashof number = $g\rho_0^2\beta(T_1 - T_0)L^3/\mu^2$
g = acceleration due to gravity
k = ratio of specific heats = c_p/c_v

L = characteristic length, lift force
M = Mach number = V/a
P_r = Prandtl number = $v/\alpha = \mu c_p/\kappa$
p = pressure
R = gas constant
Re = Reynolds number = LV_0/v
r = position vector
T = absolute temperature
t = time
V = velocity vector
$( )_0$ = free stream value
$( )^*$ = dimensionless variable
$( )^†$ = dimensionless variable

α = thermal diffusivity = $\kappa / \rho c_p$

β = volume coefficient of thermal expansion, bulk modulus

Θ = dimensionless temperature

κ = thermal conductivity

μ = absolute viscosity

ν = kinematic viscosity

Π = dimensionless group

ρ = density

Φ = dissipation function

ψ = gravitational potential

Chapter 5

Boundary Layer Flow and Flow in Pipes and Ducts

5.1 INTRODUCTION

In this chapter we will study some important types of incompressible flow where the viscous effects are important. Some flow situations can be analyzed rather easily, and in fact exact solutions are readily obtainable. Such is the case for fully developed viscous laminar flow in a pipe or channel. There the nonlinear inertia effects are not present and the equation of motion reduces to a balance between pressure gradient and viscous forces. However, other types of flow involve inertia effects which produce nonlinearities in the equation of motion, and solutions generally require approximations or numerical techniques.

One of the most important advances in fluid mechanics was contributed by Prandtl in 1904. He suggested that fluid motion around objects could be divided into two regions: a thin region close to the object where frictional effects are important, and an outer region where friction may be neglected. This chapter is concerned primarily with the region where friction is important (the boundary layer), while Chapter 6 is concerned with flows of negligible friction (potential flow). Frictional forces in a fluid come about because of viscosity and/or turbulence. For any given flow configuration at sufficiently low Reynolds numbers the flow is generally laminar and viscosity gives rise to frictional forces (and the flow is known as "viscous"), but as the Reynolds number increases the flow passes through a region of "transition" and eventually becomes turbulent. In turbulent flow frictional forces are generated by the turbulent fluctuations and usually dominate the viscous effects. In this chapter we will study viscous flow and discuss some empirical relationships for turbulent flow, particularly in reference to boundary layer flow.

In this chapter we will look at the basic character of flow with friction. We will examine the methods of determining (1) the size of the boundary layer, (2) the resulting velocity distribution, (3) the pressure distribution, and (4) the force of a fluid on a solid surface.

It is convenient to split the main topics in terms of "external flows" and "internal flows". External flows occur around solid objects, and internal flows occur inside objects such as pipes and channels. While the differential equations describing these flows are essentially the same, the boundary conditions are different so that the resulting flows are entirely different. We first discuss boundary layer flow, then internal flow (such as flow in pipes and channels). As we will learn, the equations that describe boundary layer flow generally have no exact closed form analytical solutions and require either analytical power series solutions or numerical analysis. However, solutions can often be approximated by closed form expressions. On the other hand, internal viscous flow may be described exactly by simple analytical expressions for many flow situations.

The less complicated flows will be studied first. This means that laminar flows will be considered first, then turbulent flows. While some space in this chapter is devoted to turbulent flow, Chapter 9 considers this subject in much greater depth.

Before considering the details of the equations and solutions of boundary layer flows, we will make some general observations of engineering importance concerning the physical character of the boundary layer which is illustrated in Fig. 5-1.

There is no precise dividing line between the potential flow region where friction is negligible and the boundary layer, but it is customary to define the boundary layer as that region where the fluid velocity (parallel to the surface) is less than 99% of the free stream velocity which is described by potential flow theory. The thickness of the boundary layer, δ, grows along a surface (over which fluid is flowing) from the leading edge. At the leading edge of a flat plate the thickness is zero, but on the front of a blunt body such as a cylinder there is a finite thickness even at the stagnation point.

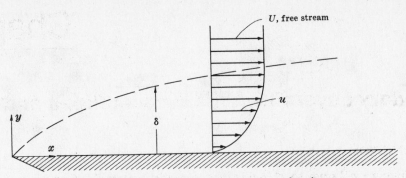

Fig. 5-1. Velocity distribution in the boundary layer.

The flow within the boundary layer begins as laminar flow, but as the layer grows along a surface a transition region occurs and the flow in the boundary layer may become turbulent if the surface is long enough. The laminar-transition-turbulence sequence occurs in all flows, if the surface is long enough, regardless of whether the free stream is laminar or turbulent, but as the degree of turbulence in the free stream increases, the transition from laminar to turbulent flow in the boundary layer occurs earlier, that is, closer to the leading edge.

It can be shown that even in flows where the pressure varies along the surface, such as flow over a curved surface, the pressure variation normal to the surface is negligible through the boundary layer. It is assumed then that the pressure distribution in the boundary layer is imposed by the potential flow free stream pressure gradient just outside the boundary layer. In many problems the boundary layer is so thin that the potential flow solution may be computed neglecting the boundary layer entirely and that solution used to find the pressure distribution for the boundary layer calculations. This approach is employed in aerodynamics to find the flow over streamlined bodies such as airfoils (see Chapter 6).

The shape of the velocity profile and the rate of increase of the boundary layer thickness, δ, depend on the pressure gradient, $\partial p/\partial x$. For example, if the pressure increases in the direction of flow, the boundary layer thickness increases rapidly and the velocity profiles will appear as shown in Fig. 5-2. If this adverse pressure gradient is large enough, then separation will occur followed by a region of reversed flow. (The separation point is defined as the point where $\partial u/\partial y|_{y=0} = 0$). If the pressure decreases in the direction of flow, the boundary layer thickness increases slowly. u is the velocity parallel to the wall (in the x direction) and y is the coordinate normal to the wall.

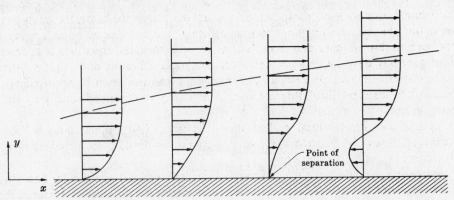

Fig. 5-2. Velocity profiles for flow on a flat plate where $\partial p/\partial x > 0$.

The effect of a pressure gradient is very important in establishing the flows in diffusers and nozzles and around objects. The diffuser, illustrated in Fig. 5-3, has a positive pressure gradient. Thus the boundary layer grows rapidly and if the diffuser angle of divergence is too large, separation will occur. This separation results in a diffuser of poor performance since the resulting pressure recovery would not

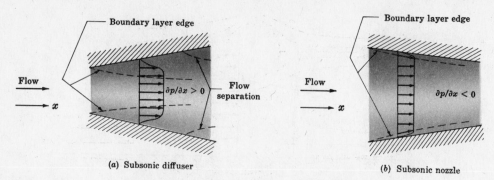

(a) Subsonic diffuser   (b) Subsonic nozzle

Fig. 5-3. Comparison of diffuser and nozzle flows.

be as large as that of a diffuser operating without separation. The design of diffusers is one of compromise of length and angle. If the angle is too large, separation will occur and if the angle is too small, an excessive length is required to obtain a given pressure, resulting in large friction losses.

The nozzle, in contrast, involves flow with a decreasing pressure in the direction of flow (favorable pressure gradient). As a result, the boundary layer remains relatively small as shown in Fig. 5-3. Separation is not a problem in nozzle flows and the design problems for nozzles are somewhat simpler.

The potential flow solution for the flow past an object as shown in Fig. 5-4 predicts a decreasing pressure over the front portion of the body and an increasing pressure over the rear portion. Again, there results a relatively thin boundary layer over the front portion and a thick boundary layer and possible separation over the rear portion. If the rear of the body is too "blunt," separation will occur because the pressure gradient $\partial p/\partial x$ becomes too large. By gently "fairing" or "streamlining" the rear of the body as shown in the tear drop shape of Fig. 5-5, separation can be prevented. At the point of separation the flow breaks away from the surface, creating a wake. After the separation point the flow is actually reversed along the surface, giving rise to vortices or eddies in the wake. The structure of the wake depends critically on the Reynolds number of the free stream flow (based on a characteristic dimension of the object). At a very low Reynolds number (Re ≪ 1) the flow is often called very viscous or "creeping flow," and in fact the boundary is so thick that viscous effects are felt far out into the main flow and there is essentially no potential flow region. Under such circumstances, neither is there a definite wake (Fig. 5-6a). However, the flow pattern is not symmetrical, front to back. Since there is drag on the object, a simple momentum balance on a control volume surrounding the object proves this statement. As the Reynolds number increases a pair of bound vortices appears in the wake (Fig. 5-6b, d). Eventually with higher Reynolds numbers the vortices form and shed alternately from side to side forming a "von-Karman vortex street." Such periodic behavior even in the presence of a steady main

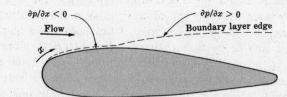

Fig. 5-4. External flow effect of pressure gradient on boundary layer growth.

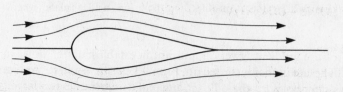

Fig. 5-5. Streamlined flow over a teardrop shape.

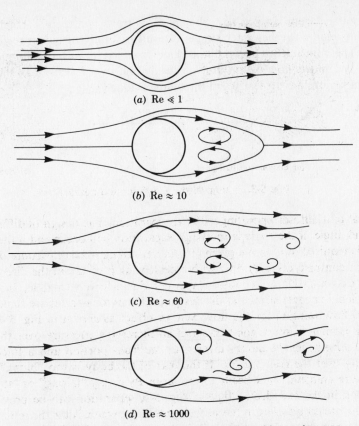

(a) Re ≪ 1

(b) Re ≈ 10

(c) Re ≈ 60

(d) Re ≈ 1000

Fig. 5-6. Flow over a cylinder at different Reynolds numbers.

flow may seem strange but is a very important phenomenon. Such periodic shedding gives rise to a periodic force on the object, and if coupling occurs with the mechanical system of the object itself a self-sustained oscillation may result. Resonance conditions may cause catastrophic effects.

The actual position of separation is difficult to compute analytically because of the interaction of the wake and potential flow region. The wake itself changes the potential flow pattern upstream and the corresponding pressure gradient along the surface. The separation point depends on the pressure gradient along the surface and the turbulence level in the boundary layer. The location of the separation point moves toward the rear (or trailing) edge as the turbulence level increases. The level of turbulence in the boundary layer is affected by the roughness of the surface and the level of turbulence in the free stream outside the boundary layer. Later we will discuss the effect of separation and the boundary layer on drag. In many flows one finds that the boundary layer is so small over the entire body that a solution neglecting viscosity completely (i.e. a potential flow solution) gives accurate results for the pressure distribution. Such is the case for flow over streamlined airfoils, for example. However, if the body is blunt in the rear and the wake becomes appreciable because of boundary layer thickening or separation, the potential flow solution is incorrect except over the front portion of the body where the boundary is thin.

5.2 EXTERNAL FLOWS—BOUNDARY LAYERS

Flow Over a Flat Plate

The flow over a flat plate is characteristic of external flows in general. This flow is shown in Fig. 5-1. The boundary layer thickness is zero at the leading edge and increases with distance along the plate surface. The early portion of the boundary layer is laminar but may be followed by a transition region where the flow changes from laminar to turbulent. This transition region actually consists of bursts of turbulence which spread until they intermingle to result in a fully turbulent region as shown in the top view of Fig. 5-7. These turbulence bursts do not have fixed locations but continually move. Thus we see that while the fully laminar flow region may be steady and two-dimensional, the transition region and the fully turbulent regions are unsteady and three-dimensional.

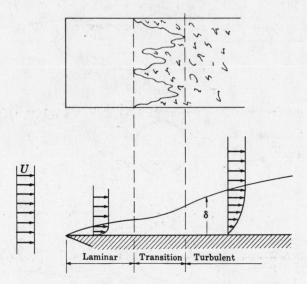

Fig. 5-7.   Boundary layer flow over a flat plate.

The exact solution to the appropriate equations describing the laminar boundary layer is difficult and only a few simple problems can be treated easily. The solution to the flow over a flat plate cannot be expressed in closed form and an infinite series expression, known as the Blasius solution, is necessary.

Several approximate methods for treating laminar boundary layer flow have been developed. Below we will discuss one, the momentum-integral method which is of great importance in many types of boundary layer calculations.

The chief results of the Blasius solution and integral method for the flat plate are then discussed and compared.

von Karman Momentum-Integral Equation

Consider the control volume for flow over a flat plate as shown in Fig. 5-8. We will write the momentum equation for this control volume for the x direction; see equation (3.9). We have

$$F_{s_x} = \int_{\text{C.S.}} \rho V_x \mathbf{V} \cdot d\mathbf{A}$$

The surface forces are

$$F_{s_x} = pW\delta - \left(p + \frac{\partial p}{\partial x}\,dx\right)(\delta + d\delta)W - \tau_0 W\,dx + \left(p + \epsilon\,\frac{\partial p}{\partial x}\,dx\right)d\delta$$

where W is the width of the plate, δ is the boundary layer thickness, and τ_0 is the shear stress on the wall defined as $\tau_0 = \mu\,\partial u/\partial y\,|_{y=0}$. Neglecting second order terms and simplifying,

$$F_{s_x} = \left[-\delta\,\frac{\partial p}{\partial x}\,dx - \tau_0\,dx\right]W$$

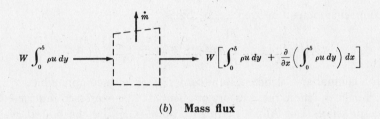

(a)   **Forces on control volume**

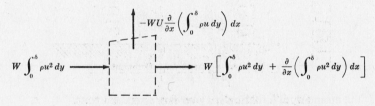

(b)   **Mass flux**

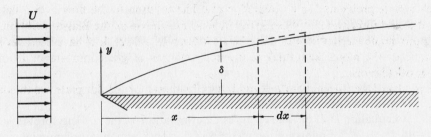

(c)   **Momentum flux**

Fig. 5.8.   Development of boundary layer integral equations. W is the width of the plate (in the z direction).

Before we determine the momentum flux terms, we will consider the mass flux. As shown in Fig. 5-6(b) there will be mass flux across three faces. The mass flux $\dot{m}$ across the top surface then is

$$\dot{m} = W \int_0^\delta \rho u \, dy - W \left[\int_0^\delta \rho u \, dy + \frac{\partial}{\partial x} \left(\int_0^\delta \rho u \, dy \right) dx \right]$$

$$= -W \frac{\partial}{\partial x} \left(\int_0^\delta \rho u \, dy \right) dx$$

where u is the x component of velocity. The x momentum flux $\dot{M}_x$ of the fluid crossing the top surface is

$$\dot{M}_x = \dot{m} U = -W U \frac{\partial}{\partial x} \left(\int_0^\delta \rho u \, dy \right) dx$$

where U is the free stream velocity (in the x direction), and the total momentum flux is

$$\int_{\text{c.s.}} \rho V_x \mathbf{V} \cdot d\mathbf{A} = W\left[\int_0^\delta \rho u^2 \, dy + \frac{\partial}{\partial x}\left(\int_0^\delta \rho u^2 \, dy\right) dx\right]$$

$$- WU \frac{\partial}{\partial x}\left(\int_0^\delta \rho u \, dy\right) dx - W\int_0^\delta \rho u^2 \, dy$$

Simplifying, $\quad \int_{\text{c.s.}} \rho V_x \mathbf{V} \cdot d\mathbf{A} = W\frac{\partial}{\partial x}\left[\int_0^\delta \rho u(u - U) \, dy\right] dx - W\frac{\partial U}{\partial x}\int_0^\delta \rho u \, dy$

So that finally the complete momentum equation becomes

$$\tau_0 + \delta\frac{\partial p}{\partial x} = \frac{d}{dx}\int_0^\delta \rho u(U - u) \, dy - \frac{\partial U}{\partial x}\int_0^\delta \rho u \, dy \tag{5.1}$$

which is valid for both laminar and turbulent flow, compressible or incompressible.

By using the Bernoulli equation, the pressure gradient may be expressed in terms of the free stream velocity U for incompressible flow. Along a stream line just outside the boundary layer (neglecting effects due to elevation change)

$$\frac{U^2}{2} + \frac{p}{\rho} = \text{constant}$$

Differentiating $\qquad U\frac{dU}{dx} + \frac{1}{\rho}\frac{dp}{dx} = 0$

Combining with (5.1) the momentum equation may be expressed as

$$\rho\frac{d}{dx}\int_0^\delta (U - u)u \, dy + \rho\frac{dU}{dx}\int_0^\delta (U - u) \, dy = \tau_0 \tag{5.2}$$

where the ρ has been removed from the derivatives to emphasize that this form is valid for incompressible flow only. If the external pressure gradient is zero (and hence U is a constant), the equation becomes

$$\tau_0 = \frac{d}{dx}\int_0^\delta \rho u(U - u) \, dy \tag{5.3}$$

The solution to equation (5.1) or (5.2) requires determination of the velocity profile $u(x, y)$ and the boundary layer thickness $\delta(x)$. Following the method of T. von Karman, a polynomial expression may be assumed for u in incompressible laminar flow. The assumption is implied that the ratio u/U is a function of only (y/δ) which indeed will be validated. A polynomial of any order may be assumed and the boundary conditions on u at $y = 0$, δ used to find the coefficients.

As an example let us choose a third degree polynomial velocity distribution as

$$u = a + by + cy^2 + dy^3$$

There results

$$u/U = \tfrac{3}{2}(y/\delta) - \tfrac{1}{2}(y/\delta)^3 \tag{5.4}$$

The coefficients are determined by satisfying the boundary conditions that $u = 0$ at $y = 0$, $u = U$ at $y = \delta$, and $\partial u/\partial y = 0$ at $y = \delta$. Further, the equation of motion, $\partial^2 u/\partial y^2 = 0$, at $y = 0$ is used to complete the determination. These conditions are approximate in that the boundary layer thickness is assumed to be defined by $u = U$ instead of $u = 0.99U$ but are consistent with the polynomial approximation. Then we determine τ_0 as

$$\tau_0 = \mu\frac{\partial u}{\partial y}\bigg|_{y=0} = \tfrac{3}{2}\mu\left(\frac{U}{\delta}\right) \tag{5.5}$$

where μ is the fluid viscosity. Substituting the velocity distribution of equation (5.4) into equation (5.3) and integrating, we have

$$\frac{3}{2}\mu\left(\frac{U}{\delta}\right) = \frac{39}{280}\rho U^2 \frac{d\delta}{dx}$$

Separating variables and integrating,

$$\delta/x = 4.64/\sqrt{Ux/\nu} = 4.64/\sqrt{\mathrm{Re}_x} \tag{5.6}$$

where ν is the kinematic viscosity and Re_x is the Reynolds number based on the length x. Now we can combine equations (5.5) and (5.6) to obtain

$$\tau_0 = 0.323\rho U^2/\sqrt{\mathrm{Re}_x} \tag{5.7}$$

The skin friction coefficient C_f is defined as $C_f = \tau_0/(\frac{1}{2}\rho U^2)$, so that we find the approximate value of C_f as

$$C_f = 0.646\sqrt{\nu/Ux} = 0.646/\sqrt{\mathrm{Re}_x} \tag{5.8}$$

An average value of τ_0, denoted as $\bar{\tau}_0$, and $\bar{C}_f$, an average value of C_f over a plate of length L, may be defined as

$$\bar{\tau}_0 = \frac{\int_0^L \tau_0\, dx}{L} = \frac{0.646\rho U^2}{\sqrt{\mathrm{Re}_l}} = 2\tau_0 \Big|_{x=L} \tag{5.9}$$

$$\bar{C}_f = \frac{\bar{\tau}_0}{\frac{1}{2}\rho U^2} = \frac{1.292}{\sqrt{\mathrm{Re}_l}} \tag{5.10}$$

The values of C_f and τ_0 are based on the third degree polynomial approximation for the velocity profile. In the next section we will find a more exact value of C_f and compare the two.

The same procedure may be used to determine the rate of boundary layer growth for turbulent flow. However, in turbulent flow it is necessary to use an empirical velocity distribution and wall shear stress. Blasius found that for smooth surfaces,

$$\tau_0 = 0.0225\rho U^2(\nu/U\delta)^{1/4}$$

The power law velocity distribution discussed below will be used, where

$$u/U = (y/\delta)^{1/7}$$

Substituting the above shear stress and velocity distribution into equation (5.3),

$$\delta/x = 0.376(Ux/\nu)^{-1/5} = 0.376(\mathrm{Re}_x)^{-1/5} \tag{5.11}$$

where it is assumed that the boundary layer is turbulent from the leading edge. The shear stress is in terms of the distance x:

$$\tau_0 = 0.0286\rho U^2(\nu/Ux)^{1/5} = 0.0286\rho U^2(\mathrm{Re}_x)^{-1/5} \tag{5.12}$$

Prandtl Boundary Layer Equations and the Blasius Solution

If we consider incompressible laminar flow at a large Reynolds number over a flat plate, the Navier-Stokes equations reduce to

$$u\frac{\partial u}{\partial x} + v\frac{\partial u}{\partial y} = -\frac{1}{\rho}\frac{\partial p}{\partial x} + \nu\frac{\partial^2 u}{\partial y^2} \tag{5.13}$$

where

$$\frac{\partial p}{\partial y} \ll \frac{\partial p}{\partial x} \tag{5.14}$$

and u and v are the x and y components of velocity respectively and the flow is assumed to be incompressible with a constant viscosity. These equations may be derived by an order of magnitude analysis of the complete Navier-Stokes equations. The reader may refer to the references for the details.

These equations are generally valid in a cartesian coordinate system even for a curved surface (i.e., cross section of a cylinder with z axis) such as, for example, the surface of an airfoil if the boundary layer thickness δ is much smaller than the radius of curvature. This is generally the case in practical situations. The extension to two-dimensional boundary layer flow will not be discussed here. It is rather complicated but necessary for the flow description over three-dimensional bodies.

The pressure gradient $\partial p / \partial y$ in the direction normal to the plate is neglected compared to $\partial p / \partial x$ and hence the pressure in the boundary layer is imposed by the free stream flow outside the boundary layer. Usually $p(x)$ is found by solving the potential (inviscid flow) over the body outside the boundary layer assuming the boundary layer is of negligible thickness and then evaluating the resulting pressure along the surface of the body. This procedure cannot be used if the boundary layer separates since the potential flow then becomes influenced by the position of separation and the shape of the wake. The flows then become coupled in a complex manner and often experiments must be performed to completely describe the flow.

The continuity equation is

$$\frac{\partial u}{\partial x} + \frac{\partial v}{\partial y} = 0 \tag{5.15}$$

Thus the problem reduces to two equations and two unknowns, u and v. The pressure gradient $\partial p / \partial x$ is determined from the potential flow outside the boundary layer where the viscous effects can be neglected. This problem is taken up in the next chapter.

Consider flow over a flat plate with zero pressure gradient. The equations for this flow are

$$u \frac{\partial u}{\partial x} + v \frac{\partial u}{\partial y} = v \frac{\partial^2 u}{\partial y^2}, \qquad \frac{\partial u}{\partial x} + \frac{\partial u}{\partial y} = 0$$

with boundary conditions: $u = v = 0$ at $y = 0$; $u = U$ at $y = \infty$. It is assumed that the velocity curves at different axial locations are similar in shape (which is indeed the situation here) and thus may be written as

$$u/U = g(y/\delta)$$

where $g(\ )$ is a functional notation.

Further, we let

$$\eta = y \sqrt{U/vx}, \qquad u = \partial \psi / \partial y, \qquad v = -\partial \psi / \partial x, \qquad \psi = \sqrt{vxU} f(\eta)$$

where ψ is known as the stream function defined in Chapter 3 and f is an unknown function to be found. Then in terms of the stream function ψ, the momentum equation becomes

$$\frac{\partial \psi}{\partial y} \frac{\partial^2 \psi}{\partial y \, \partial x} - \frac{\partial \psi}{\partial x} \frac{\partial^2 \psi}{\partial y^2} = v \frac{\partial^3 \psi}{\partial y^3}$$

And in terms of f we have the ordinary differential equation

$$f \frac{d^2 f}{d\eta^2} + 2 \frac{d^3 f}{d\eta^3} = 0 \tag{5.16}$$

and the boundary conditions: $f = f' = 0$ at $\eta = 0$; $f' = 1$ at $\eta = \infty$. This equation was solved by Blasius (see reference 5) by a series expansion. The result is

$$f = \sum_{n=0}^{\infty} \left(-\frac{1}{2} \right)^n \frac{\alpha^{n+1} C_n}{(3n+2)!} \eta^{(3n+2)}$$

where $\alpha = 0.3320$ and the first few values of C_n are

$$C_0 = 1$$
$$C_1 = 1$$
$$C_2 = 11$$
$$C_3 = 375$$
$$C_4 = 27,897$$
$$C_5 = 3,817,137$$

More recently equation (*5.16*) (and extensions of it to include pressure gradients) have been solved extensively by digital computers. A plot u/U is shown in Fig. 5-9.

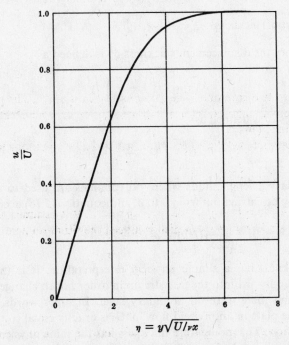

Fig. 5-9. Velocity distribution for laminar flow along a flat plate with zero pressure gradient.

Since the skin friction coefficient is defined by $C_f = \tau_0/(\frac{1}{2}\rho U^2)$ where $\tau_0 = \mu \, \partial u/\partial y|_{y=0}$ we find C_f as

$$C_f = 2\sqrt{v/Ux} \left.\frac{d^2 f}{d\eta^2}\right|_{\eta=0}$$

From the Blasius solution, $d^2f/d\eta^2|_{\eta=0} = 0.332$ and we have

$$C_f = 0.664\sqrt{v/Ux} = 0.664/\sqrt{\mathrm{Re}_x} \qquad (5.17)$$

Comparing this value of C_f to the one obtained previously by the integral method, we see that the third degree polynomial approximation leads to an error in C_f of only about 2.7%, which points out the value of the approximate methods.

Integral Equations

The integral form of the boundary layer equation (5.1) for incompressible flow may also be derived by integrating the Prandtl equation across the boundary layer as follows:

$$\rho \int_0^\delta \left(u \frac{\partial u}{\partial x} + v \frac{\partial u}{\partial y} \right) dy = - \int_0^\delta \frac{\partial p}{\partial x} \, dy + \mu \int_0^\delta \frac{\partial^2 u}{\partial y^2} \, dy$$

By using continuity and the definition of $\tau_0 = \mu(\partial u / \partial y)|_{y=0}$ a straightforward bit of algebra yields the result

$$\rho \frac{d}{dx} \int_0^\delta u(U - u) \, dy - \rho \frac{dU}{dx} \int_0^\delta u \, dy = \delta \frac{\partial p}{\partial x} + \tau_0 \qquad (5.18)$$

which is identical to (5.1) with ρ constant.

Displacement and Momentum Thickness

In boundary layer theory the displacement thickness δ^* is defined as

$$\delta^* = \int_0^\infty \left(1 - \frac{u}{U} \right) dy \qquad (5.19)$$

and the momentum thickness θ is defined as

$$\theta = \int_0^\infty \left(1 - \frac{u}{U} \right) \frac{u}{U} \, dy \qquad (5.20)$$

In the approximate boundary layer methods where polynomials are used to approximate the velocity profile, the integrations may be carried out from 0 to δ, and equation (5.18) becomes

$$\frac{d}{dx} (U^2 \theta) + \rho U \frac{dU}{dx} \delta^* = \tau_0 \qquad (5.21)$$

The displacement thickness has a simple physical interpretation. It is the distance that the wall would have to be displaced outward into the free stream in order not to change the flow field if the fluid were completely inviscid and there were no boundary layer. In other words, the flow reduction that would be due to moving the plate in an inviscid flow, $\delta^* U$, is exactly equal to the flow defect due to the boundary layer in the real flow. The momentum thickness has the same physical significance except that it is based on momentum flux instead of mass flux.

Turbulent Boundary Layers

There is no completely analytical solution for the mean velocity distribution in turbulent flows even for such simple situations as flow over a flat plate and fully-developed pipe flow. The purpose of this section will be to present the semi-empirical methods for describing mean velocity and pressure distributions in turbulent flows. In all of the methods that follow, the results will apply only to two-dimensional flow.

Power Laws

The power laws come from Blasius' resistance formula for smooth straight pipes. They apply as well, however, to other channel flows and two-dimensional boundary layers. We shall assume that the walls are smooth and that the pressure gradient is negligible.

If we assume that the skin friction coefficient, C_f, can be expressed as a power function of Reynolds number, (based on δ), we have

$$C_f = \frac{\text{constant}}{(U\delta/v)^m} \tag{5.22}$$

where U is the centerline velocity for pipe flow and the free stream velocity for boundary layer flow. δ is the radius for pipe flow and boundary layer thickness for flow over a flat plate. Then from the definition of C_f and equation (5.22),

$$U/u_\tau = (\text{constant})(u_\tau \delta/v)^{m/(2-m)} \tag{5.23}$$

where we have introduced the friction velocity u_τ defined as $\sqrt{\tau_0/\rho}$. We assume that the velocity, u, at any distance from the wall, y, may be expressed by an equation similar to (5.23), obtaining

$$u/u_\tau = (\text{constant})(u_\tau y/v)^{m/(2-m)} \tag{5.24}$$

If we further assume that the mean velocity profiles are similar and combine equations (5.23) and (5.24), we obtain

$$u/U = (y/\delta)^{m/(2-m)} \tag{5.25}$$

It is found that by taking $m = \frac{1}{4}$, equation (5.25) expresses the variation of friction coefficients in pipes for $3000 < U\delta/v < 70{,}000$. This then is the advantage of introducing the power law distribution. Ideally one would hope for no change in the exponent m. And, since it varies but weakly with Reynolds number, it is found to be useful in many engineering calculations.

For pipe flow the equations become

$$\left. \begin{array}{l} C_f = 0.0466(U\delta/v)^{-1/4} \\ u/u_\tau = 8.74(u_\tau y/v)^{1/7} \\ u/U = (y/\delta)^{1/7} \end{array} \right\} \quad \begin{array}{l} \text{Pipe flow} \\ 3000 < \text{Re} < 70{,}000 \end{array}$$

For increasing values of Reynolds number the exponent m for the above equation decreases.

Law of the Wall

If we assume that there is a region close to the wall where viscosity is important and that the wall shear stress is the important constraint on the flow (i.e., pressure gradient can be neglected), we may write the following functional relationship for the velocity distribution:

$$u = F(\tau_0, y, \mu, \rho)$$

By dimensional analysis we have

$$u/u_\tau = f(yu_\tau/v) \tag{5.26}$$

Equation (5.26) is the expression for the "law of the wall." It says that if one plots yu_τ/v vs. u/u_τ for many different flows there will be a single curve. This has been verified experimentally for a variety of flows. For example, the results hold well for smooth surfaces and moderate pressure gradients even at large distances from the wall (but not for the outermost portion of the boundary layer). The effect of large adverse pressure gradients is to cause deviations from the law of the wall to occur closer to the wall. Some experimental results are shown in Fig. 5-10.

Velocity Defect Law

We now assume that the viscosity is no longer important for the outer portion of the boundary layer. We assume that the velocity defect $(U - u)$ depends on the shear stress at the wall and the

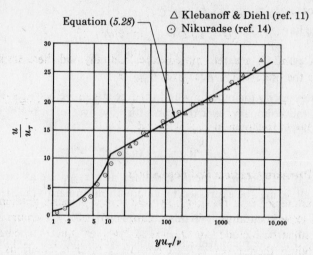

Fig. 5-10. Velocity distribution for the law of the wall. u_τ is the friction velocity, $\sqrt{\tau_0/\rho}$.

distance δ to which the effect has diffused from the wall. Then

$$U - u = G(\tau_0, y, \delta, \rho)$$

and by dimensional analysis we have

$$(U - u)/u_\tau = g(y/\delta) \tag{5.27}$$

The velocity defect law is satisfied experimentally for zero pressure gradient flows for the outer portion of the boundary layer. It is independent of wall roughness. The results are shown in Fig. 5-11.

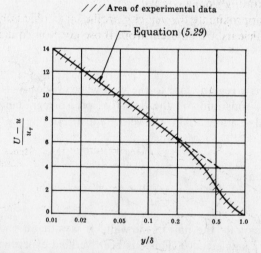

Fig. 5-11. Velocity distribution for velocity defect law.

Logarithmic Forms of Law of the Wall and Velocity Defect Law

It has been observed that there is an intermediate region of overlap where both the law of the wall and the velocity defect law hold simultaneously. In this region of overlap where both laws are valid it can be shown that the functions f and g mentioned above must be logarithmic. That is, equation (5.26) becomes

$$u/u_\tau = C_1 \ln (yu_\tau/v) + C_2 \tag{5.28}$$

and equation (5.27) becomes

$$(U - u)/u_\tau = C_3 \ln (y/\delta) + C_4 \qquad (5.29)$$

The constants C_1, C_2, C_3 and C_4 are determined experimentally and there are slight variations in these quantities as reported in the literature. Clauser (ref. 6) gives

$$C_1 = 2.44, \qquad C_2 = 4.9, \qquad C_3 = -2.44, \qquad C_4 = 2.5$$

Comparison with experiment is shown in Fig. 5-10 and 5-11.

Boundary Layer with a Pressure Gradient and Separation

In general a body immersed in a flowing flow will have a pressure gradient along its surface. This pressure may be found experimentally, or in some cases analytically if separation does not occur. The pressure gradient term must be included in the boundary layer equations, complicating their solution. Numerical methods lending themselves well to boundary layer calculations are used extensively in aerodynamics and other practical applications. We will look at a few examples, using the approximate technique so that the physical behavior does not become obscured by the analysis.

A laminar boundary layer invariably begins on the leading edge or nose of an object immersed in a flow. The boundary layer will become turbulent if given sufficient length to grow. In general as Re_x approaches a critical value downstream (as x increases) the flow undergoes a transition to turbulence. This critical value depends on the surface roughness and turbulence level in the face stream. If a sufficiently large adverse pressure gradient exists, the boundary layer will separate. At the separation point $\tau_0 = 0$ (hence $\partial u/\partial y = 0$) and flow reversal along the surface occurs which feeds the vortices in the wake. Increased turbulence generally delays separation, pushing the separation point further toward the trailing edge. Streamlining the body tends to reduce the adverse pressure gradient and if effected properly prevents separation entirely. The effect of this on drag is discussed in the next section. Figure 5-12 shows a boundary layer separating from a circular cylinder.

As an example, we will approximate the velocity profile by a cubic polynomial, but because there is a pressure gradient the coefficients are different from those given in equation (5.4). The condition that

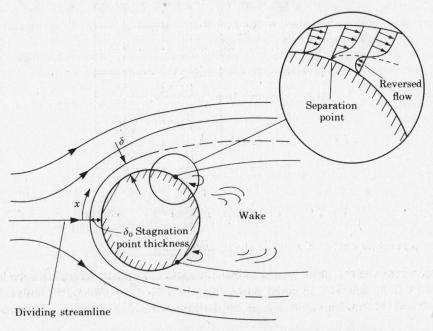

Fig. 5-12. Separation from a cylinder showing stagnation.

$\partial^2 u/\partial y^2 = 0$ at $y = 0$ is no longer valid and instead the condition (which is simply the equation of motion)

$$\mu \frac{\partial^2 u}{\partial y^2} = \frac{dp}{dx} = -\rho U \frac{dU}{dx}$$

at $y = 0$ must be used. Evaluation of the coefficients gives

$$\frac{u}{U} = \left(\frac{3}{2} + \frac{\lambda}{2}\right)\frac{y}{\delta} - \lambda\left(\frac{y}{\delta}\right)^2 - \left(\frac{1}{2} - \frac{\lambda}{2}\right)\left(\frac{y}{\delta}\right)^3 \qquad (5.30)$$

where

$$\lambda = \frac{\delta^2 \rho}{2\mu} \frac{dU}{dx}$$

The wall shear $\mu(\partial u/\partial y)|_{y=0}$ becomes

$$\tau_0 = \mu\left(\frac{3}{2} + \frac{\lambda}{2}\right)\frac{U}{\delta}$$

At the point of separation τ_0 is zero and the above expression gives the value of λ at the separation point.

$$\lambda = -3 = \frac{\delta^2 \rho}{2\mu} \frac{dU}{dx} \qquad (5.31)$$

This is an approximate value for λ at separation. More exact calculations yield a somewhat different result. U may be determined from the potential flow solution but δ must be found from the boundary layer calculations themselves. Hence a problem arises in practice since the wake formation is strongly coupled through the boundary layer to the potential flow. In most instances complex numerical iterative techniques must be used or some experimental data in the form of $p(x)$ are required.

Another problem of interest in aerodynamics is the boundary layer behavior in the vicinity of a stagnation point (Fig. 5-12). At the leading edge or nose of an object the surface is blunt and normal to the free stream flow, unlike a flat plate with a thin sharp leading edge. The flow stagnates just at some point on the nose, dividing the stream into flow over the top and bottom surfaces. Just at this stagnation point δ is not zero but has a finite thickness which can be calculated. Moreover, unlike on the flat plate where τ_0 approaches an infinite value of the leading edge, τ_0 at the stagnation point is zero since $\partial u/\partial y = 0$ there.

For simplicity we will use a quadratic velocity profile approximation. We assume

$$\frac{u}{U} = a + bx + cx^2$$

and evaluate a, b, and c from the conditions $u = 0$, $\partial u/\partial y = 0$ at $y = \delta$ and $u = 0$ at $y = 0$. It may be noted that the pressure gradient does not enter into the determination of the quadratic profile and hence a quadratic profile is not adequate for determination of separation. It does, however, give a reasonable description of the boundary layer in the vicinity of the stagnation point. The result is

$$\frac{u}{U} = 2(y/\delta) - (y/\delta)^2 \qquad (5.32)$$

It is convenient to expand the potential flow (which will be discussed in Chapter 6) at the surface $U(x)|_{y=0}$ in a power series of x, the distance along the surface measured from the stagnation point.

$$U(x)|_{y=0} = U_\infty(Ax + Bx^2 + \cdots) \qquad (5.33)$$

where U_∞ is the far free stream velocity with respect to the object (i.e., speed of the airplane, for example). Retaining only the leading term $U_\infty Ax$ (which is valid for small x near the stagnation point), substituting (5.32) and (5.33) into (5.3), and integrating over the boundary layer there results

$$\frac{x}{15}\frac{d\delta}{dx} = \frac{\mu}{A\rho U_\infty \delta} - \frac{9}{30}\delta \qquad (5.34)$$

which is valid only in the vicinity of the stagnation point. The value of δ_0 (δ at $x = 0$), the stagnation point boundary layer thickness, may be found immediately since $d\delta/dx = 0$ at $x = 0$ and the right hand side of (5.34) must be zero at $x = 0$ so that

$$\delta_0 = \sqrt{\frac{30\,\mu}{9 A \rho U_\infty}}$$

$$= 1.83\sqrt{\frac{\mu}{A\rho U_\infty}} \qquad (5.35)$$

A cubic velocity profile gives a value of $\delta_0 = 2.4\sqrt{\mu/A\rho U_\infty}$ which is the same as the exact solution to two significant figures.

5.3 EXTERNAL FLOWS—LIFT AND DRAG

Whenever an object is placed in a moving fluid (or moves through a stationary fluid) it will experience a force in the direction of the motion of the fluid relative to the object (drag force, D) and it may experience a force normal to the flow direction (lift force, L). These forces may be expressed as

$$D = C_D(\rho V^2/2)A \qquad (5.36)$$

$$L = C_L(\rho V^2/2)A \qquad (5.37)$$

Here A is the characteristic area which is usually the surface area or the projected area normal to the flow direction. Equations (5.36) and (5.37) define the drag coefficient C_D and the lift coefficient C_L. In all except a few cases these coefficients must be experimentally determined and they generally depend on the Reynolds number.

The drag and lift forces are caused by the sum of the tangential and normal forces at the surface of the body. The drag due to tangential stresses is called friction, skin friction or viscous drag. This kind of drag is most important where the surface area parallel to the flow direction is large compared to the projected area normal to the flow. For example, skin friction drag accounts for all of the drag on a flat plate aligned with the flow.

The drag due to normal stresses is called form or pressure drag. Pressure drag is more important and often dominant for bluff bodies. In Chapter 6 it is seen that if the fluid passing the object were frictionless, and thus there were no boundary layer, the drag force would be zero. However, the fluid is not frictionless and a boundary layer exists. The boundary layer grows more rapidly for an adverse pressure gradient and if the pressure gradient is large enough, separation may occur. The large boundary layer or wake over the rear portion of the body results in a lower pressure than would be obtained for frictionless flow. This reduced pressure on the rear portion of the body, then, results in a net force in the direction of flow. This is shown in Fig. 5-13.

We see that in order to reduce pressure drag it is necessary to reduce the magnitude of the adverse pressure gradient over the rear portion of the body and to prevent separation if possible. This means that there should be a long gradual taper over the rear portion of the body. However, if the body is too long, the gain by reducing the pressure drag may be offset by the increase in skin friction drag. The problem of the design of a body for minimum drag is one of compromise between skin friction drag and pressure drag.

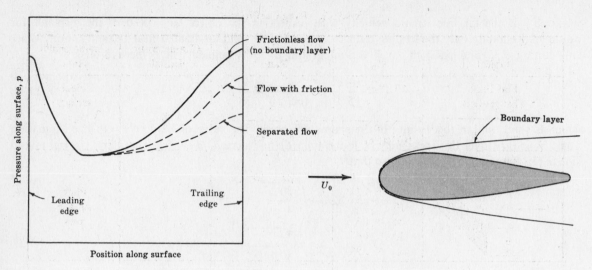

Fig. 5-13. Effect of boundary layer on pressure distribution along the surface of an object.

Drag coefficients for bodies of given shapes depend primarily on the Reynolds number. And, although the free stream turbulence intensity and surface roughness usually have only minor effects on drag, under certain conditions (particularly for non-streamlined bodies where boundary layer separation occurs) they may be very important. Table 5.1 gives the drag coefficients for several different body shapes. For a more complete listing of drag coefficients the reader may refer to Hoerner (ref. 9).

Lift is explained, in principle, by the frictionless flow analysis of Chapter 6.

We will consider the flow past a circular cylinder in order to demonstrate some of the ideas previously expressed concerning the causes of drag. The drag coefficient for a circular cylinder is shown in Fig. 5-14. The flow is shown in Fig. 5-15.

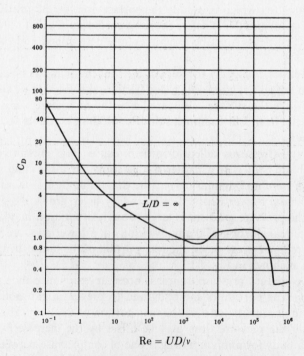

Fig. 5-14. Drag coefficients for circular cylinders.

Table 5.1.   Drag Coefficients

Object	C_D	Reynolds No. Range	Characteristic Length	Characteristic Area
Flat Plate (Tangential)	$1.33(\text{Re})^{-1/2}$	laminar	L	Plate surface area
	$0.074(\text{Re})^{-1/5}$	$\text{Re} < 10^7$		
Flat Plate (Normal)	L/d 1 1.18 5 1.2 10 1.3 20 1.5 30 1.6 ∞ 1.95	$\text{Re} > 10^3$	d	Plate surface area
Circular Disk (Normal)	1.17	$\text{Re} > 10^3$	d	
Sphere	$24(\text{Re})^{1/2}$	$\text{Re} < 1$	d	Projected area
	0.47	$10^3 < \text{Re} < 3 \times 10^5$		
	0.2	$\text{Re} > 3 \times 10^5$		
Hollow Hemisphere	0.34	$10^4 < \text{Re} < 10^6$	d	Projected area
	1.42	$10^4 < \text{Re} < 10^6$		
Solid Hemisphere	0.42	$10^4 < \text{Re} < 10^6$	d	Projected area
	1.17	$10^4 < \text{Re} < 10^6$		
Circular Cylinder	L/d 1 0.63 5 0.8 10 0.83 20 0.93 30 1.0 ∞ 1.2	$10^3 < \text{Re} < 10^5$	d	Projected area
Square Cylinder	2.0	$3.5(10)^4$	d	Projected area

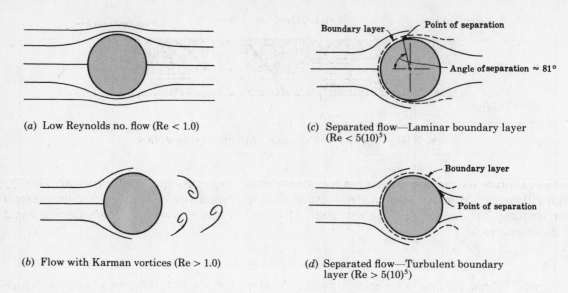

(a) Low Reynolds no. flow (Re < 1.0)

(b) Flow with Karman vortices (Re > 1.0)

(c) Separated flow—Laminar boundary layer (Re < 5(10)^5)

(d) Separated flow—Turbulent boundary layer (Re > 5(10)^5)

Fig. 5-15.   Flow past a circular cylinder.

The character of the flow is determined by the Reynolds number. For example, for a very low Reynolds number (Re < 1) the flow does not separate and skin friction is important. For the very low Reynolds number the drag is proportional to the velocity and the drag coefficient decreases as the Reynolds number increases. As the Reynolds number is increased, the flow will tend to separate. The separation takes place in a periodic way in the form of shedding of Karman vortices. A further increase in Reynolds number will result in a fully separated flow. The laminar boundary layer over the front portion of the cylinder is thin because of the favorable pressure gradient. However, an adverse pressure gradient exists over the rear portion, resulting in a rapid growth of the boundary layer and separation. For a laminar boundary layer the point of separation is located at 81° from the stagnation point. Pressure drag is much greater than skin friction and the drag coefficient is relatively constant for this case.

The boundary layer becomes turbulent at a certain Reynolds number. As a result, there is a sudden decrease in drag coefficient because the point of separation is delayed, resulting in a lower pressure drag and thus lower total drag.

In situations where separation is unavoidable as, for example, flow over a circular cylinder or sphere, the profile (or pressure drag) usually dominates the friction drag at moderate to high Reynolds numbers. In order to minimize the profile drag, separation may be delayed (decreasing the wake size) by increasing the turbulence level in the boundary layer. An effective way to increase the turbulence is to roughen the surface. For example, golf balls are "dimpled" to decrease their drag and greatly increase their flight range. Similarly, baseballs are roughened by their seams and course covers.

5.4   INTERNAL FLOW

Entrance Flows

Internal flows differ from external flows in that in the entry region of an internal flow there is a boundary layer and a uniform free stream which accelerates according to the rate of growth of the boundary layer. A second, and more important, difference exists when the flow becomes fully developed. Here the velocity varies over the entire channel and there is no free stream or well defined boundary layer.

Consider laminar flow in the entry region of a tube as shown in Fig. 5-16. The velocity is uniform at the entrance. The boundary layer grows with distance from the entrance until the flow becomes fully

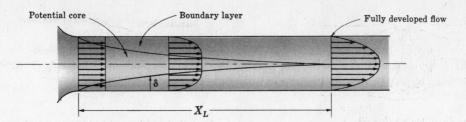

Fig. 5-16.   Flow in the entry region of a pipe for laminar flow.

developed. From the continuity equation it is seen that the frictionless core must accelerate. Then by writing Bernoulli's equation along a streamline in this free stream region it is seen that the pressure must decrease. The laminar development length X_L for the flow to become fully developed was found by Boussinesq to be

$$X_L = 0.03 \mathrm{Re} D$$

Fig. 5-17 shows the flow in the entry region for the case where the Reynolds number is large enough for the flow to be turbulent (Re > 2300). There are several ways of setting criteria for fully developed flow. For example, one may define fully developed flow on the basis of pressure drop, mean velocity distribution or turbulence quantities. The actual lengths for these are substantially different. The pressure gradient generally takes on the fully developed value after three or four diameters of entrance length. The mean velocity requires from 30 to 60 diameters of entrance length before it becomes fully developed. And the turbulence quantities require a much greater length. Strictly speaking, the criterion for establishing fully developed flow should be that the rate of change of all mean quantities (except pressure) with respect to the coordinate in the flow direction is zero. However, the criterion used most frequently in the literature is the point where the mean velocity profiles are not changing with distance in the flow direction.

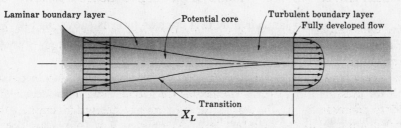

Fig. 5-17.   Flow in the entry region of a pipe for turbulent flow.

Fully Developed Flows

Transition

Flow in a pipe may be well ordered and smooth (laminar) or it may assume a chaotic fluctuating motion (turbulent) superimposed on the mean flow. The character of the flow is determined by wall roughness and Reynolds number. This is demonstrated by the classic Reynolds experiment. A stream of dye is introduced into the flow in a glass tube. For small values of flow rate the dye forms a smooth line. As the flow is increased, a point is reached at which the dye breaks up into jagged patterns indicating turbulent motion. The Reynolds number for the transition from laminar to turbulent flow is approximately 2300. However, for special conditions transition has been known to occur at Reynolds numbers as high as 40,000.

Laminar Flow

Consider fully developed laminar flow between parallel walls as shown in Fig. 5-18. The velocity is a function of y only (so that $\partial u/\partial y$ may be replaced by du/dy) and will be a maximum at the center and zero at the walls. Also, the velocity distribution will be symmetric about the y axis.

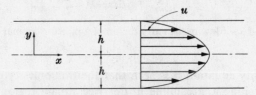

Fig. 5-18. Fully developed laminar flow between parallel walls.

The equation of motion for the x direction for this case becomes

$$0 = -dp/dx + \mu(d^2u/dy^2)$$

Integrating once gives

$$\mu(du/dy) = (dp/dx)y + C_1$$

and using the condition that at $y = 0$, $du/dy = 0$, we have $C_1 = 0$. Since $dp/dx =$ constant for fully developed flow and $\tau = \mu(du/dy)$, we have

$$\tau = (dp/dx)y$$

Thus the shear stress is a linear function of y. This result applies for turbulent flow also.

If we integrate again and apply the condition that $u = 0$ at $y = h$, we have

$$u = \frac{1}{2\mu}\frac{dp}{dx}(y^2 - h^2) \tag{5.38}$$

Next consider fully developed laminar flow in a circular tube as shown in Fig. 5-19. Such flow is known as Poiseuille flow. Again by applying the equation of motion and boundary conditions and integrating directly, we obtain

$$0 = -\frac{dp}{dx} + \mu\left(\frac{d^2u}{dr^2}\right)$$

with the boundary condtions $u = 0, r = R$, and $\partial u/\partial r = 0, r = 0$ with the result for u

$$u = \frac{1}{4\mu}\frac{dp}{dx}(r^2 - R^2) \tag{5.39}$$

The flow rate Q is obtained by integrating the velocity over the cross section of the tube

$$Q = \int_0^R 2\pi r u\,dr = -\frac{\pi R^4}{8\mu}\frac{dp}{dx} \tag{5.40}$$

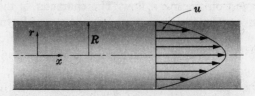

Fig. 5-19. Fully developed laminar flow in a pipe.

For flow in the positive x direction (positive Q) the pressure gradient must be negative (i.e., the pressure decreases in the x direction). The pressure gradient balances the retarding friction forces. Consider a tube of length Δx in the x direction. As an alternative to beginning with the equation of motion we may consider an overall force balance on the fluid in the tube of radius R as

$$\left(P \Big|_x - P \Big|_{x+\Delta x} \right)\pi R^2 + 2\pi R\,\Delta x\left(\mu\,\frac{du}{dr} \right)_{r=R} = 0$$

which yields

$$\frac{dp}{dx} = \frac{\mu}{R}\frac{du}{dr}\Big|_{r=R}$$

If a tube of arbitrary radius r is chosen the force balance gives

$$\frac{dp}{dx} = \frac{\mu}{r}\frac{du}{dr}$$

which is just the first integral of the equation of motion. A subsequent integration would yield the velocity profile (5.39).

Couette Flow

Couette flow is the flow between parallel (or near parallel) surfaces where one of the surfaces is moving laterally in its own plane. If both surfaces are stationary the flow reduces to Poiseuille flow discussed in the previous section. Referring to Fig. 5-20 it is convenient to take the coordinate system attached to the bottom stationary plate. (Both plates are assumed very large in the direction out of the page.) The plates are separated a distance h. As in Poiseuille flow the equation of motion is again

$$0 = -\frac{dp}{dx} + \mu\,\frac{d^2u}{dy^2}$$

but now the boundary conditions are $u = 0$, $y = 0$ and $u = U$, $y = h$. Integrating twice and applying the boundary conditions we obtain

$$u = \frac{1}{2\mu}\frac{dp}{dx}(y^2 - hy) + \frac{Uy}{h}$$

which we recognize as the superposition of a linear profile (due to the motion of the top surface) and a quadratic profile due to the pressure gradient. The quadratic profile here looks a bit different from that of (5.38) but that is because we have defined h differently and taken our origin for y on the bottom plate instead of the centerline. The reader should verify that both expressions are equivalent.

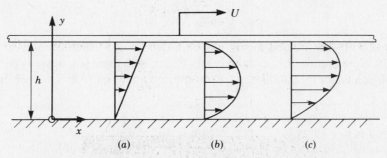

Fig. 5-20. Couette flow. (a) shows a linear profile when no pressure gradient exists, (b) shows the flow (Poiseuille) when $U = 0$ but a pressure gradient is imposed, $dp/dx < 0$, and (c) shows general Couette flow which is a linear superposition of flows (a) and (b).

Integration of the velocity profile across the fluid gives the flow rate per unit depth (into the page)

$$Q = \int_0^h u \, dy = -\frac{h^3}{12\mu}\frac{dp}{dx} + \frac{Uh}{2} \tag{5.41}$$

Friction Factor and Head Loss

Pressure losses will occur in internal flows as a result of friction. These losses, which are important to the engineer, may occur in straight pipes or ducts (major losses) or in sudden expansions, valves, elbows, etc. (minor losses).

The energy equation for a control volume between two points in a flow channel is

$$V_1^2/2 + p_1/\rho + gz_1 = V_2^2/2 + p_2/\rho + gz_2 + u_2 - u_1 - q$$

or
$$V_1^2/2g + p_1/\rho g + z_1 = V_2^2/2g + p_2/\rho g + z_2 + H_L \tag{5.42}$$

where $H_L = (u_2 - u_1 - q)/g$ is the head loss and u here is the specific internal energy. This H_L term actually represents the decrease (loss) in mechanical energy between points 1 and 2 and in general contains both major and minor losses.

We will now consider the methods of determining the losses. There is no way of determining the losses for turbulent flow by a purely analytical method. Thus the results are highly empirical in nature.

First, let us look at the method of determining the major losses. Our analysis will be restricted to fully-developed, incompressible turbulent flow in a constant diameter tube.

By observing equation (5.42) we see that pressure changes result from velocity changes, elevation changes and friction losses. For the constant area, incompressible case being considered, we have $V_1 = V_2$ and we assume $z_1 = z_2$. Thus equation (5.42) becomes $H_L = (p_1 - p_2/\rho g)$.

The pressure change is known to depend on (1) pipe diameter D, (2) average velocity V, (3) length L, (4) viscosity μ, (5) density ρ, and (6) wall roughness ϵ. Thus

$$\Delta p = F(D, V, L, \mu, \rho, \epsilon)$$

From dimensional analysis we get four dimensionless parameters, or,

$$\frac{\Delta p}{\rho V^2} = G(\rho V D/\mu, \epsilon/D, L/D)$$

Experiments show that two of the parameters may be combined to give

$$\frac{\Delta p}{\frac{1}{2}\rho V^2}\left(\frac{D}{L}\right) = f(\rho V D/\mu, \epsilon/D)$$

or
$$\frac{\Delta p}{\rho g} = H_L = \left(\frac{L}{D}\right)\frac{V^2}{2g} f \tag{5.43}$$

where f = friction factor. The friction factor has been determined experimentally and the results are shown in Fig. 5-21. Note that for Reynolds numbers below 2000 there is a single curve because the flow is laminar. The friction factor for laminar flow is determined analytically as (see Problem 5.6)

$$f = 64/\text{Re} \tag{5.44}$$

The friction factor f is known as the Darcy friction factor and may be defined in terms of wall shear stress, τ_0, as $f = 8\tau_0/\rho V^2$. This can be verified by considering the static equilibrium of a control volume of fluid by balancing the pressure forces with the shear forces.

Next we will consider minor losses. In accounting for losses in elbows, valves and expansions one must resort to experiment. It is customary to write such losses in the form

$$H_L = KV^2/2g \tag{5.45}$$

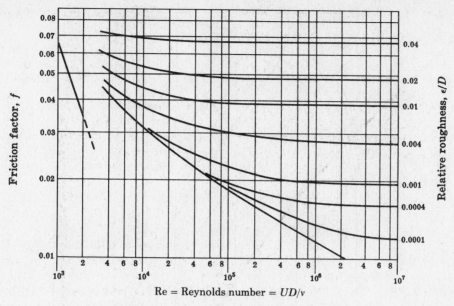

Fig. 5-21.   Friction factors for flow in pipes.

where K is a friction loss coefficient for various types of minor losses and is given for commercial pipe fittings in handbooks. Table 5.2 lists some approximate K values.

Table 5.2.   Head Loss Coefficients for Minor Losses, K

Valves, Fittings and Piping	K
Globe valve (wide open)	10.0
Gate valve (wide open)	0.19
90° elbow	0.90
45° elbow	0.42
Sharp-edged entrance to circular pipe	0.50
Rounded entrance to circular pipe	0.25
Sudden expansion	$(1 - A_1/A_2)^{2*}$

*A_1 = upstream area, A_2 = downstream area

Velocity Distributions for Turbulent Flow

Typical velocity distributions for fully developed flow in a pipe are shown in Fig. 5-22. The velocity is approximated by the power law velocity as was previously used for a flat plate. We have

$$u/u_{\max} = (y/R)^{1/n} \tag{5.46}$$

where y is the distance from the pipe wall measured toward the center.

The exponent $1/n$ varies weakly with Reynolds numbers from 1/6 to 1/10 for Reynolds numbers ranging from 4×10^3 to 3×10^6.

The logarithmic form of the law of the wall may be used to approximate the velocity distribution except for the region not close to the centerline. Here we have

$$u/u_\tau = 2.44 \ln (yu_\tau/v) + 4.9 \tag{5.47}$$

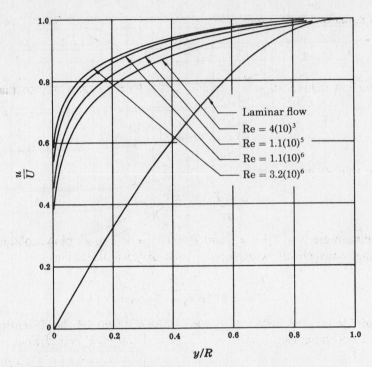

Fig. 5-22. Velocity distribution for fully developed flow in smooth pipes (ref. 14).

5.5 THERMAL ASPECTS OF VISCOUS FLOW

We have concentrated on incompressible flow in the previous sections. If the flow were compressible the density would depend on pressure and temperature and the energy equation would have to be satisfied along with the equations of motion and continuity, greatly complicating the calculations. Often compressible flow may be approximated as isothermal or isentropic, which allows simplification. Such approximations are discussed in subsequent chapters. Moreover, even if the flow is incompressible the temperature may be coupled into the descriptive equations if viscosity variations with temperature become important.

For the present we will not further discuss these problems where energy and motion are coupled and must be solved simultaneously. However, the temperature distribution in incompressible constant viscosity flow may be found directly once the velocity profile is found, since velocity and pressure are uncoupled from the temperature and energy considerations. This temperature distribution in the fluid is often needed, particularly in order to find heat transfer characteristics in flowing fluid. Convection film coefficients are found by determining the temperature profile in the boundary layer over a surface where the surface and free stream are at different temperatures.

As a simple example we will find the temperature profile in fully developed (laminar) Couette flow where the two plates are held at different temperatures. The heat generated by viscous shear (viscous dissipation) will be taken into account, although in many practical situations this dissipation is negligible. Some examples of situations where the dissipation is important are in high shear flows (such as thin lubricant films), in boundary layers in high speed (supersonic) flight, and in large volume rate pipe flows. The oil flowing in the Alaskan pipeline (which is insulated) is maintained at a temperature considerably above ambient by frictional dissipation.

Consider the fully developed steady Couette flow (without a pressure gradient) shown in Fig. 5-23. The linear velocity profile is simply

$$\frac{u}{U} = \frac{y}{h}$$

The general energy equation (*3.67*) simplifies to

$$0 = \kappa \frac{\partial^2 T}{\partial y^2} + \Phi$$

The dissipation function reduces to $\mu(\partial u/\partial y)^2$ and u is a function of y only so that the appropriate energy equation is

$$0 = \kappa \frac{d^2 T}{dy^2} + \mu \left(\frac{du}{dy}\right)^2$$

and using the linear velocity profile we obtain

$$\frac{d^2 T}{dy^2} = -\frac{\mu}{\kappa}\left(\frac{U}{h}\right)^2 \tag{5.48}$$

The boundary conditions are $T = T_0$, $y = 0$ and $T = T_1$, $y = h$. The solution is obtained by integrating directly for T and may conveniently be expressed in non-dimensional form as

$$\frac{T - T_0}{T_1 - T_0} = \frac{y}{h} + \frac{\mu U^2}{2\kappa(T_1 - T_0)}\left(\frac{y}{h}\right)\left(1 - \frac{y}{h}\right) \tag{5.49}$$

The temperature profile may be further simplified by making use of the dimensionless parameters, Prandtl number and Eckert number. Remember the Prandtl number P_r is defined as

$$P_r = \frac{\mu c_p}{\kappa} = \frac{v}{\alpha} \approx \frac{\mu c_v}{\kappa}$$

where α, the thermal diffusivity, is $\kappa/\rho c_p$ (and we have made use of the fact that $c_p \approx c_v$ for a liquid). The Eckert number, E_c, is defined as

$$E_c = \frac{U^2}{c_p(\Delta T)} \approx \frac{U^2}{c_v(\Delta T)}$$

where ΔT is the characteristic temperature difference ($T_1 - T_0$). Equation (*5.48*) may be written in the dimensionless form

$$\frac{T - T_0}{T_1 - T_0} = \eta - \tfrac{1}{2}E_c P_r \dot{\eta}(1 - \eta) \tag{5.50}$$

where $\eta = y/h$. This profile is plotted in Fig. 3-23.

The product $E_c P_r$ is a dimensionless measure of dissipation, and if this term is negligible the temperature profile is, of course, a straight line.

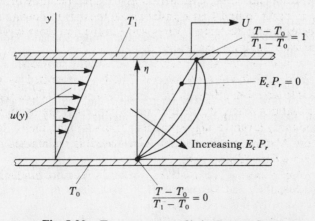

Fig. 5-23. Temperature profile in Couette flow.

Another interesting aspect of this problem is the adiabatic wall temperature. If the top plate is insulated while the bottom plate is held at T_1, the temperature will "float" thermally and its temperature will be the adiabatic wall temperature T_{ad}. T_{ad} may be found immediately by setting the heat flux into the top plate to zero and solving for the resultant wall or plate temperature. From (5.49) we set $dT/dy|_{y=h} = 0$ and solve for the resultant T_1 which is T_{ad}. The result is

$$T_{ad} - T_0 = \frac{\mu U^2}{2\kappa} \qquad (5.51)$$

or in terms of the Eckert number E_{ad} which is defined in terms of $(T_{ad} - T_0)$ we have the surprisingly simple result

$$E_{ad} P_r = 2 \qquad (5.52)$$

We will not pursue the thermal aspects of fluid flow further at this time, but the concept of heat generation by dissipation is of great importance in many aspects of fluid mechanics.

5.6 SUMMARY

In this chapter we have looked at flow with friction. It was convenient to divide the discussion into external flows and internal flows. The former was concerned with flow around objects and the latter with flow in pipes and ducts.

For external flows we obtained expressions for boundary layer growth on a flat plate, equation (5.6), and wall shear stress, equation (5.7). These were obtained from the control volume momentum integral equation where a velocity distribution was assumed.

The Blasius velocity distribution for laminar flow on a flat plate with zero pressure gradient was presented. Also the methods of describing turbulent velocity distributions were discussed. These were

$$u/U = (y/\delta)^{m/(2-m)} \qquad \text{(Power law)} \qquad (5.25)$$

$$u/u_\tau = 2.44 \ln (yu_\tau/v) + 4.9 \qquad \text{(Log form of law of wall)} \qquad (5.28)$$

$$(U - u)/u_\tau = -2.44 \ln (y/\delta) + 2.5 \qquad \text{(Log form of velocity defect law)} \qquad (5.29)$$

Methods of describing drag and lift were presented. There are two kinds of drag; one is skin friction drag which depends on the tangential stresses and the second is form drag which depends on the pressure distribution.

The pressure drag is strongly influenced by separation which may occur at some point on the rear portion on the body. The location of the point of separation is determined largely by the rate of growth of the boundary layer which in turn is determined by the external pressure gradient. Turbulence tends to delay separation and reduce the pressure drag.

Internal flows were considered in terms of entry flows and fully developed flows. The velocity distributions for fully developed laminar flow between parallel walls and in a circular pipe were derived.

Transition from laminar to turbulent flow in a circular tube occurs at a Reynolds number of 2300 nominally, although much higher values have been measured under special conditions. The shear stress for both fully developed laminar and turbulent flow was found to be a linear function of distance from the centerline.

The pressure changes for internal flow were presented in a highly empirical manner in terms of "head" or "friction loss". For flow in constant diameter pipes the results were given in terms of friction factor which depends on Reynolds number and wall roughness. For other types of losses (elbows, valves, etc.) the losses were given in terms of an empirical loss coefficient.

References

1. Anderson, D. A., Tannehill, J. C., and Pletcher, R. H., *Computational Fluid Mechanics and Heat Transfer*, Hemisphere Publishing Corp., 1984.

2. Baker, A. J., *Finite Element Computational Fluid Mechanics*, Hemisphere Publishing Corp., 1983.

3. Batchelor, G. K., *An Introduction to Fluid Dynamics*, Cambridge University Press, 1967.

4. Bird, R. B., Stewart, W. E., and Lightfoot, E. N., *Transport Phenomena*, John Wiley, 1960.

5. Blasius, H., "Grenzschichten in Flussigkeiten mit kleiner Reibung," *Z. Math. u. Phys.*, 56, 1, 1908, English translation NACA TM, No. 1256.

6. Clauser, F. H., *The Turbulent Boundary Layer, Advances in Applied Mechanics*, Vol. 4, pp. 1–51, Academic Press, 1956.

7. Fletcher, C. A. J., *Computational Techniques for Fluid Dynamics*, Vols. I and II, Springer-Verlag, 1988.

8. Goldstein, S., *Modern Developments in Fluid Dynamics*, Vols. I and II, Oxford University Press, 1938; also Dover Publications.

9. Hoerner, S. F., *Fluid Dynamic Drag*, published by author, Midland Park, New Jersey, 1958.

10. Hughes, W. F., *An Introduction to Viscous Flow*, Hemisphere Publishing Corp., 1979.

11. Klebanoff, P. S., and Diehl, F. W., *Some Features of Artificially Thickened Fully Developed Turbulent Boundary Layers with Zero Pressure Gradient*, NACA Report 1110, 1952.

12. Lai, W. M., Rubin, D., and Krempl, E., *Introduction to Continuum Mechanics*, Pergamon Press, 1974.

13. Li, W. H., and Lam, S. H., *Principles of Fluid Mechanics*, Addison-Wesley, 1964.

14. Nikuradse, J., *Gesetzmässigkeiten der turbulenten Strömung in glatten Rohren*, Forschungs-Arb. Ing.-Wesen, 356, 1932.

15. Schlichting, H., *Boundary Layer Theory*, 7th ed., McGraw-Hill, 1986.

16. Shames, I. H., *Mechanics of Fluids*, McGraw-Hill, 1962.

17. Shapiro, A. H., *Shape and Flow*, Anchor Books, Doubleday, 1961.

18. White, F. M., *Viscous Fluid Flow*, McGraw-Hill, 1974.

Solved Problems

5.1. Air at 70°F, 14.7 psi and 50 fps free stream velocity, flows over a flat plate. Determine the boundary layer thickness at a point 5 ft from the leading edge.

The kinematic viscosity for air at these conditions is $v = 1.8 \times 10^{-4}$ ft^2/sec. The Reynolds number based on the plate length is

$$R_{e_x} = Ux/v = 50(5)/(1.8 \times 10^{-4}) = 1.39 \times 10^6$$

Now we will determine the boundary layer thickness assuming the boundary layer is laminar. Using equation (5.5),

$$\delta/x = 4.64/\sqrt{R_{e_x}}, \qquad \delta/5 = 4.64/(1.39 \times 10^6)^{1/2}, \qquad \delta = 0.0197 \text{ ft} = 0.236 \text{ in.}$$

If we assume the boundary layer is turbulent from the leading edge,

$$\delta/x = 0.376/(R_{e_x})^{1/5}, \qquad \delta/5 = 0.376/(13.9 \times 10^5)^{1/5}, \qquad \delta = 0.111 \text{ ft} = 1.33 \text{ in.}$$

We can see that there is a large difference in the answer depending on whether we assume laminar or turbulent flow. And neither answer is correct since the boundary layer is neither completely laminar or turbulent over the entire length. The length for transition to occur has been found experimentally and is given approximately when $R_{e_x} = 3.2 \times 10^5$. Thus the transition length X_T for this problem is

$$X_T = (v/U)R_{e_x} = [(1.8 \times 10^{-4})/50](3.2 \times 10^5) = 1.15 \text{ ft}$$

Thus the answer assuming a turbulent boundary layer would be closer to the correct value.

5.2. Work Problem 5.1 taking into account that part of the boundary layer is laminar and the rest is turbulent.

The boundary layer thickness at the point of transition is

$$\delta_{transition} = \frac{1.15(4.64)}{[3.2 \times 10^5]^{1/2}} = 0.00944 \text{ ft} = 0.113 \text{ in.}$$

At this point the turbulent boundary layer starts. If the boundary layer were turbulent from the leading edge to this point, the length X required would be given as

$$\frac{\delta}{X} = \frac{0.376}{(UX/\nu)^{1/5}} \quad \text{or} \quad X^{4/5} = \frac{\delta(U/\nu)^{1/5}}{0.376}$$

so that

$$X = \left(\frac{\delta}{0.376}\right)^{5/4}\left(\frac{U}{\nu}\right)^{1/4} = \left(\frac{0.944 \times 10^{-2}}{0.376}\right)^{5/4}\left(\frac{50}{1.8 \times 10^{-4}}\right)^{1/4} = 0.230 \text{ ft}$$

This means that the total length of the equivalent turbulent boundary layer would be

$$X = (5.0 - 1.15) + 0.23 = 4.08 \text{ ft}$$

and

$$\delta = \frac{4.08(0.376)}{[50(4.08)/(1.8 \times 10^{-4})]^{1/5}} = \frac{1.53}{[11.3 \times 10^5]^{1/5}} = 0.0942 \text{ ft} = 1.13 \text{ in.}$$

5.3.   Determine the total drag force for Problem 5.1.

Assuming that the boundary layer is laminar, from equation (5.6) we have

$$\tau_0 = \frac{0.323\rho U^2}{(U/\nu)^{1/2}x^{1/2}} = \frac{0.323(0.00237)(50)^2 x^{-1/2}}{[50/(1.8 \times 10^{-4})]^{1/2}} = (3.62 \times 10^{-3})x^{-1/2}$$

The drag is

$$D = \int_A \tau_0 \, dA = W \int_0^5 \tau_0 \, dx$$

so that

$$\frac{D}{W} = 3.62 \times 10^{-3} \int_0^5 x^{-1/2} \, dx = 0.0163 \text{ lb}_f/\text{ft}$$

Assuming that the flow is turbulent,

$$\tau_0 = \frac{0.0286\rho U^2}{(U/\nu)^{1/5}x^{1/5}} = \frac{0.0286(0.00237)(50)^2 x^{-1/5}}{[50/(1.8 \times 10^{-4})]^{1/5}} = 0.0139x^{-1/5}$$

and

$$\frac{D}{W} = 0.0139 \int_0^5 x^{-1/5} \, dx = 0.0627 \text{ lb}_f/\text{ft}$$

5.4.   How would the boundary layer thickness be affected in Problem 5.1 if the plate were curved as shown in Fig. 5-24?

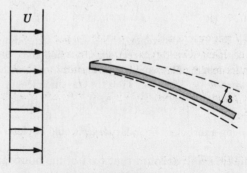

Fig. 5-24

Along the top surface there would be an adverse pressure gradient. This means that the boundary layer would grow more rapidly than for the flat plate case where the pressure gradient is zero. The opposite is true along the bottom surface; here we have a favorable pressure gradient and a more slowly growing boundary layer. Thus

$$\delta_{top} > \delta_{flat\ plate} = 1.13 \text{ in.}$$

$$\delta_{bottom} < \delta_{flat\ plate} = 1.13 \text{ in.}$$

5.5. Determine the total force exerted by an 80 mph wind on a 10 ft × 40 ft billboard. The wind is blowing at right angles to the billboard.

Assume standard air with $v = 1.8 \times 10^{-4}$ ft^2/sec and $\rho = 0.00237$ slug/ft^3, so that

$$Re = UW/v = 117(10)/(1.8 \times 10^{-4}) = 6.50 \times 10^6$$

where W is the width, 10 ft, and the velocity of the air is 80 mi/hr = 117 ft/sec. From Table 5.1, we find $C_D = 1.2$ and (5.36) gives

$$D = C_D(\tfrac{1}{2}\rho U^2)A = 1.2(\tfrac{1}{2})(0.00237)(117)^2(400) = 7850 \text{ lb}_f$$

5.6. Derive an expression for the friction factor for fully developed flow in a circular tube of length L where the Reynolds number is less than 2000.

The velocity distribution as given by equation (5.39) for laminar flow is

$$u = \frac{1}{4\mu}\frac{dp}{dx}(r^2 - R^2)$$

The volume flow rate, Q, is

$$Q = \int u\, dA = \int_0^R 2\pi u r\, dr$$

Integrating,

$$Q = -\frac{\pi R^4}{8\mu}\frac{dp}{dx}$$

and if V_{av} is the average velocity,

$$V_{av} = \frac{Q}{A} = \frac{Q}{\pi R^2} = -\frac{R^2}{8\mu}\frac{dp}{dx}$$

Then

$$\frac{dp}{dx} = -\frac{\Delta p}{L} = -\frac{8\mu V_{av}}{R^2}$$

From equation (5.43), we have

$$f = \frac{\Delta p}{L}\frac{4R}{\rho V_{av}^2}$$

and hence

$$f = \frac{8\mu V_{av}}{R^2}\frac{4R}{\rho V_{av}^2} = \frac{32\mu}{\rho R V_{av}} = \frac{64}{(\rho V_{av} D/\mu)} = \frac{64}{Re}$$

5.7. Air at 70°F and 14.7 psi enters a 1.0 in. diameter pipe (Fig. 5-25) with a uniform velocity and a Reynolds number of 1000. Determine the decrease in pressure in going from the entrance to 100 in. downstream from the entrance. The entrance length L_e is given by $L_e/D = 0.0288\, R_e$.

For air under the given conditions, $v = 1.8 \times 10^{-4}$ ft^2/sec, $\rho = 0.00237$ slug/ft^3. The entrance length L_e is

$$L_e = 0.030(1000)(1.0) = 30 \text{ in.}$$

The uniform velocity at the entrance is

$$U_0 = 1000v/D = 1000(1.8 \times 10^{-4})(1.0/12) = 2.16 \text{ ft/sec}$$

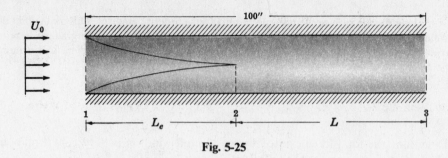

Fig. 5-25

From Problem 5.6 we see that the average velocity is

$$U_0 = V_{av} = -\frac{R^2}{8\mu}\frac{dp}{dx}$$

and the maximum velocity for fully developed flow is ($r = 0$)

$$U_{max} = -\frac{R^2}{4\mu}\frac{dp}{dx}$$

Thus $U_{max} = 2U_0 = 2(2.16) = 4.32 \text{ ft/sec}$

We note that in going from section 1 to section 2 along the centerline the flow is frictionless. Thus we can apply Bernoulli's equation to determine the pressure difference between those two points.

$$p_1 - p_2 = \tfrac{1}{2}\rho(U_{max}^2 - U_1^2) = \tfrac{1}{2}(0.00237)[(4.32)^2 - (2.16)^2] = 0.0166 \text{ lb}_f/\text{ft}^2$$

From Problem 5.6 we have (between points 2 and 3)

$$p_2 - p_3 = \Delta p = \frac{8\mu U_1 L}{R^2} = \frac{8(1.8 \times 10^{-4})(0.00237)(2.16)(70/12)}{(0.5/12)^2} = 0.0250 \text{ lb}_f/\text{ft}^2$$

Then the total pressure drop is

$$p_1 - p_3 = (p_1 - p_2) + (p_2 - p_3) = 0.0166 + 0.0250 = 0.0416 \text{ lb}_f/\text{ft}^2$$

5.8. Determine the flow rate for the system shown in Fig. 5-26. The fluid is water and the pipe is hydrodynamically smooth. Neglect all losses except through the pipe.

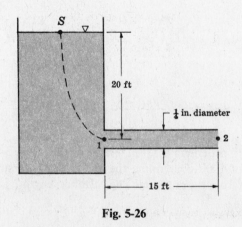

Fig. 5-26

First we write the energy equation (5.42) between points 1 and 2.

$$V_1^2/2g + p_1/\rho g + z_1 = V_2^2/2g + p_2/\rho g + z_2 + H_L$$

where $H_L = f(L/D)(V^2/2g)$

Since $V_1 = V_2$ and $z_1 = z_2$, we have

$$(p_1 - p_2)/\rho g = f(L/D)(V^2/2g) = H_L$$

Next we write Bernoulli's equation between the free surface and point 1 along the dotted line as shown.

$$p_S/\rho g + V_S^2/2g + z_S = p_1/\rho g + V_1^2/2g + z_1$$

We note also that $p_2 = p_{\text{atm}} = p_S$. Then we can combine the two equations to obtain

$$f(L/D)(V_1^2/2g) = H_L = (z_S - z_1) - V_1^2/2g$$

$$V_1^2 = \frac{2g(z_S - z_1)}{f(L/D) + 1} = \frac{64.4(20)}{f[15/(1/48)] + 1} = \frac{1290}{f(720) + 1}$$

The friction factor depends on the Reynolds number which depends on the velocity. Thus we must solve the problem by trial and error. The Reynolds number Re is given by

$$\text{Re} = V_1 D/\nu = V_1(1/48)/(1 \times 10^{-5}) = 2080 V_1$$

The table below shows the trial and error solution. We assume a value for V_1 and then calculate V_1.

V_1 (assumed)	Re	f	V_1 (calculated)
10	$2.08(10)^4$	0.026	8.10
7	$1.46(10)^4$	0.028	7.70
7.9	$1.64(10)^4$	0.0272	7.92

Thus $V_1 = 7.92$ ft/sec and $Q = AV = AV_1 = \frac{1}{4}\pi D^2 V_1 = \frac{1}{4}\pi(1/48)^2(7.92) = 0.00270$ ft^3/sec.

5.9.   Determine the flow rate for the system shown in Fig. 5-27. The fluid is water and the pipe is hydrodynamically smooth. First find the flow rate neglecting minor losses, then find the flow rate with the minor losses included.

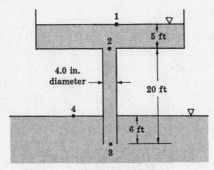

Fig. 5-27

First we write Bernoulli's equation between points 1 and 2.

$$\frac{p_1}{\rho g} + \frac{V_1^2}{2g} + z_1 = \frac{p_2}{\rho g} + \frac{V_2^2}{\rho g} + z_2$$

or
$$\frac{p_2 - p_1}{\rho g} = (z_1 - z_2) - \frac{V_2^2}{2g} \qquad (a)$$

Bernoulli's equation between points 3 and 4 gives

$$\frac{p_3}{\rho g} + \frac{V_3^2}{2g} + z_3 = \frac{p_4}{\rho g} + \frac{V_4^2}{2g} + z_4$$

or
$$\frac{p_3 - p_4}{\rho g} = (z_4 - z_3) - \frac{V_3^2}{2g} \qquad (b)$$

Next write the energy equation between 2 and 3 where head losses occur:

$$\frac{p_2}{\rho g} + \frac{V_2^2}{2g} + z_2 = \frac{p_3}{\rho g} + \frac{V_3^2}{2g} + z_3 + H_L$$

which becomes

$$\frac{p_2 - p_3}{\rho g} = (z_3 - z_2) + \frac{fL}{D}\frac{V^2}{2g} \tag{c}$$

where $V_2 = V_3 = V$ and $p_1 = p_4 = p_{atm}$. Combining equations (a) and (b) gives

$$\frac{p_2 - p_3}{\rho g} = (z_1 - z_2) - (z_4 - z_3)$$

Now combining this equation with equation (c) yields

$$(z_3 - z_2) + \frac{fL}{D}\frac{V^2}{2g} = (z_1 - z_2) - (z_4 - z_3)$$

so that

$$V^2 = \frac{2g(z_1 - z_4)}{fL/D} = \frac{64.4(19)}{f(20/\frac{1}{3})} = \frac{20.4}{f}$$

Since both f and V are unknown we must work simultaneously with the friction factor diagram to obtain a solution. $Re = VD/\nu = 3.33(10)^4 V$. We use the method of the previous problem:

V (assumed)	Re	f	V (calculated)
10	$3.33(10)^5$	0.0143	37.9
40	$1.33(10)^6$	0.0114	42.3
43	$1.43(10)^6$	0.0112	43.0

and we arrive at $V = 43.0$ ft/sec and $Q = \frac{1}{4}\pi D^2 V = 3.75$ ft^3/sec, where we have neglected minor losses. Minor losses would occur at the entrance and exit of the straight pipe. This would have the effect of altering the pressures as indicated by Bernoulli's equation in the previous calculations. We may account for these losses by modifying equation (c) as follows:

$$\frac{p_2 - p_3}{\rho g} = (z_3 - z_2) + \frac{fL}{D}\frac{V^2}{2g} + K_1 \frac{V^2}{2g} + K_2 \frac{V^2}{2g}$$

and we obtain by combining with equations (a) and (b) as before,

$$\frac{fL}{D}\frac{V^2}{2g} + K_1 \frac{V^2}{2g} + K_2 \frac{V^2}{2g} = z_1 - z_4$$

or

$$V^2 = \frac{2g(z_1 - z_4)}{fL/D + K_1 + K_2} = \frac{64.4(19)}{f(20/\frac{1}{3}) + 0.5 + 1.0} = \frac{1220}{f(60) + 1.5}$$

where K_1 = entrance loss coefficient = 0.50 and K_2 = exit loss coefficient = 1.0, obtained from Table 5.2. Then by trial and error we determine f and V as before.

V (assumed)	Re	f	V (calculated)
10	$3.33(10)^5$	0.0143	22.8
24	$8.0(10)^5$	0.0122	23.4
23.3	$7.76(10)^5$	0.0122	23.4

We find $V = 23.4$ ft/sec and $Q = \frac{1}{4}\pi D^2 V = \frac{1}{4}\pi(4/12)^2(23.4) = 2.04$ ft^3/sec.

5.10. Water flows from a large reservoir and discharges into the atmosphere as shown in Fig. 5-28. Determine the volume flow rate.

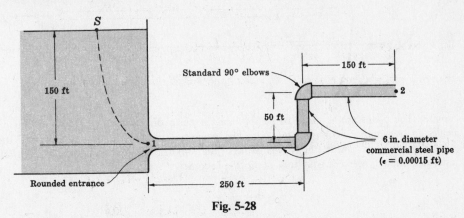

Fig. 5-28

First write Bernoulli's equation between the free surface and the entrance. By continuity, $V_1 = V_2 = V$.

$$\frac{p_1 - p_S}{\rho g} = (z_S - z_1) - \frac{V_1^2}{2g}$$

Next write the energy equation between sections 1 and 2:

$$\frac{p_1 - p_2}{\rho g} = (z_2 - z_1) + \frac{fL}{D}\frac{V^2}{2g} + K_1\frac{V^2}{2g} + 2K_2\frac{V^2}{2g}$$

where K_1 = loss coefficient for rounded entrance = 0.25, K_2 = loss coefficient for elbow = 0.90. Combining the two equations gives

$$(fL/D + K_1 + 2K_2 + 1)\frac{V^2}{2g} = (z_S - z_2)$$

or
$$V^2 = \frac{2g(z_S - z_2)}{fL/D + K_1 + 2K_2 + 1} = \frac{64.4(100)}{f(450/\frac{1}{2}) + 0.25 + 2(0.90) + 1} = \frac{6440}{f(900) + 3.05}$$

Since both f and V are unknown, we must work simultaneously with the friction factor diagram to obtain a solution. Re is given as

$$\text{Re} = VD/\nu = V(\tfrac{1}{2})/(1 \times 10^{-5}) = (5 \times 10^4)V \qquad \text{and} \qquad \epsilon/D = 0.00015/\tfrac{1}{2} = 0.00030$$

V (assumed)	Re	f	V (calculated)
10	$5(10)^5$	0.0165	361
370	$1.85(10)^7$	0.0150	390
390	$1.95(10)^7$	0.0150	390

We find $V = 390$ ft/sec and $Q = AV = \frac{1}{4}\pi D^2 V = \frac{1}{4}\pi(\frac{1}{2})^2(390) = 76.6$ ft³/sec.

5.11. Discuss the fully developed viscous laminar flow between two parallel plates, one of which is in motion as shown in Fig. 5-29. Such flow is known as Couette flow. The top plate has velocity U with respect to the bottom plate. The pressure at the inlet of the flow region is p_2 and the pressure at the outlet is p_1.

We assume that the length L of the plate is much larger than the spacing h, so that the entrance or development effects may be neglected. The flow is essentially one dimensional then, with variations in velocity only in the y direction across the channel. A coordinate system is attached to the top plate which is

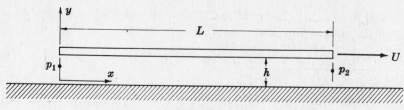

Fig. 5-29

moving. The plates are assumed to be very long in the z direction so that flow in that direction may be neglected. Assuming incompressible flow, the equations of motion may be obtained directly from the general equation of motion (the Navier-Stokes equation) or one may take a small element of fluid and write a momentum balance. The equations of motion become:

$$0 = -\frac{\partial p}{\partial x} + \mu \frac{\partial^2 u}{\partial y^2}, \qquad 0 = -\frac{\partial p}{\partial y}$$

where μ is the fluid viscosity, and u and v are respectively the x and y components of velocity measured relative to the top plate. The coordinate system is attached to and moves with the top plate, but the origin is made to coincide with the bottom plate for convenience. Then, relative to the top plate and coordinate system the bottom surface has velocity $-U$.

Since there is no variation in the x direction, the continuity equation $\partial u/\partial x + \partial v/\partial y = 0$ tells us that the y component of velocity, v, must be zero and hence $\partial p/\partial y$ must be zero and the pressure must be only a function of x. Integration of the equations of motion along with the boundary conditions that $u = -U$ at $y = 0$ and $u = 0$ at $y = h$ gives

$$u = \frac{1}{2\mu}\frac{dp}{dx}(y^2 - hy) + U[(y/h) - 1]$$

Since the pressure p is a function only of x, and u is a function only of y, the total flow rate $Q = \int_0^h u\, dy$ may be evaluated and the pressure drop found as

$$Q = -\frac{h^3}{12\mu}\frac{dp}{dx} - \frac{Uh}{2} \qquad p_2 - p_1 = -\frac{12L\mu}{h^3}\left(\frac{Uh}{2} + Q\right)$$

The above equations tell us that the pressure gradient is a constant and relates the pressure drop to the flow rate Q, which is the flow rate relative to the top plate since our coordinate system is attached to that plate. A positive flow is a flow in the positive x direction. If the pressure gradient is zero, the velocity profile is linear and we have a simple shear flow.

We could just as well have taken the origin on the moving plate and let the y axis point downward. Then the differential equations would have been the same, but the boundary conditions would have been: $y = 0$, $U = 0$; $y = h$, $u = -U$. The velocities are again measured with respect to the top plate and again we find Q as above.

5.12. The lubrication of bearings is a problem in fluid mechanics. The oil or lubricant between the bearing and slider is a viscous fluid and the determination of the pressure distribution, load capacity, friction, etc., may be accomplished by studying the laminar viscous flow of the lubricant. Most bearings operate in the laminar range, with a very small Reynolds number. The spacing between the bearing and slider (the flow gap) is much smaller than the length of the slider so that the flow becomes fully developed throughout most of the gap. Because the Reynolds number is so small, the inertia of the fluid is negligible compared to the pressure and viscous forces. The treatment of a lubrication problem is similar to the Couette flow problem, Problem 5.11.

Consider a step bearing as shown in Fig. 5-30. The spacings h_1 and h_2 are much less than L_1 and L_2, and the width of the bearing in the z direction is assumed to be very large so that

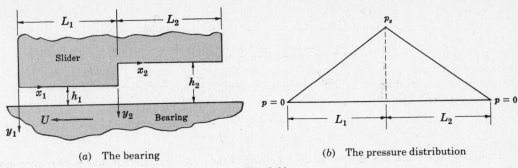

(a) The bearing   (b) The pressure distribution

Fig. 5-30

leakage in the z direction may be neglected. For a slider velocity U and a fluid of viscosity μ, find the pressure distribution in the bearing.

Each section, the L_1 and L_2 sections, may be treated just as the Couette problem. We attach a coordinate system to the slider as shown and let the stationary bearing move with velocity U in the negative x direction with respect to the slider. In each section then at $y = 0$, $u = 0$ and at $y = h$, $u = -U$. The pressure at the inlet, $x_2 = L_2$, and at the outlet, $x_1 = 0$, is zero gage. The pressure at the step, $x_1 = L_1$ or $x_2 = 0$, is unknown and must be found by using the continuity condition that $Q_1 = Q_2$. If we use the expressions developed in the previous problem we have

$$Q_1 = -\frac{h_1^3}{12\mu}\frac{p_s}{L_1} - \frac{Uh_1}{2} = Q_2 = \frac{h_2^3}{12\mu}\frac{p_s}{L_2} - \frac{Uh_2}{2}$$

The pressure in each section varies linearly with x and the pressure distribution forms a triangle as shown in the figure. The peak pressure p_s is found by equating Q_1 to Q_2, and the result is

$$p_s = \frac{6\mu U(h_2 - h_1)}{(h_1^3/L_1) + (h_2^3/L_2)}$$

which completely specifies the pressure as a function of x. The total load carrying capacity is simply the area under the pressure distribution curve. The load capacity W per unit z width is

$$W = (p_s/2)(L_1 + L_2)$$

Physically, if the load were changed the spacing h_1 and h_2 would change, the slider rising or sinking as the load is decreased or increased respectively.

Once the velocity is known throughout the flow film, the friction can be found by integrating the shear stress along either the slider or the bearing. Do this calculation as an exercise. You will find that the shear stresses along the top and bottom surfaces are not quite equal, but if the pressure force against the step is included in the calculation of the friction of the top plate or slider, the two calculations give the same result.

5.13. A hydrostatic thrust bearing is shown in Fig. 5-31. Two circular disks (of radius b) are spaced a distance h apart (in practice h is of the order of 0.001 to 0.010 inch) and the diameter may be several inches. Pressurized liquid lubricant is supplied to the stator through a cup or recess region of radius a where the pressure is maintained at a high pressure p_0. The tap disk rotates at speed ω and supports a load W. The lubricant leaks radially between the disks to p_b. Find the pressure distribution in the lubricant and the load carrying capacity W.

Since $h \ll b$ we assume the Reynolds number $\mathrm{Re} \ll 1$ and neglect inertia terms in the equation of motion. The radial equation of motion becomes

$$0 = -\frac{\partial p}{\partial r} + \mu \frac{\partial^2 u}{\partial z^2}$$

The tangential equation of motion is

$$\frac{\partial^2 v}{\partial z^2} = 0$$

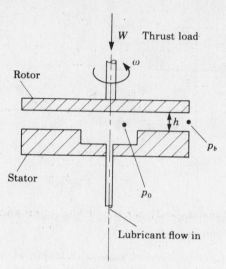

Fig. 5-31

The boundary conditions are $u = 0$, $y = 0$, h, and $v = 0$, $y = 0$; $v = r\omega$, $y = h$. Direct integration given.

$$v = r\omega\left(\frac{z}{h}\right)$$

$$u = \frac{1}{2\mu}\frac{dP}{dr}(z^2 - hz)$$

The flow rate Q is

$$Q = \int_0^k 2\pi r u \, dz = -\frac{\pi r h^3}{6\mu}\frac{dp}{dr}$$

which is a first order equation which may be integrated for $p(r)$ and Q with the two boundary conditions $p = p_0$, $r = a$, and $p = p_b$, $r = b$.

$$Q = \frac{\pi h^3 (p_0 - p_b)}{6\mu \ln (b/a)} \qquad \frac{p_0 - p}{p_0 - p_b} = \frac{\ln (r/a)}{\ln (b/a)}$$

and the total load W is

$$W = (p_0 - p_b)\pi a^2 + \int_a^b (p - p_b)2\pi r \, dr = \frac{\pi(b^2 - a^2)(p_0 - p_b)}{2 \ln (b/a)}$$

5.14. Two parallel flat circular disks of radius a are immersed in a shallow oil bath (Fig. 5-32). They are held a distance h apart where $h \ll a$. Suddenly the disks are pulled apart with a constant velocity V. The acceleration time is small compared to times of interest, and we can assume V is achieved rapidly. Assume the velocity is slow enough and h small enough that inertia effects may be neglected and the radial flow between the disks is similar to that of Problem 5.13. Find the critical velocity V such that cavitation occurs if it is exceeded.

Referring to Problem 5.13 the radial velocity profile is

$$u = \frac{1}{2\mu}\frac{dp}{dr}(z^2 - hz)$$

Consider a cylindrical constant volume of radius r between the disks. Continuity applied to this cylinder yields

$$\int_0^h 2\pi r u \, dz = -\pi r^2 V - \frac{\pi r h^3}{6\mu}\frac{dp}{dr} = \pi r^2 V$$

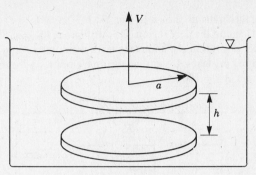

Fig. 5-32

with the boundary conditions $P = P_0$ at $r = a$ where P_0 is the nominal hydrostatic pressure at the edge of the disks. Integrating,

$$p = p_0 - \frac{3\mu V}{h^3}(a^2 - r^2)$$

The minimum pressure occurs at $r = 0$.

$$p = p_0 - \frac{3\mu V a^2}{h^3}$$

and if cavitation occurs at near zero absolute (or more exactly at the vapor pressure of the liquid which is small compared to atmospheric) the critical value of V, V_{cr}, is

$$V_{cr} = \frac{p_0 h^3}{3\mu a^2}$$

If V_{cr} is exceeded, the fluid cavitates.

Supplementary Problems

5.15. Determine the boundary layer thickness 10 ft from the leading edge of a flat plate for the flow of air (70°F and 14.7 psi) where the free stream velocity is 60 fps. Assume laminar flow.

5.16. Determine the shear stress 10 ft from the leading edge for the conditions of Problem 5.15.

5.17. Determine the total drag force and drag coefficient for Problem 5.15 if the plate width is 3.0 ft.

5.18. Work Problem 5.15 assuming that the boundary layer is turbulent.

5.19. Work Problem 5.16 assuming that the boundary layer is turbulent.

5.20. Work Problem 5.17 assuming that the boundary layer is turbulent.

5.21. Derive an expression for the drag coefficient for flow over a flat plate. Express the results in terms of a Reynolds number based on the length of the plate.

5.22. A truck hauling 5 ft diameter 30 ft long thin wall tubing is traveling at 60 mph. Determine the power (hp), resulting from the drag, required for a single tube which is well above the cab of the truck.

5.23. What is the boundary layer thickness at the point where the air is leaving the tubing in Problem 5.22?

5.24. Air flows between parallel walls as shown in Fig. 5-33. The velocity V_1 is uniform and equal to 100 fps at the entrance (section 1) and in the core region. Downstream a distance of 900 in. the velocity varies over the entire width. The velocity varies in the boundary layer region according to $V = V_c(y/\delta)^{1/7}$ where $\delta = 0.1\sqrt{x}$ and δ and x are measured in inches. Determine the acceleration on the axis of symmetry for $0 \leqq x \leqq 900$ in. Evaluate the acceleration at $x = 100$ in.

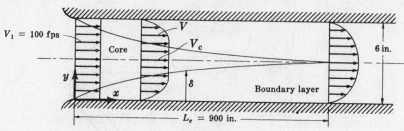

Fig. 5-33

5.25. Assume a laminar flow over a flat plate given by

$$u = Uy/\delta \qquad \text{for} \qquad 0 \leqq y \leqq \delta$$

$$u = U \qquad \text{for} \qquad y > \delta$$

where $\delta = 0.1\sqrt{x}$ and $x = $ distance from the leading edge of the plate and y is the distance measured perpendicular to the plate. All dimensions are measured in feet. If the plate is 1 ft wide and 25 ft long, find the drag force.

5.26. Air is flowing between the parallel flat plates as shown in Fig. 5-34. The velocity is uniform at the entrance, having a magnitude of 10 fps. The boundary layer thickness is $\delta = \frac{1}{10}\sqrt{x}$. The width of the plates, W, is much larger than the distance between the plates, h, and therefore end effects may be neglected. The velocity distribution in the boundary layer is given by $u/U = (y/\delta)^2$ where U is the core velocity and y is the coordinate measured from the wall of either plate. Determine the pressure at a point 25 ft from the entrance. δ and x are measured in feet.

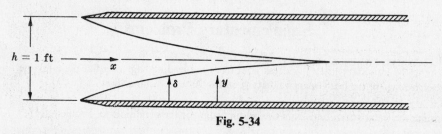

Fig. 5-34

5.27. In Problem 5.26 find the pressure at a position downstream located at twice the distance required for the flow to become fully developed.

5.28. In Problem 5.26 determine the total force of the fluid on the walls between the entrance and the position where the flow becomes fully developed.

5.29. Work Problem 5.26 for a pipe rather than parallel plates.

5.30. Work Problem 5.28 for a pipe rather than parallel plates.

5.31. It has been stated that a ball with a roughened surface will travel greater distances than a ball with a smooth surface, for the same initial velocity. Explain.

5.32. Determine an approximate value for the initial velocity of a golf ball which travels 250 yards on the fly. Assume that the ball has a velocity of 20 fps when it hits the ground.

5.33. Determine the total force on a $\frac{1}{8}$ in. diameter 4 ft car radio antenna if the car is traveling at a speed of 60 mph.

5.34. Determine the power required as a result of the force on the antenna in Problem 5.33.

5.35. A cubical weight of 5 lb falls between two parallel walls, as shown in Fig. 5-35, at a velocity of $\frac{1}{2}$ fps. Determine the viscosity of the oil.

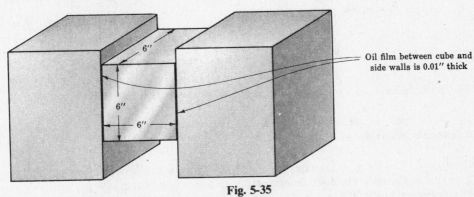

Fig. 5-35

5.36. A rotating sleeve bearing, between the shaft and the bearing housing, is used in a turbosuper-charger in order to minimize the relative velocity between moving parts.

   The dimensions of the turbosupercharger bearing are shown in inches on the sketch, Fig. 5-36. The shaft rotates at a speed of 60,000 revolutions per minute. The oil used in this bearing has a viscosity of $\mu = 1 \times 10^{-4}$ lb-sec/ft^2. Calculate the angular speed of the sleeve bearing. The various clearances C_1, C_2, C_3 between the sleeve bearing and the housing and the shaft are indicated in the figure.

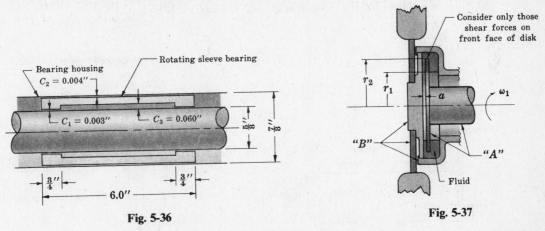

Fig. 5-36   **Fig. 5-37**

5.37. In Fig. 5-37 is shown a fan drive designed such that there is relative motion between the drive shaft and disk assembly "A", and the housing and fan assembly "B". The angular velocities are ω_1 and ω_2 respectively $(\omega_1 > \omega_2)$. Torque T is transmitted between the two parts by means of a fluid having viscosity μ. Determine the clearance a required for the given conditions. Express a in terms of $\omega_1, \omega_2, \mu, r_1, r_2$ and T.

5.38. Water that is at 60°F ($\nu = 1.217 \times 10^{-5}$ ft^2/sec) is flowing through a $\frac{1}{8}$ inch inside diameter smooth tube as shown in Fig. 5-38. Considering friction losses between sections 1 and 2 only, find the velocity at the outlet (section 2).

5.39. Air is flowing in a galvanized iron pipe ($\epsilon = .0005$ ft) having a 3.5 ft diameter at a rate of 12,000 ft^3/min. The length of the pipe is 800 ft. The air temperature is 60°F. What is the difference in elevation between inlet and outlet if the static pressure change is zero?

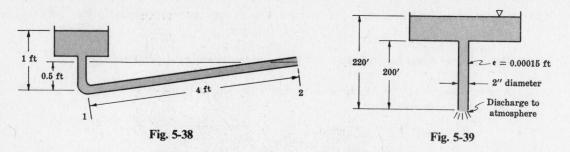

Fig. 5-38   **Fig. 5-39**

5.40. Calculate the average velocity of water at 70°F for the pipe as shown in Fig. 5-39.

5.41. Determine the length of pipe which will produce the maximum flow rate in Problem 5.40.

5.42. Two water reservoirs are connected by 100 ft of 2 in. diameter smooth pipe. What is the flow rate when the difference in elevation is 25 ft?

5.43. Consider air flowing at an average velocity of 100 fps in a 6 in. smooth pipe between the two plenum chambers as shown in Fig. 5-40. Determine the difference in pressure, $p_1 - p_2$, assuming no minor losses. The air temperature is 70°F.

5.44. Determine the pressure difference in Problem 5.43 if the minor losses are included.

5.45. A straight-sided open glass tank for washing photographic prints has a siphon tube in one side which drains off the water. The operation is as follows. Referring to Fig. 5-41 a stream of water flows into the tank until the water reaches the top of the siphon tube; then it is cut off. The water that has filled the tank runs through the siphon until the end of the siphon is uncovered, then the cycle begins again with more fresh water flowing into the tank. How long does the drainage part of the cycle take?

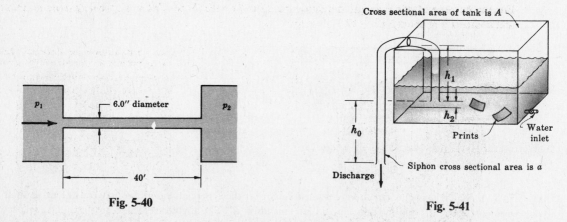

Fig. 5-40   **Fig. 5-41**

5.46. Water is flowing in the piping system shown in Fig. 5-42. For a flow rate in the 8 in. pipe of 4.0 ft³/sec up to the reservoir, calculate the flow rates in the other pipes and the necessary pump horsepower. The friction factors f, pipe lengths and reservoir elevations are given.

5.47. The 200 inch Mt. Palomar telescope is mounted on a yoke mechanism which allows the lens to be focused on a desired star or nebula. The yoke can be preset so that it will track a star if its path is known.

Due to the exceptionally heavy construction of the supporting frame and yoke, the bearing design was extremely important. The total weight of the telescope is in excess of 1,000,000 pounds. This weight is supported by two sets of bearings at the north and south ends of the main rotating frame. The semicircular

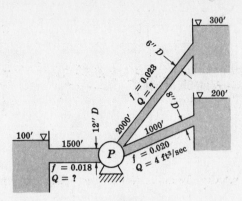

Fig. 5-42

horseshoe shaped yoke is supported by four pad type hydrostatic bearings. Each pad must sustain a load of 164,000 lb under conditions of zero velocity. Hence the oil must be forced into the pads under pressure.

Fig. 5-43 and Fig. 5-44 show the yoke and pad detail. Each pad is 28 in. square and has four deep recessed regions 7 in. square. The oil is forced up through small tubes into the center of these recessed regions, the pressure over each recess being constant at the inlet value. The oil then flows slowly through the thin gap between the pad surface and the yoke surface, out to a reservoir at atmospheric pressure where it is recirculated through the pump. Assume that the oil film in the flow gap is 0.005 in. Find the following quantities:

1. Necessary oil flow rate
2. Oil inlet pressure
3. Capacity of the oil pump
4. Coefficient of friction between the yoke and bearing
5. The necessary motor capacity to rotate the yoke at a rate of one revolution per 24 hours for tracking purposes.

For the purpose of this calculation assume the oil to be SAE 20 which has a specific gravity of about 0.8 and a viscosity of about 3.85×10^{-6} lb-sec/in^2.

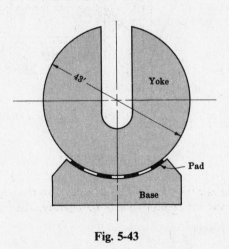

Fig. 5-43

Oil flows over the surfaces surrounding the recesses. The surface mates to the yoke with 0.005″ clearance.

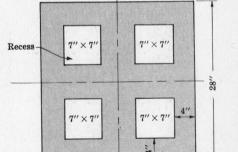

Pad detail

Fig. 5-44

5.48. A hydraulic dashpot or damper shown in Fig. 5-45 consists of an outer cylinder of inside diameter D_c into which is placed a solid cylinder of diameter D which is nearly equal to D_c. The ends of the outer casing cylinder are closed except for a small hole which allows the shaft to be attached to the solid inner cylinder or plunger. The casing is filled with oil and the shaft is sealed so that oil cannot leak out around it. The

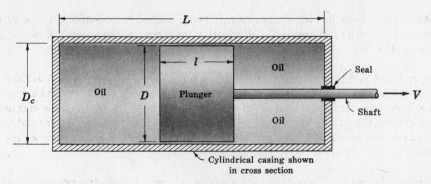

Fig. 5-45

outer casing may be fastened rigidly to a mounting and the shaft fastened to a machine member which is to be damped. Automobile shock absorbers work on this principle.

The small clearance between the two cylinders impedes the flow of oil from the front of the plunger to the rear, or vice versa, so that the plunger can only move very slowly. As the plunger moves, the oil must flow around it through the small passage created by the clearance.

Find the resisting force of the dashpot as a function of the relevant parameters and the velocity V at which the plunger is moved.

Hint. Neglect the inertia of the fluid in the clearance space. Treat the flow in this region as a lubrication type film. The pressure difference between the front and rear of the plunger determines the flow rate between the two regions. The force on the plunger is the sum of the pressure forces and the viscous shear.

5.49. Repeat Problem 5.13 for air as a lubricant. Assume the temperature is constant (isothermal flow) and that air is an ideal gas with $p = \rho RT$ where R is the gas constant and T the absolute temperature.

Ans. $\dfrac{p_0^2 - p^2}{p_0^2 - p_b^2} = \dfrac{\ln(r/a)}{\ln(b/a)}$

5.50. Figure 5-46 shows a partial stepped journal bearing that has been proposed as a practical load-supporting device. The bearing is very long and covers a 180° arc, with the step located at the midpoint or 90°. During operation the journal is aligned so that its center is coincident with the centers of the two bearing arcs. If the pressure at the inlet and outlet is atmospheric, find the load W and its direction. The flow is laminar, and assume the flow is fully developed throughout with no inlet or outlet losses.

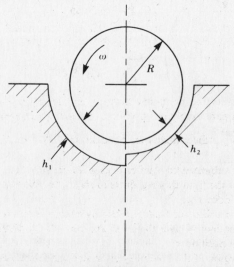

Fig. 5-46

5.51. In the above problem find the frictional torque on the journal. Is the frictional torque on the stepped stator equal (in magnitude) and opposite in direction to the torque on the journal? If not, why? *Hint*: Consider the pressure acting on the step.

5.52. Rework Problem 3.16 but include friction in the pipe. Assume $D = 6$ inches, $h = 120$ ft, the length of the pipe is 100 ft, and the pipe is smooth. (The lake is deep so that $L < h$.) *Hint*: Assume a flow rate, find the power output of the turbine, and plot power vs. flow rate Q. If $Q = 0$ there is maximum torque but no speed of the turbine and the power is zero. At maximum speed the torque is zero and the fluid flows freely but the power is zero. Construct a curve of power output vs. flow rate and obtain a value for the maximum power.   *Ans.*   51 hp

5.53. A submersible pump, 1/2 horsepower at 80% rated efficiency, is used to pump water from a deep well into a storage tank (see Fig. 5-47). The water level in the well is constant at 100 feet below the ground level, and the pump is located at a depth of 200 feet. A 1-inch ID pipe (with friction factor $f = 0.01$) connects the pump to a storage tank at ground level. The tank is located at a distance of 100 feet from the well head. The tank is fitted with an air chamber so that the pressure in the tank is constant at 60 psig.

At what rate can the pump deliver water to the tank (in gallons per minute)?

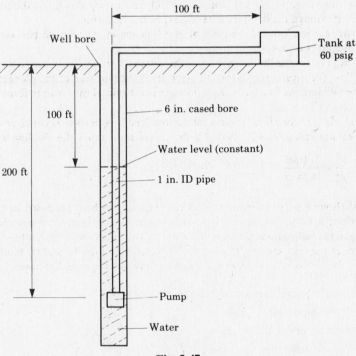

Fig. 5-47

5.54. Determine the behavior of the boundary layer in the vicinity of the stagnation point assuming a cubic velocity profile and compare to the results given in the text for a quadratic velocity profile.

Ans. $\delta_0 = 2.4\sqrt{\mu/A\rho U_\infty}$, where A is given by equation (5.33).

5.55. A porous flat plate is subjected to suction along its entire length on its underside, which results in a constant velocity (v_0) of the fluid flowing through the plate. The suction velocity is small compared to the constant free-stream velocity U_0.

At a sufficiently great distance from the leading edge, it is found that the suction stabilizes the boundary layer growth. The boundary layer thickness and the velocity distribution within the boundary layer are independent of the axial position x.

In this stabilized region of the boundary layer, find the velocity distribution and the shear stress on the plate.

NOMENCLATURE FOR CHAPTER 5

A = area

C_D = drag coefficient

C_f = skin friction coefficient

C_L = lift coefficient

c_p = specific heat at constant pressure

c_v = specific heat at constant volume

D = drag force and diameter

E_c = Eckert number

f = friction factor, functional notation

F_s = surface force

F_{s_x} = surface force in the x direction

h = distance or half distance between parallel plates

H_L = head loss

K = loss coefficient for minor losses

L = length

m = exponent for power law velocity distribution, mass

$\dot{m}$ = mass flux

$\dot{M}$ = momentum flux

n = exponent for power law velocity distribution

P_r = Prandtl number

p = pressure

q = heat transfer per unit mass of flowing fluid

Q = volume flow rate

r = radial coordinate

R = radius of pipe

Re = Reynolds number

Re_x = length Reynolds number = Ux/v

T = temperature

u = velocity in x direction, internal energy per unit mass

u_τ = wall friction velocity, $\sqrt{\tau_0/\rho}$

U = free stream or maximum velocity

v = velocity in r direction or y direction

V = velocity

$\mathbf{V}$ = velocity vector

V_x = velocity in x direction

W = width

x = coordinate

X_L = entrance length for the flow to become fully developed

y = coordinate

z = elevation

δ = boundary layer thickness

δ^* = displacement thickness

ϵ = average roughness height

η = similarity parameter
θ = momentum thickness
μ = viscosity
v = kinematic viscosity
ρ = density
τ = shear stress
τ_0 = shear stress at the wall
Φ = dissipation function
ψ = stream function

Chapter 6

Incompressible Potential Flow

6.1 POTENTIAL FLOW THEORY

In Chapter 1 we mentioned that outside the boundary layer the flow is frictionless and irrotational and hence is known as potential flow. By potential flow we mean that the velocity is derivable from a scalar velocity potential ϕ as

$$\mathbf{V} = -\nabla\phi \tag{6.1}$$

Mathematically, the necessary and sufficient condition that a vector field which is single valued (in this case $\mathbf{V}$) can be derived from a scalar potential function ϕ is that the curl of the vector be zero (except at singular points). The significance of the curl of the velocity vector (which is called vorticity $\boldsymbol{\omega}$) was discussed in Chapter 3. If the vorticity is zero the flow is "irrotational" since the rotation or angular velocity of any infinitesimal element of the fluid is zero.

In this chapter we will be concerned with incompressible two-dimensional potential flow theory which is valid for subsonic flow where the Mach number M is less than about 0.3. In Chapter 8 we will discuss potential compressible flow where M approaches unity or exceeds unity for supersonic flow. If the flow is two-dimensional and incompressible, simplifications are possible which allow the use of complex variable theory to "map" the flow by means of conformal transformations.

One might ask, if a real fluid has viscosity how can the flow ever be irrotational or potential. The answer is that outside the boundary layer, which forms near a surface, the effects of viscosity (or friction) are quite small compared to the effects of inertia and pressure (as long as the Reynolds number is large compared to unity). Except for flow in pipes and channels where the boundary grows to fill the entire pipe or external "creeping" flow ($\text{Re} \ll 1$) where the boundary layer may extend far from the object there will generally be regions of flow which are essentially irrotational. In practice the boundary layer over streamlined objects (where there is no separation) grows to a thickness usually much smaller than the object itself and can be neglected in determining the potential flow outside the boundary layer. That is, as far as the potential flow is concerned the boundary layer only serves to increase the effective size of the object but only by a negligible amount. For example, the flow of air over an airplane is a problem in potential flow. Once the potential flow solution is obtained the velocity and pressure on the "surface" of the object may be found. This "surface" value of velocity is used as the boundary condition at the outer edge of the boundary layer flow and the pressure is simply imposed on the boundary layer.

One of the most important applications of potential flow theory is in aerodynamics. We confine ourselves to two-dimensional flow in this chapter. The extension to three dimensions is not difficult in principle but the mathematical techniques such as the introduction of a stream function become much more complex.

Remember, from Chapter 3, that the rotation $\boldsymbol{\omega}$ (or vorticity as it is often called) in a fluid is defined as

$$\boldsymbol{\omega} = \nabla \times \mathbf{V} \tag{6.2}$$

and the angular velocity of an infinitesimal element of fluid $\boldsymbol{\Omega}$ is related to the vorticity as

$$\boldsymbol{\omega} = 2\boldsymbol{\Omega} \tag{6.3}$$

We can illustrate this concept geometrically by examining Fig. 6-1. Restricting the analysis to one cartesian component for simplicity, the z component of $\nabla \times \mathbf{V}$ is

$$\omega_z = (\nabla \times V)_z = \frac{\partial v}{\partial x} - \frac{\partial u}{\partial y}$$

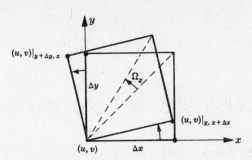

Fig. 6-1. Rotation of a fluid element.

and from Fig. 6-1 we see that ω_z is twice the average value of the z component of the angular velocity of the element $\Delta x \, \Delta y$. The Δx line has angular velocity $(v|_{x+\Delta x} - v|_x)/\Delta x$ and the Δy line has angular velocity $-(u|_{y+\Delta y} - u|_y)/\Delta y$ which when averaged gives $\frac{1}{2}(\partial v/\partial x - \partial u/\partial y)$, the average angular velocity of the square $\Delta x \, \Delta y$.

If we now consider any two-dimensional flow (in the xy plane so that only ω_z exists), we can use Stokes' theorem to relate the rotation to a line integral as follows (see Fig. 6-2).

$$\int_A \boldsymbol{\omega} \cdot d\mathbf{A} = \int \nabla \times \mathbf{V} \cdot d\mathbf{A} = \oint \mathbf{V} \cdot d\mathbf{l} \qquad (6.4)$$

which states that the area sum of the rotation over a given area is equal to the integral of the velocity (along the curve) integrated around the curve bounding the area. This line integral is called circulation, which is denoted as Γ. For any closed curve, then, the circulation is given by

$$\Gamma = \oint \mathbf{V} \cdot d\mathbf{l} = \int \boldsymbol{\omega} \cdot d\mathbf{A} \qquad (6.5)$$

We can obtain a physical picture of rotation in a fluid as follows. Imagine a tiny cross floating on a fluid surface (see Fig. 6-3). If the fluid is truly irrotational, the cross moves along always parallel to itself since the fluid nowhere has any angular velocity. In actuality, of course, the cross has finite length and so may actually begin to rotate even though the fluid is irrotational. However, in the limit as the length of the test cross becomes infinitesimal it would not rotate in an irrotational flow. In contrast to this behavior, imagine a cross floating in a viscous shear flow which has rotation. Obviously the cross rotates as it moves along with the fluid, since the shearing behavior of the fluid tends to move the ends of the straw with different velocities.

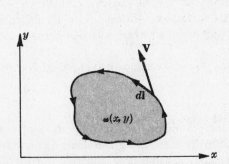

Fig. 6-2. Rotation and circulation. $\boldsymbol{\omega}\,(x, y)$ may exist throughout the area and is related to the line integral $\oint \mathbf{V} \cdot d\mathbf{l}$.

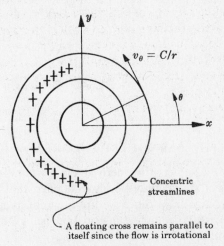

A floating cross remains parallel to itself since the flow is irrotational

Fig. 6-3. The potential vortex.

A simple example of irrotational flow is the potential vortex, of which the common whirlpool or tornado is a close approximation. In the potential vortex the velocity (which has only an angular component v_θ) is given by

$$v_\theta = C/r \tag{6.6}$$

and hence the velocity potential ϕ is

$$\phi = -C\theta \tag{6.7}$$

so that $-\nabla\phi$ is $(C/r)\,\hat{\theta}$ where C is a constant and $\hat{\theta}$ is a unit vector in the θ direction. The flow is irrotational since

$$\nabla \times \mathbf{V} = \frac{1}{r}\frac{\partial}{\partial r}\left(r \cdot \frac{C}{r}\right) = 0$$

However, at $r = 0$ there is a singularity. The circulation Γ about any contour not enclosing the origin is zero, but if the contour encloses the origin there is a finite value of Γ which is due to the rotation at the origin. Referring to Fig. 6-4, let us form Γ_1 about contour C_1.

$$\Gamma_1 = \int_0^{2\pi} v_\theta \bigg|_{r=b} b\, d\theta = 2\pi C$$

Any other contour such as C_2 enclosing the origin also gives $2\pi C$ for the circulation. The velocity potential can be written then as $\phi = -\Gamma\theta/2\pi$. Any contour such as C_3 will give zero circulation. The constant $C = \Gamma/2\pi$ is known as the strength of the vortex. In reality a vortex cannot have infinite velocity at its center and the central core of the vortex rotates as a rigid body of diameter a and angular velocity Ω. From Stokes' theorem then, $\frac{1}{2}a\Omega^2 = 2\pi C$. In nature, the tropical hurricane or typhoon is an example of a vortex. The eye of the hurricane is a central core of comparative calm where the fluid rotates approximately as a rigid body.

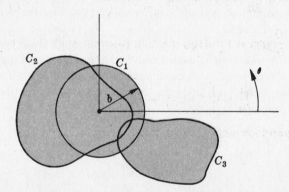

Fig. 6-4. Contours of integration about the potential vortex.

6.2 THE BERNOULLI THEOREM

In Chapter 3 we derived the Bernoulli equation by integrating the inviscid equation of motion along a streamline. We also obtained the same relationship by integrating the energy equation for frictionless flow. Since we are now concerned only with incompressible flow, the energy equation is unnecessary, complete information being obtainable from the equation of motion and the continuity equation.

In irrotational flow with conservative body forces (derivable from a potential as $\mathbf{B} = -\nabla\psi$), the Bernoulli equation holds between any two points in the flow (not necessarily on the same streamline).

Let $d\mathbf{r}$ be an element of distance in the flow field (not necessarily along a streamline). Then for steady incompressible irrotational flow,

$$\rho \int \nabla\left(\frac{V^2}{2}\right) \cdot d\mathbf{r} - \rho \int \mathbf{V} \times (\nabla \times \mathbf{V}) \cdot d\mathbf{r} = -\int \nabla p \cdot d\mathbf{r} - \rho \int \nabla\psi \cdot d\mathbf{r} \qquad (6.8)$$

But in irrotational flow $\nabla \times \mathbf{V} = 0$ so that

$$\frac{V^2}{2} + \frac{p}{\rho} + \psi = \text{constant} \qquad (6.9)$$

throughout the flow. It can further be shown that if a fluid is frictionless (inviscid) and the body forces are irrotational (conservative), then the flow is necessarily irrotational (except perhaps at singular points such as the center of the potential vortex).

The Bernoulli equation (6.9) is useful for the calculation of the pressure throughout the flow field once the velocity $\mathbf{V}$ is known. The determination of the velocity $\mathbf{V}$ is an important problem in potential flow and subsonic aerodynamics and will be addressed in the following sections.

6.3 THE KELVIN VORTEX THEOREM AND VORTEX MOTION

The circulation Γ about any contour always composed of the same fluid particles (a fluid line) is constant in an inviscid fluid with only conservative body forces. Although the vorticity under such conditions is generally zero, circulation can exist because of local regions of vorticity or singular points of vorticity.

By multiplying the equation of motion by $d\mathbf{l}$ (an element of fluid line) and integrating around a contour, we obtain $\left(\text{remembering that } \dfrac{D}{Dt}(\mathbf{V} \cdot d\mathbf{l}) = \mathbf{V} \cdot \dfrac{D}{Dt}(d\mathbf{l}) + d\mathbf{l} \cdot \dfrac{D\mathbf{V}}{Dt}\right)$

$$\frac{D\Gamma}{Dt} = \oint \frac{D}{Dt}(\mathbf{V} \cdot d\mathbf{l}) = -\oint \left(\frac{\nabla p}{\rho} + \nabla\psi\right) \cdot d\mathbf{l} + \oint \mathbf{V} \cdot \frac{D}{Dt}(d\mathbf{l}) \qquad (6.10)$$

But since $d\mathbf{l} = d\mathbf{r}$ and $\mathbf{V} \cdot \dfrac{D}{Dt}(d\mathbf{r}) = \mathbf{V} \cdot d\mathbf{V} = d(V^2/2)$, (where $\mathbf{r}$ is the Lagrangian position vector of a fluid particle).

$$\frac{D\Gamma}{Dt} = \oint \left[-dp/\rho - d\psi - d(V^2/2)\right] = 0$$

since the terms in the integrand are single valued. Hence

$$\frac{D\Gamma}{Dt} = 0 \qquad (6.11)$$

which means physically that Γ about any fluid line contour remains constant in time as we move along with the fluid, a very important result. Note that (6.11) is valid for both incompressible or compressible fluids.

We have only sketched the proof of the Kelvin theorem here, and the reader is referred to the references for a more detailed proof. It should be emphasized here that $d\mathbf{l}$ is an element of a fluid line (always composed of the same fluid molecules) and moves along with the fluid and may distort and change its orientation. The closed circuit around which $d\mathbf{l}$ is integrated is always composed of the same particles of fluid and always forms a complete closed loop in two or three dimensions. Kelvin's theorem may also be derived by integrating the vorticity $\boldsymbol{\omega}$ over the surface formed by the loop and by using Stokes' theorem to transform to a line integral. The element $d\mathbf{l}$ is not to be confused with an element $d\mathbf{r}$ (fixed in the coordinate space and which may be written in terms of differentials of the Eulerian coordinates).

Consider a three-dimensional vortex, such as a tornado, whose core forms a line of perhaps irregular configuration. This line, which is the continuous locus of the centers of the two-dimensional vortices, is called a *vortex line*. More precisely, a *vortex line* is a line in the fluid such that the tangent to it at each point is in the direction of the vorticity vector at that point. In a rotational fluid an infinite set of vortex lines exist, but only a single line exists along the centers of the potential vortices if the vorticity is due to potential vortex motion.

A *vortex tube* is a tube which is the locus of vortex lines drawn through every point of a closed curve.

A *vortex filament* is a vortex tube of infinitesimal cross sectional area. We can think of the solid core of a tornado as being a vortex tube, and as the core becomes smaller the tube becomes infinitesimal in cross sectional area and becomes a vortex line. See Fig. 6-5.

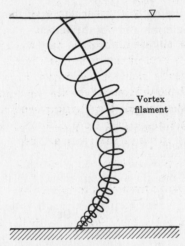

Vortex
filament

Fig. 6-5. A vortex filament associated with a potential vortex in
a fluid. In a tornado, the core is of finite dimensions,
enclosed in a vortex tube.

A *vortex filament* and a *vortex line* are identical except that we associate strength with the filament. The vortex tube or filament can be associated with strength defined as the integral $\int \omega \cdot d\mathbf{A}$ over the cross section. This strength must be constant along the vortex tube. For a filament, $\omega \cdot d\mathbf{A}$ is an intensity, vorticity per unit area, and must be constant along the filament. Hence filaments must form closed rings (vortex rings), like a smoke ring, or end at the fluid boundaries. A vortex filament cannot just end in the fluid unless viscosity is present to dissipate the vorticity.

Our discussion has not been confined to steady state and we must remember that all these lines and filaments move along with the fluid.

6.4 THE VELOCITY POTENTIAL AND STREAM FUNCTION

The condition of irrotationality is a necessary and sufficient one for the velocity to be derivable from a scalar velocity potential ϕ as

$$\mathbf{V} = -\nabla\phi \qquad (6.12)$$

which in cartesian coordinates is

$$u = -\partial\phi/\partial x, \qquad v = -\partial\phi/\partial y, \qquad w = -\partial\phi/\partial z$$

A velocity potential may be defined for general three-dimensional flow of a compressible fluid. However, we will reserve such a general discussion for Chapter 8 and confine our study here to two-dimensional steady incompressible flow. The assumption of incompressible flow is valid for subsonic aerodynamics (Mach number $M < 0.3$, approximately) and the restriction to two dimensions allows a tractable mathematical analysis, although we can discuss, at least qualitatively here, three-dimensional effects. Although the velocity potential can be defined for any irrotational flow, the term "potential flow" is often taken to mean two-dimensional incompressible flow unless otherwise indicated.

Under the conditions of incompressibility, the velocity in terms of the potential ϕ may be substituted into the continuity equation $\nabla \cdot \mathbf{V} = 0$ to yield the condition that ϕ is harmonic (satisfies Laplace's equation):

$$\nabla^2 \phi = 0 \qquad (6.13)$$

which in cartesian coordinates is

$$\frac{\partial^2 \phi}{\partial x^2} + \frac{\partial^2 \phi}{\partial y^2} + \frac{\partial^2 \phi}{\partial z^2} = 0$$

Another important function, the stream function ψ, may be defined for any two-dimensional flow field regardless of whether the flow is irrotational or not, and incompressible or compressible, although we will restrict ourselves here to steady incompressible flow. In a two-dimensional flow, the lines of constant ψ are the streamlines, and the difference between the numerical values of two streamlines is equal to the flow rate between the streamlines. The physical significance of the stream function may be seen by referring to Fig. 6-6. Going along a path from ψ_1 to ψ_2, the flow is considered positive from right to left. ψ is defined in terms of $\mathbf{V}$ in cartesian coordinates as

$$u = -\partial \psi / \partial y, \qquad v = \partial \psi / \partial x \qquad (6.14)$$

The flow rate between ψ_1 and ψ_2 (positive from right to left) is

$$Q_{12} = \int_1^2 (v \, dx - u \, dy) = \int_1^2 \left(\frac{\partial \psi}{\partial x} \, dx + \frac{\partial \psi}{\partial y} \, dy \right) = \int_1^2 d\psi = \psi_2 - \psi_1 \qquad (6.15)$$

The integral is independent of the path so long as it connects the two streamlines. Physically, we insist that ψ be single-valued so that $\oint d\psi = 0$ about any closed contour unless the contour encloses a singularity such as a source or sink, in which case ψ may not be single-valued unless we restrict its domain to, say, $0 < \theta < 2\pi$.

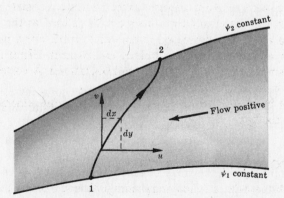

Fig. 6-6. The streamlines and stream function ψ.

For two-dimensional flow, then, we have from the condition of irrotationality, $\nabla \times \mathbf{V} = 0$ or $\partial v / \partial x - \partial u / \partial y = 0$, which combined with the definition of ψ gives for any steady two-dimensional incompressible potential flow,

$$\frac{\partial^2 \psi}{\partial x^2} + \frac{\partial^2 \psi}{\partial y^2} = 0$$

so that ψ is harmonic (i.e. satisfies Laplace's equation) and in any coordinate system

$$\nabla^2 \psi = 0 \tag{6.16}$$

Further,

$$u = -\frac{\partial \phi}{\partial x} = -\frac{\partial \psi}{\partial y}, \qquad v = -\frac{\partial \phi}{\partial y} = \frac{\partial \psi}{\partial x} \tag{6.17}$$

which are known as the Cauchy-Riemann conditions.

In polar coordinates r and θ, we can write the fundamental relationships as

$$v_r = -\frac{\partial \phi}{\partial r} = -\frac{1}{r}\frac{\partial \psi}{\partial \theta}, \qquad v_\theta = -\frac{1}{r}\frac{\partial \phi}{\partial \theta} = \frac{\partial \psi}{\partial r} \tag{6.18}$$

and of course we obtain $\nabla^2 \phi = \nabla^2 \psi = 0$.

An important consequence of the fact that ϕ and ψ are harmonic and satisfy the Cauchy-Riemann conditions is that lines of constant ϕ and ψ are orthogonal. We can easily demonstrate this fact by proving that

$$\left.\frac{\partial y}{\partial x}\right|_{\phi = \text{constant}} = -\left.\frac{\partial x}{\partial y}\right|_{\psi = \text{constant}}$$

Along a constant ϕ line,

$$d\phi = \frac{\partial \phi}{\partial x}\, dx + \frac{\partial \phi}{\partial y}\, dy = 0$$

$$= -u\, dx - v\, dy$$

and along a constant ψ line,

$$d\psi = \frac{\partial \psi}{\partial x}\, dx + \frac{\partial \psi}{\partial y}\, dy = 0$$

$$= v\, dx - u\, dy$$

so that from $d\phi = 0$ we obtain $dy/dx = -u/v$ and from $d\psi = 0$ we have $dy/dx = v/u$, and thus

$$\left.\frac{dy}{dx}\right|_{\phi = \text{constant}} = -\left.\frac{dx}{dy}\right|_{\psi = \text{constant}}$$

which is the mathematical statement that constant ϕ lines and constant ψ lines form an orthogonal network. Since these lines are perpendicular and satisfy the same differential equations, the role of ϕ and ψ may be interchanged to represent different flows.

The lines of constant ϕ and constant ψ form a mesh or network of so-called curvilinear squares. In a uniform flow the lines are all straight, but in general the streamlines and potential lines are curved. Since no fluid crosses constant ψ lines, they may be taken as solid boundaries. That is, a solid boundary may replace a streamline (ψ line) without changing the flow pattern.

Since the equations $\nabla^2 \phi = \nabla^2 \psi = 0$ are linear, we may superpose solutions for different flows and add directly the values of ϕ and ψ at every point in space to obtain the new values of ϕ and ψ, which represent physically the direct superposition of the various flows.

For example, we may superpose the flows created by a source, or sink, or potential vortex onto a uniform flow. In the next section we will discuss some simple two-dimensional flow patterns and some of the methods of superposing these simple flows to create more complex flow fields.

After the ϕ and ψ lines are determined for a given flow, the velocity components are then known and the pressure may be found from Bernoulli's equation.

6.5 SOME SIMPLE FLOW PATTERNS

In this section we will discuss some simple flows and their ϕ and ψ functions. In the next section we will outline some methods of solutions. However, once the simple flows are understood, many more complicated flows can be synthesized merely by superposing these simple solutions.

Uniform Flow

Assume the flow fills all space and is uniform; the velocity is $U_0\hat{x}$ parallel to the x axis, where $\hat{x}$ is a unit vector in the x direction. The only velocity component is u, so that $-\partial\phi/\partial x = U_0 = $ constant. Hence $\phi = -\int U_0\,dx + f(y) = -U_0 x + C_1$ since v is zero and ϕ must be independent of y. The constant C_1 is arbitrary and we take it as zero. The stream function is found from $U_0 = -\partial\psi/\partial y$ so that by similar reasoning,

$$\psi = -U_0 y + C_2 = -U_0 y$$

The flow rate between any two lines of constant ψ ($y = $ constant) is given then by

$$\psi_2 - \psi_1 = \psi\Big|_{y=y_2} - \psi\Big|_{y=y_1} = Q_{12} = -U_0(y_2 - y_1)$$

a negative number for positive flow from right to left, since the flow here is of course from left to right. Hence for uniform flow $U_0\hat{x}$ parallel to the x axis,

$$\phi = -U_0 x, \qquad \psi = -U_0 y \tag{6.19}$$

The ϕ and ψ lines are shown in Fig. 6-7.

Sources and Sinks

A point source or sink is a singularity from which the ψ lines radiate and about which the ϕ lines form concentric circles. For a source of flow rate Q, the radial velocity v_r is $Q/2\pi r$ and the angular

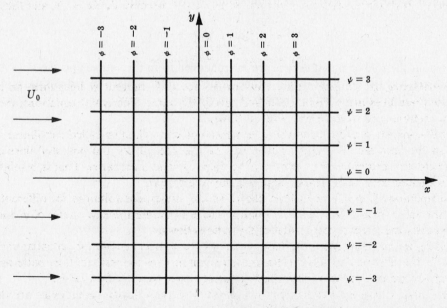

Fig. 6-7. Uniform flow parallel to the x axis.

velocity v_θ is zero. Q is the source strength and is physically the total flow rate per unit depth of fluid. Hence since $v_r = -\dfrac{\partial \phi}{\partial r} = -\dfrac{1}{r}\dfrac{\partial \psi}{\partial \theta}$,

$$\psi = -\frac{Q}{2\pi}\theta, \qquad \phi = -\frac{Q}{2\pi}\ln r \qquad (6.20)$$

For a sink (6.20) is still valid but Q is negative, so that v_r is negative and the flow is inward. Of course any arbitrary constant may be added to ϕ or ψ without changing the velocities.

The flow pattern is shown in Fig. 6-8. At the origin, of course, $\phi \to \infty$ as $r \to 0$, which is no surprise since, in reality, we must always have a finite area, not a point, into which the fluid flows.

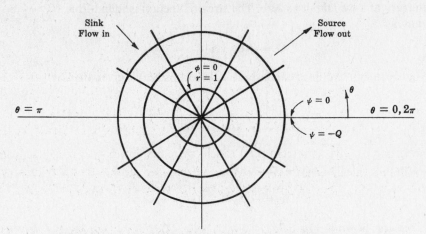

Fig. 6-8. The source and sink. If at $\theta = 2\pi$ we say $\psi = -Q$, then Q is a positive number for a source and negative for a sink.

Potential Vortex

We have discussed the potential vortex earlier; now we can discuss it in terms of ϕ and ψ. For the vortex, we can integrate $v_\theta = C/r = \Gamma/2\pi r$ to find ϕ and ψ. We obtain

$$\phi = -\frac{\Gamma}{2\pi}\theta, \qquad \psi = \frac{\Gamma}{2\pi}\ln r \qquad (6.21)$$

We may note that the role of ϕ and ψ here are interchanged from the source and the sink. And, indeed, in Fig. 6-9 we see that the constant ϕ and ψ lines form a similar pattern of radial lines and concentric circles. The term $\Gamma/2\pi$ is known as the strength of the vortex.

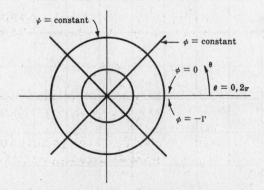

Fig. 6-9. The potential vortex and its ϕ and ψ lines.

Superposition

As an example of the superposition of two or more potential flows let us examine the flow shown in Fig. 6-10, the Rankine oval. A source and sink of equal strength are spaced equidistant from the origin on the x axis in a uniform flow $U_0 \hat{x}$. All of the fluid from the source is absorbed by the sink, and there is a definite dividing streamline between the fluid of the uniform stream and the fluid transferring from the source to the sink. This dividing streamline may be considered as the surface of the cross section of an oval-shaped cylinder. The superposition of these flows, then, will give us the external flow around an oval cylinder. By combining many sources and sinks we could obtain the approximate flow about an arbitrarily-shaped cylinder, symmetrical about the x axis. And, by using a distributed source along the x axis we can find the exact flow about such a body; however, the strength distribution function may be difficult to calculate in general, and involves the solution to an integral equation. Such methods are useful in aerodynamics.

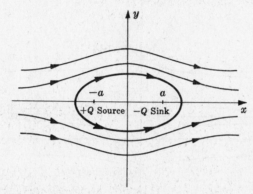

Fig. 6-10.   The flow over a Rankine oval. The oval is the dividing streamline between the fluid flowing between the source and sink and the fluid in the free stream.

Returning to the Rankine oval, we have

$$\phi = -U_0 x - (Q/2\pi) \ln r_1 + (Q/2\pi) \ln r_2$$

$$\psi = -U_0 y - (Q/2\pi)\theta_1 + (Q/2\pi)\theta_2 \qquad (6.22)$$

which may be written as

$$\phi = -U_0 x - \frac{Q}{4\pi} \ln \frac{(x+a)^2 + y^2}{(x-a)^2 + y^2}$$

$$\psi = -U_0 y - \frac{Q}{2\pi}\left(\tan^{-1}\frac{y}{x+a} - \tan^{-1}\frac{y}{x-a}\right) \qquad (6.23)$$

The Method of Images

As we have stated, a streamline may be considered as a solid boundary. If we can find a flow such that a constant ψ line coincides with a boundary, we can specify the flow along that boundary. For flow over an object, the surface of the object is a line of constant ψ.

Often it is possible by superposing flow patterns to create a constant ψ line coincident with a wall or boundary. One useful example of this method is the method of images. Consider two identical flows separated by a midplane; the midplane must have no flow across it and hence can be thought of as a solid boundary.

By the method of images we superpose flows by reflecting about a solid boundary across which there is no flow. Many rather complex flows can be synthesized by this method.

For example, consider the flow from a source (or sink) near a wall (the x axis) as shown in Fig. 6-11. We construct the flow due to a source at $y = a$ and an image source at $y = -a$. The two flows buck each other along the x axis which is a dividing streamline or wall. The functions ϕ and ψ are, obviously,

$$\phi = -\frac{Q}{4\pi} \ln \{[(y-a)^2 + x^2][(y+a)^2 + x^2]\}$$

$$\psi = -\frac{Q}{2\pi} \left(\tan^{-1} \frac{y-a}{x} + \tan^{-1} \frac{y+a}{x} \right) \tag{6.24}$$

such that at $y = 0$ the component of velocity normal to the wall, $v(y = 0)$, is zero.

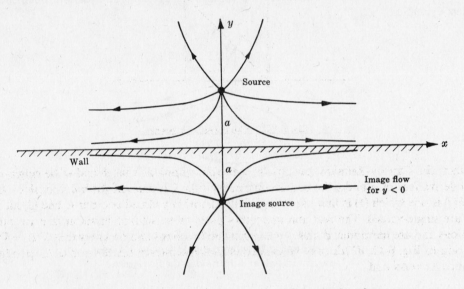

Fig. 6-11. The method of images used to create the flow from a source near a wall.

6.6 THE COMPLEX POTENTIAL

In general, problems in potential flow involve the solution of the Laplace equations $\nabla^2\phi = 0$ and $\nabla^2\psi = 0$ subject to the appropriate boundary conditions, which are usually that at infinity the flow is uniform or zero and that the fluid does not cross solid bodies over which the fluid flows. However, except for certain simple geometries for which ϕ and ψ may be easily found either by solving the harmonic equation or by direct integration of $V = -\nabla\phi$ if the velocity is known such as in the simple examples of the preceding section, the determination of ϕ and ψ is best accomplished by the use of complex variable theory and conformal transformations.

The Complex Function $F(z)$

In two dimensions, the fact that ϕ and ψ are harmonic and satisfy the Cauchy-Riemann equations is necessary and sufficient for the definition of a complex function F (called the complex potential) as

$$F = \phi + i\psi = F(z) \tag{6.25}$$

where $i = \sqrt{-1}$ and $z = x + iy$. In the complex $(\phi + i\psi)$ plane, ϕ and ψ form a rectangular coordinate grid. We consider ϕ and ψ to be functions of z, the complex variable, instead of x and y. The xy plane represents the physical flow plane.

In general,

$$z = x + iy = re^{i\theta} = r(\cos\theta + i\sin\theta)$$

See Fig. 6-12. z is a complex number with real part x and imaginary part y. F may be written as a function of z; then the real part of F is $\phi(x, y)$ and the imaginary part of F is $\psi(x, y)$.

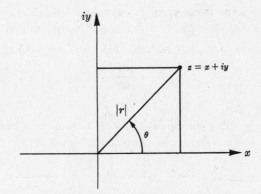

Fig. 6-12.   The complex z plane.

Together, the Cauchy-Reimann conditions with the conditions that ϕ and ψ be single-valued and all partial derivatives of ϕ and ψ be continuous, imply that F is analytic (or holomorphic). An analytic function $F(z)$ is one which (1) is finite and single-valued within a closed contour C and (2) all derivatives exist and are single-valued. The real and imaginary parts of an analytic function of z are called conjugate functions and are harmonic. ϕ and ψ are conjugate functions and we know that $\nabla^2\phi = \nabla^2\psi = 0$.

Referring to Fig. 6-12, dF/dz may be evaluated for an arbitrary Δz. If we take Δz parallel to the x axis, we have $\Delta z = \Delta x$ and

$$\frac{dF}{dz} = \frac{\partial\phi}{\partial x} + i\frac{\partial\psi}{\partial x}$$

and if we take Δz parallel to the y axis we have $\Delta z = i\,\Delta y$ and

$$\frac{dF}{dz} = -i\frac{\partial\phi}{\partial y} + \frac{\partial\psi}{\partial y}$$

Hence either expression is appropriate and both must be equal, so that we obtain the Cauchy-Riemann conditions by equating real and imaginary parts:

$$\frac{\partial\phi}{\partial x} = \frac{\partial\psi}{\partial y} \qquad\text{and}\qquad \frac{\partial\psi}{\partial x} = -\frac{\partial\phi}{\partial y}$$

The Complex Velocity

By differentiating the complex potential F, we get

$$\frac{dF}{dz} = \frac{\partial\phi}{\partial x} + i\frac{\partial\psi}{\partial x}$$

or

$$-\frac{dF}{dz} = u - iv \tag{6.26}$$

and $-dF/dz = u - iv$ is called the complex velocity. The conjugate potential $\bar{F} = (\phi - i\psi)$ may also be differentiated with respect to $\bar{z}$, $(x - iy)$ the complex conjugate variable, to give $-d\bar{F}/d\bar{z} = u + iv$. Then

$$\frac{dF}{dz} \cdot \frac{d\bar{F}}{d\bar{z}} = u^2 + v^2 = V^2 \tag{6.27a}$$

which is the square of the velocity in the fluid. It is often useful to find V^2 once $F(z)$ is known, without further calculation. To illustrate the meaning of $\bar{F}$, suppose $F = z + iz^2$, then $\bar{F} = \bar{z} - i\bar{z}^2$. If $F = z + a^2/z$, $\bar{F} = \bar{z} + a^2/\bar{z}$. $\bar{F}(\bar{z})$ means that all explicit i are changed in sign and all z are changed to $\bar{z}$. Alternatively,

$$V^2 = \left| \frac{dF}{dz} \right|^2 \tag{6.27b}$$

since $|V|$ is the modulus of $(u - iv)$.

Once V^2 is known, the Bernoulli equation may be used to find the pressure in the flow.

A stagnation point where $u = v = 0$ is found by setting $dF/dz = 0$.

Conformal Mapping

The physical plane where motion takes place is the z or (x, y) plane where $\psi = $ constant lines are curved and represent streamlines. In the F plane ϕ and ψ form a rectangular network. Now it is possible to pass from the z plane to another plane, say the $\zeta = \eta + i\xi$ plane by a transformation which preserves the orthogonal nature of ϕ and ψ. Such a transformation is known as a mapping function of the form

$$\zeta = f(z) \tag{6.28}$$

It can be shown that an infinitesimal triangle in the z plane maps into a similar infinitesimal triangle in the ζ plane with preservation of the angles and similarity. Such transformations are used in map making; the Mercator projection is a conformal mapping of the Earth onto a flat surface. By choosing appropriate functions of the form (6.28), we can construct flow patterns about complex shapes if we know the flow pattern $F(z)$ for a simple shape. Then by (6.28) we can arrive at $F(\zeta)$ which allows description in the ζ plane of a more complex flow.

For example, consider the mapping shown in Fig. 6-13 below. The upper half of the ζ plane may be mapped into the sector shown in the z plane by the transformation

$$\zeta = z^\alpha \tag{6.29}$$

where the origin $0 \to 0'$ is excluded.

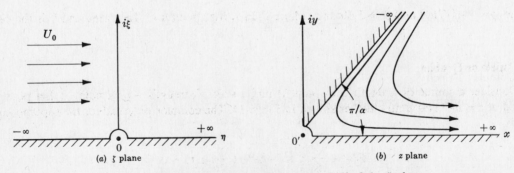

Fig. 6-13. Conformal mapping of the upper half of the ζ plane into the sector π/α in the z plane.

If we were to consider a uniform flow from left to right in the ζ plane,

$$F = -U_0 \zeta = -U_0 \eta - iU_0 \xi = \phi + i\psi \tag{6.30}$$

then this flow would be

$$F = -U_0 \zeta = -U_0 z^\alpha \tag{6.31}$$

in the z plane and be as shown in Fig. 6-13(b), representing flow in a corner.

6.7 THE COMPLEX POTENTIAL FOR SOME SIMPLE FLOWS

The method of complex variables is one of the most powerful tools in potential flow theory and forms the basis of subsonic aerodynamics. By successive transformations from a simple flow pattern into a more complex one it is often possible to construct the flow around objects such as cylinders, airfoils, etc. In this section we list a few important complex potentials and describe their flow patterns. Furthermore, complex potentials can be superposed, just as ϕ and ψ were, to generate various patterns.

The Uniform Flow Field

As we have previously stated,

$$F = -U_0 z = -U_0(x + iy) = \phi + i\psi \tag{6.32}$$

is the complex potential for a uniform flow U_0 parallel to the x axis. Equating real and imaginary parts, $\phi = -U_0 x$ and $\psi = -U_0 y$, which we have already learned previously and which has been shown in Fig. 6-7.

Sources and Sinks

For a source of strength Q, the complex potential is

$$F = -(Q/2\pi) \ln z = -(Q/2\pi) \ln re^{i\theta} \tag{6.33}$$

It is convenient to represent z as $re^{i\theta}$ here so that we can separate F into real and imaginary parts: $F = -(Q/2\pi)(\ln r + i\theta)$. Hence we obtain $\phi = -(Q/2\pi) \ln r$ and $\psi = -(Q/2\pi)\theta$ as shown in Fig. 6-8. A sink is identical except that Q is negative, or if we define Q as the strength of the sink,

$$F = (Q/2\pi) \ln z \tag{6.34}$$

Potential Vortex

For a potential vortex, as shown in Fig. 6-9,

$$F = i \frac{\Gamma}{2\pi} \ln z = i \frac{\Gamma}{2\pi} \ln re^{i\theta} \tag{6.35}$$

so that $\phi = -(\Gamma/2\pi)\theta$ and $\psi = \Gamma/2\pi \ln r$, where $\Gamma/2\pi$ is the strength of the vortex and Γ is the circulation.

The Dipole or Doublet

Consider a source of strength Q at point A and a sink of strength $-Q$ at point B. Let point A be located at $z = ae^{i\alpha}$ and B at $-ae^{i\alpha}$ as shown in Fig. 6-14. The complex potential for the superposed flow is then

$$F = -\frac{Q}{2\pi} \ln (z - ae^{i\alpha}) + \frac{Q}{2\pi} \ln (z + ae^{i\alpha}) \tag{6.36}$$

The streamlines are circles passing through A and B.

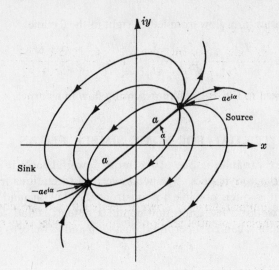

Fig. 6-14. A source and sink.

Now as the points A and B approach each other, the limiting flow as $A \to B$ $(a \to 0)$ is known as a doublet or dipole. The complex potential for this flow (as $a \to 0$ and A and B coincide) is

$$F = \frac{me^{i\alpha}}{z} \qquad (6.37)$$

where $m = Qa/\pi = $ constant, even as $a \to 0$. As $a \to 0$, $Q \to \infty$ and $\lim_{a \to 0} Qa/\pi = m$, the strength of the dipole. The flow patterns are nested circles for streamlines and velocity potential as shown in Fig. 6-15. These patterns are similar to the electrostatic dipole field configuration of a positive and a negative charge, and to the radiation pattern from a dipole radio antenna.

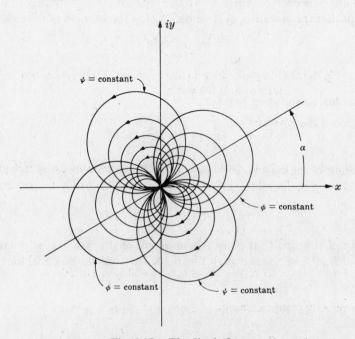

Fig. 6-15. The dipole flow.

Separating (6.37) into ϕ and ψ, we get

$$\phi = \frac{m(x \cos \alpha + y \sin \alpha)}{(x^2 + y^2)}, \qquad \psi = \frac{m(x \sin \alpha - y \cos \alpha)}{(x^2 + y^2)} \qquad (6.38a)$$

which represent circles tangent to the origin. If $\alpha = 0$ we get

$$\phi = \frac{mx}{x^2 + y^2}, \qquad \psi = -\frac{my}{x^2 + y^2} \qquad (6.38b)$$

Streaming Motion Past a Circular Cylinder

Referring to Fig. 6-16, a uniform flow U_0 flows in the positive x direction. A cylinder of radius a is located at the origin. The complex potential for flow about the cylinder is given by

$$F = -U_0(z + a^2/z) \qquad (6.39)$$

The stream function is obtained from (6.39) as

$$\psi = -U_0\left(r \sin \theta - \frac{a^2}{r} \sin \theta\right) = -U_0 y(1 - a^2/r^2) = -U_0 y + \frac{a^2 U_0 y}{x^2 + y^2} \qquad (6.40)$$

which shows that the flow over a cylinder is represented as the superposition of a uniform flow over a dipole of strength $-a^2 U_0$. (The negative sign merely interchanges the role of source and sink in Fig. 6-15.)

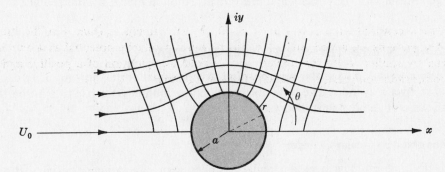

Fig. 6-16. Flow around a circular cylinder. The figure is symmetric about the x axis.

The velocity potential ϕ is given by

$$\phi = -U_0 x(1 + a^2/r^2) \qquad (6.41)$$

At $r = a$, the contour of the cylinder must coincide with a streamline. And indeed, at $r = a$, $\psi = 0$.

The velocity field may be found from $-dF/dz = u - iv$ which may be put into polar coordinate form,

$$\frac{dF}{dz} = -U_0 + \frac{U_0 a^2}{z^2} = -U_0 + \frac{U_0 a^2 e^{-2i\theta}}{r^2}$$

$$= U_0\left(\frac{a^2}{r^2} \cos 2\theta - 1\right) + iU_0 \frac{a^2}{r^2} \sin 2\theta$$

so that the cartesian velocity components are

$$u = -U_0\left(\frac{a^2}{r^2} \cos 2\theta - 1\right), \qquad v = -U_0 \frac{a^2}{r^2} \sin 2\theta \qquad (6.42)$$

and the polar coordinate velocity components v_r and v_θ are then

$$v_r = U_0(1 - a^2/r^2) \cos \theta, \qquad v_\theta = -v_0(1 + a^2/r^2) \sin \theta \qquad (6.43)$$

and hence

$$V^2 = u^2 + v^2 = v_r^2 + v_\theta^2 = \frac{dF}{dz} \cdot \frac{d\bar{F}}{d\bar{z}} = U_0^2\left[1 + \frac{a^4}{r^4} + \frac{2a^2}{r^2}(\sin^2 \theta - \cos^2 \theta)\right]$$

The pressure around the cylinder can now be found from Bernoulli's equation

$$\frac{p}{\rho} + \frac{V^2}{2} = \text{constant} = \frac{p_\infty}{\rho} + \frac{U_0^2}{2} = \frac{p_0}{\rho}$$

where p_0 is the stagnation pressure at any point in the flow where V is zero and p_∞ is the free stream pressure where the velocity is U_0. On the surface of the cylinder, at $r = a$,

$$p\bigg|_{r=a} = p_0/\rho - 2U_0^2(1 + \sin^2 \theta - \cos^2 \theta)$$

The maximum velocity is $2U_0$ and occurs on the top and bottom of the cylinder ($\theta = \pi/2, 3\pi/2$), and there the pressure is a minimum. Since the pressure is symmetrical about the x and y axes, there is no net force on the cylinder. In actuality, of course, separation would occur on the rear of the cylinder and drag would be present. However, if the cylinder is deformed so that the rear is drawn out to a point, separation may be prevented and the potential flow solution is quite good except for the boundary layer which gives rise to skin friction. In the next section we will discuss airfoil theory where potential flow theory gives good results for the velocity and pressure distribution over a streamlined body.

6.8 CIRCULATION AND THE JOUKOWSKI THEOREM

In this section we will discuss the superposition of a potential vortex flow (of finite circulation) on the uniform flow over a cylinder. The resultant pressure distribution gives rise to a lift on the cylinder, and forms the basis for aerodynamic theory.

The Circulation about a Circular Cylinder

Consider the uniform flow over a circular cylinder with circulation. The circulation could be brought about by rotation of the cylinder such that if there were no uniform flow, the peripheral velocity of the cylinder of radius a would be the corresponding tangential velocity of the potential vortex at radius a. The total complex potential is then

$$F = -U_0(z + a^2/z) + i\frac{\Gamma}{2\pi} \ln z/a \qquad (6.44)$$

so that the cylinder is part of the line $\psi = 0$. To determine the form of the streamlines ($\psi = \text{constant}$) we find the complex velocity $-dF/dz$.

$$\frac{dF}{dz} = U_0(a^2/z - 1) + i\frac{\Gamma}{2\pi z}$$

and the stagnation points are given by solving for z the equation

$$U_0(a^2/z - 1) + i\frac{\Gamma}{2\pi z} = 0$$

so that

$$z\bigg|_{v=0} = a\left(i\frac{\Gamma}{4\pi a U_0} \pm \sqrt{1 - \frac{\Gamma^2}{16\pi^2 a^2 U_0^2}}\right) \qquad (6.45)$$

There are three cases: $\Gamma^2 < (4\pi a U_0)^2$, $\Gamma^2 = (4\pi a U_0)^2$, $\Gamma^2 > (4\pi a U_0)^2$. In Fig. 6-17 we plot the flow for each case.

There is a lift L on the cylinder, and we develop below a general formula for this lift and show that it is $L = -i\rho U_0 \Gamma$; that is, it is a force in the y direction of $-\rho U_0 \Gamma$.

The value of Γ here depends on the rotary speed of the cylinder, and that is why a baseball, golf ball, or ping-pong ball curves when it has spin. For the flow shown in Fig. 6-17 the value of Γ is negative and the lift is in the $+y$ direction.

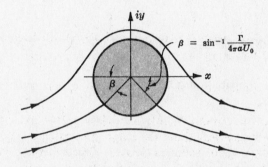

(a) $\Gamma^2 < (4\pi a U_0)^2$. The two stagnation points lie on bottom of the cylinder.

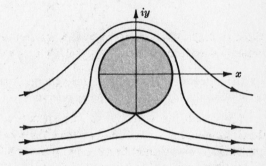

(b) $\Gamma^2 = (4\pi a U_0)^2$. One stagnation point lies on the bottom center.

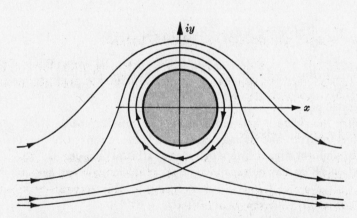

(c) $\Gamma^2 > (4\pi a U_0)^2$. The single stagnation point moves down.

Fig. 6-17.   The streamlines for flow with circulation over a circular cylinder.

The Blasius Theorem and the Theorem of Kutta and Joukowski

We will derive now the general forces and moments acting on a cylinder in potential flow and will apply the results to the calculation of lift on an airfoil. Referring to Fig. 6-18, we integrate the pressure around the cylinder to find the total complex force $X + iY$ and moment M acting on the cylinder (per unit length). From Bernoulli's equation, $p = p_0 - \frac{1}{2}\rho V^2$ where p_0 is a constant. Then,

$$p = p_0 - \tfrac{1}{2}\rho V^2 = p_0 - \tfrac{1}{2}\rho \, \frac{dF}{dz} \cdot \frac{d\bar{F}}{d\bar{z}}$$

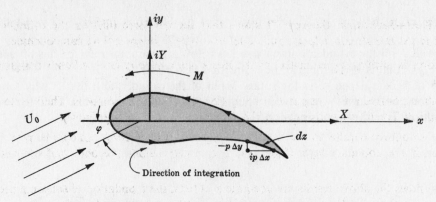

Fig. 6-18. Flow over a cylinder with lift and drag indicated.

and the integral of p_0 around the cylinder makes no contribution. On the cylinder, $\psi = $ constant ($d\psi = 0$), so that $d\bar{F} = dF$. We have, referring to Fig. 6-18,

$$X - iY = -\oint_C (p\, dy + ip\, dx) = -\oint_C ip\, d\bar{z}$$

$$= \oint_C \tfrac{1}{2} i\rho \, \frac{dF}{dz} \cdot \frac{d\bar{F}}{d\bar{z}} \cdot d\bar{z} = \tfrac{1}{2} i\rho \oint_C \left(\frac{dF}{dz}\right)^2 dz \tag{6.46}$$

where we have replaced $d\bar{F}$ by dF, canceled the $d\bar{z}$ terms, and multiplied and divided by dz. The moment M is given by

$$M = \oint_C p(x\, dx + y\, dy) = \mathrm{Re} \oint_C pz\, d\bar{z}$$

$$= \mathrm{Re} \oint_C -\tfrac{1}{2}\rho \, \frac{dF}{dz} \cdot \frac{d\bar{F}}{d\bar{z}} \cdot z\, d\bar{z} = \mathrm{Re}\left\{ -\tfrac{1}{2}\rho \oint_C z\left(\frac{dF}{dz}\right)^2 dz \right\} \tag{6.47}$$

where Re indicates the real part.

Once we know $F(z)$, then the lift and moment can be found immediately. Further, by expanding $F(z)$, explicit expressions for X, Y and M may be found. These are given by the theorem of Kutta and Joukowski for flow of a uniform stream over a cylinder of arbitrary shape.

The complex function dF/dz may be expanded (for $|z|$ sufficiently large) as

$$-\frac{dF}{dz} = U_0 e^{-i\varphi} + A/z + B/z^2 + \cdots \tag{6.48}$$

because clearly as $z \to \infty$, $-dF/dz$, the complex velocity, becomes U_0 which is the free stream value of the velocity which in general makes an angle φ with the x axis as shown in Fig. 6-18. Then F may be determined as

$$F = -U_0 e^{-i\varphi} z - A \ln z + B/z + \cdots \tag{6.49}$$

We see that the second term is exactly the complex potential due to circulation. Hence $A = -i\Gamma/2\pi$. Thus

$$\left(\frac{dF}{dz}\right)^2 = U_0^2 e^{-2i\varphi} - i\, \frac{\Gamma U_0 e^{-i\varphi}}{\pi z} - \frac{\Gamma^2 - 8\pi^2 B U_0 e^{-i\varphi}}{4\pi^2 z^2} + \cdots$$

and from the Blasius theorem and the theory of complex integration we obtain

$$X - iY = \tfrac{1}{2} i\rho \left[2\pi i \left(-i\, \frac{\Gamma U_0 e^{-i\varphi}}{\pi z} \right) \right] = i\rho \Gamma U_0 e^{-i\varphi} \tag{6.50}$$

which is the Kutta-Joukowski theorem. It states that the net force (lift) on the cylinder is directed perpendicular to the free stream velocity and equal to $-\rho U_0 \Gamma$. In general we can conclude:

1. The force is always perpendicular to the free stream velocity (and is hence designated as a lift force).

2. The lift is positive if Γ is negative (for a positive U_0 in the x direction). There is no lift without circulation. The lift depends only on Γ (which may depend on the profile).

3. There is no force parallel to the direction of U_0 (drag force). Drag can only be produced by skin friction in the boundary layer if the flow is potential around the body and there is no separation.

4. In real flows the above results are accurate and form the foundation of subsonic aerodynamics, so long as the cylinder is streamlined. If separation occurs, a drag force due to a pressure decrease in the wake occurs but is not predicted by potential flow. Flow over a circular cylinder always separates (except in very slow viscous flow) and the potential flow is not very accurate there.

In summary, then, lift is $-\rho U_0 \Gamma$, and the task that remains is to determine Γ for a given cylinder such as an airfoil, Fig. 6-19. The angle of attack α is the angle between the free stream velocity and a fixed chord line on the profile. It is defined so that L and hence Γ are zero for $\alpha = 0$. The value of Γ tends to increase for certain shapes as α is increased until stall occurs and separation occurs. Then a drastic reduction in lift suddenly occurs as Γ drops.

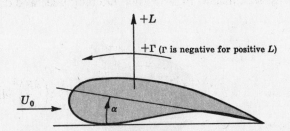

Fig. 6-19. Lift L on a cylinder. The angle of attack is α. When $\alpha = 0$, $L = 0$, by definition of α.

6.9   AIRFOIL THEORY

We will now apply the results of the previous section to the calculation of lift for some simple shapes. In particular, we are interested in the airfoil shape and the physical cause of circulation. As we pointed out, a rotating circular cylinder generates circulation; but what about an airfoil which certainly isn't rotating? The cause of circulation there is explained by the Joukowski hypothesis which very simply states that infinite velocities are inadmissible in real flows.

The Airfoil

As we have seen, lift is due to circulation; and there is, indeed, circulation about an airfoil. Referring to Fig. 6-20, the Kutta-Joukowski hypothesis states that the singular point at the trailing edge, where the velocity must be infinite by potential flow theory, is impossible in a real flow and the flow will adjust itself so that the stagnation point moves to the trailing edge and hence removes the singular point. The value of circulation, Γ, about the airfoil is just that amount necessary to shift the stagnation point to the trailing edge. Clearly Γ must be negative to effect this transfer, and hence the lift $L = -\rho U_0 \Gamma$ is positive as shown.

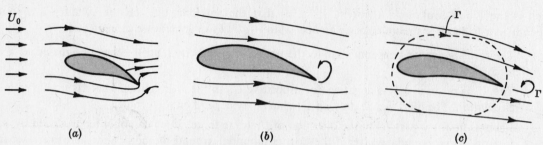

Fig. 6-20. Circulation and its formation about an airfoil. The stagnation point moves to the rear as the flow progresses and this transfer determines Γ. Fig. (a), (b) and (c) show successive stages in the startup of the airfoil.

As the flow starts up, the potential flow is as shown in Fig. 6-20(a). The stagnation point moves to the rear and a small vortex is formed which sheds off and is lost downstream. By Kelvin's theorem the total circulation in the fluid must be constant in time and hence a Γ equal but of opposite sign to the shed vortex is created about the airfoil. The total circulation in the fluid remains zero but the value about the airfoil is a negative number. The shed vortex falls far behind and gets left on the airfield. (A large control contour including the airfoil and the airfield gives zero circulation.)

The formation of this trailing vortex can easily be seen by dragging a spoon through a cup of coffee. Draw the spoon, inclined at an angle of attack through the coffee, and you will see the trailing vortex form and shed off. (See Fig. 6-21.)

Fig. 6-21. A spoon in a coffee cup generating a trailing vortex. The curved spoon approximates a cambered airfoil.

Calculation of the Potential Flow and Lift

The flow around any given airfoil may be described by the appropriate complex potential F. The simplest airfoil is the Joukowski airfoil which may be obtained from the flow about a circular cylinder by a single conformal transformation. Referring to Fig. 6-22, flow about the cylinder in the ζ plane maps into the flow in the z plane by the Joukowski transformation. The transformation from ζ to z here is

$$z = \zeta + l^2/\zeta \tag{6.51}$$

and the flow about the displaced cylinder (center at C) for flow at an apparent angle of attack φ is

$$F = -U_0\left[(\zeta - be^{i\theta})e^{-i\varphi} + \frac{a^2}{(\zeta - be^{i\theta})e^{-i\varphi}}\right] \tag{6.52}$$

If $b = 0$ (C at the origin) the Joukowski transformation maps the cylinder into an ellipse centered at the origin with major axis along the x axis. The point $\zeta = l$ transforms into $z = 2l$ and is the trailing edge where dF/dz has a singularity which must be removed by the superposition of an appropriate amount of circulation.

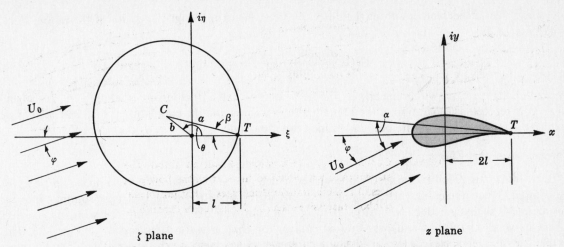

Fig. 6-22. The Joukowski transformation. The apparent angle of attack is φ. L is $\perp$ to U_0. The absolute angle of attack is α (to be determined). When $\alpha = 0$, $L = 0$ by definition of α.

To equation (6.52) we must add the complex potential due to circulation, so that the velocity at the trailing edge can be made finite. The total $F(\zeta)$ about the cylinder must then be

$$F(\zeta) = -U_0\left[e^{-i\varphi}(\zeta - be^{i\theta}) + \frac{a^2 e^{i\varphi}}{(\zeta - be^{i\theta})}\right] + i\,\frac{\Gamma}{2\pi}\ln\left(\frac{\zeta - be^{i\theta}}{a}\right) \qquad (6.53)$$

The complex velocity in the z plane at point T, the trailing edge (where $z = 2l$ and $\zeta = l$), is

$$\left.\frac{dF}{dz}\right|_T = \left.\frac{dF}{d\zeta}\cdot\frac{d\zeta}{dz}\right|_{\text{point }T}$$

and $dz/d\zeta = 1 - l^2/\zeta^2 = 0$ at the trailing edge so that $d\zeta/dz$ is singular (i.e., approaches ∞) at the trailing edge. Hence the value of $dF/d\zeta$ must be zero at $z = -2l$ for dF/dz to be finite there, and this condition determines the value of Γ and hence the lift on the airfoil.

From (6.53) we find $dF/d\zeta = 0$ as

$$\frac{dF}{d\zeta} = -U_0\left(e^{-i\varphi} - \frac{a^2 e^{i\varphi}}{(\zeta - be^{i\theta})^2}\right) + i\,\frac{\Gamma}{2\pi}\frac{1}{(\zeta - be^{i\theta})} = 0 \qquad (6.54)$$

and from Fig. 6-22, $(\zeta - be^{i\theta}) = [\zeta - l + ae^{i(-\beta)}]$ so that at $\zeta = l$, (6.54) gives

$$-U_0(1 - e^{-2i(-\beta)+2i\varphi}) + i\,\frac{\Gamma}{2\pi a}e^{-i(-\beta)+i\varphi} = 0 \qquad (6.55)$$

which may be solved for Γ. We can equate either the real or imaginary parts of (6.55) and get the same answer for Γ.

$$\Gamma = -4\pi a U_0 \sin(\beta + \varphi) \qquad (6.56)$$

and we see that Γ is a negative number which depends on the free stream velocity, apparent angle of attack φ, and the parameters a and β which determine the size and camber of the airfoil. Since $L = -\rho U_0 \Gamma$ is perpendicular to the free stream, L is a positive number for values of $(\beta + \varphi)$ positive. We call the angle $(\beta + \varphi)$ the absolute angle of attack α. By definition, when $\alpha = 0$, the lift L is zero.

In aerodynamics, the term angle of attack usually refers to the angle α, not φ.

The lift coefficient C_L is defined by

$$C_L = \frac{L}{c\rho U_0^2/2} \qquad (6.57)$$

where L is the lift per unit length of the airfoil and c is the chord length or width of the airfoil. For a Joukowski airfoil $c \approx 4a$ and we get for C_L,

$$C_L = \frac{\rho U_0 (4\pi a U_0) \sin \alpha}{4a\rho U_0^2/2} \cong 2\pi\alpha \tag{6.58}$$

In practice C_L increases with α linearly as (6.58) shows for small α, then drops off rapidly when the stall angle α_S is reached as separation of the boundary layer occurs and the circulation is lost. The stall angle and exact C_L versus α curves are best determined experimentally.

Three-Dimensional Effects

The lift on an airfoil is due to the line vortex which moves with the airfoil (the *bound vortices*). As we showed earlier, these line vortices cannot just end at the tip of the wing, but must continue into the free fluid (until dissipated by viscosity). If the circulation were uniform along the wing the vortices would be shed off the tips and pass downstream in the *trailing vortices*, as shown in Fig. 6-23. This simple "horseshoe" vortex system is not quite correct, however, because the value of Γ generally varies along the wing. As a consequence, vorticity is shed off the trailing edge all along the wing, generating a vortex sheet made of a superposition of "horseshoe" vortex systems of various strength, as shown in Fig. 6-24. Alternatively, a physical picture of the vortex sheet can be obtained by imagining a plane of ball bearings extending downstream behind the wing; the air on top tends to flow inward, and the air on the bottom tends to flow outward. This difference in flow direction of the wind coming off the trailing edge generates the vortex layer or sheet. The reason for the skewing of the flow is readily understood by remembering that there is pressure differential between the top and bottom of the wing and from the center to the tips.

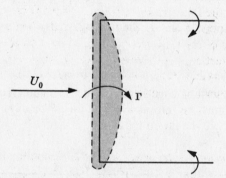

Fig. 6-23. A simple horseshoe vortex pattern over a three-dimensional wing. The top view of the wing is shown.

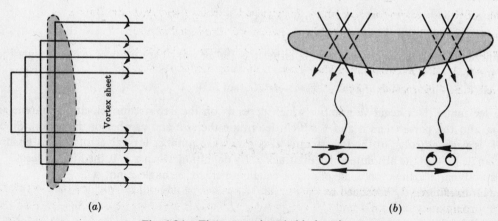

Fig. 6-24. The vortex sheet behind a wing.

Ultimately, a few wing lengths or less behind the wing, the vortex sheet forms two distinct vortex lines as shown in Fig. 6-25. This phenomenon may be observed in multi-jet aircraft (where the nacelles are wing suspended). The jet vapor trails may be seen to coalesce to form two distinct trails, one from the left wing and one from the right wing.

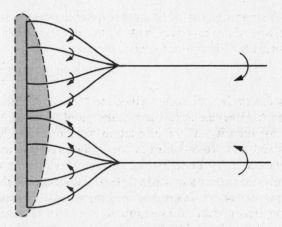

Fig. 6-25.   The vortex sheet rolls up into two distinct vortices, spaced at somewhat less than the wing span.

One last point to mention is the induced drag. Because of the vortex sheet, there is an induced "downwash" which, in effect, changes the angle of attack of the wing and causes the lift vector to point at a slight angle (to the rear) to the free stream velocity. Hence the lift vector has a component parallel to the flight direction. This component is a drag force, called the "induced" drag. The remainder of the drag, the "profile" drag, is made of the skin friction drag of the boundary layer and a "form" drag due to the shedding of the Karman vortices in the vortex sheet (i.e., the wake).

We cannot pursue the subject here, but potential flow theory and boundary layer theory together constitute the foundation of subsonic aerodynamics.

References

1. Batchelor, G. K., *An Introduction to Fluid Dynamics*, Cambridge University Press, 1967.

2. Glauert, H., *The Elements of Aerofoil and Airscrew Theory*, 2nd ed., Cambridge University Press, 1948.

3. Karamchetti, K., *Principles of Ideal-Fluid Aerodynamics*, John Wiley, 1966.

4. Kuethe, A. M., and Chow, C., *Foundations of Aerodynamics*, 4th ed., John Wiley, 1986.

5. Lamb, Sir Horace, *Hydrodynamics*, 6th ed., Cambridge University Press, 1932 (also Dover).

6. Milne-Thompson, L. M., *Theoretical Hydrodynamics*, 3rd ed., Macmillan, 1957.

7. Prandtl, L., and Tietjens, O. G., *Applied Hydro- and Aero-mechanics*, Dover, 1957.

8. Robertson, J. M., *Hydrodynamics in Theory and Application*, Prentice-Hall, 1965.

9. Shevell, R. S., *Fundamentals of Flight*, Prentice-Hall, 1983.

Solved Problems

6.1.   A tornado may be idealized as a potential vortex with a rotational "eye" or core which behaves approximately as a solid body. A rough rule of thumb is that the radius of the eye is of the order of 100 ft. How does the pressure vary along the ground around the "eye"? For a tornado with a maximum wind velocity of 100 mi/hr, what is the maximum drop in pressure? This under-pressure is partly responsible for the lifting of roofs and much of the damage inflicted by tornados.

The velocity v_θ is $\Gamma/2\pi r$ and hence

$$\Gamma = 2\pi r v_\theta = 2\pi(100 \text{ ft})(100 \times 5280/3600 \text{ ft/sec}) = 9.2 \times 10^4 \text{ ft}^2/\text{sec}$$

From Bernoulli's equation (assuming standard air at pressure p_0),

$$p - p_0 = \tfrac{1}{2}V^2\rho = -\frac{\Gamma^2\rho}{8\pi^2 r^2} = -\frac{(9.2 \times 10^4)^2(0.0023)}{8\pi^2 r^2} = -\frac{2.4 \times 10^5}{r^2} \text{ psf}$$

The negative sign indicates that the pressure is lower than the atmospheric pressure far from the tornado. At $r = 100$ ft (which is the smallest r where we can apply potential flow theory) the pressure is reduced by 24.6 psf (which is about 1.7 psi) below atmospheric.

6.2. In Fig. 6-26 a source is located at a distance a from a wall. What is the total pressure force on the wall? Behind the wall ($x > 0$) the pressure is p_0, the stagnation pressure.

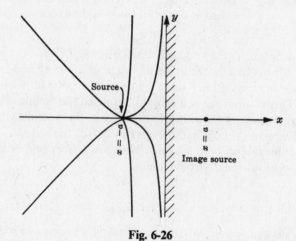

Fig. 6-26

The flow is given by assuming an image source at $x = a$ when we place the real source at $x = -a$ as shown in the figure. The complex potential for the total flow is

$$F = -\frac{Q}{2\pi}[\ln(z+a) + \ln(z-a)]$$

and

$$\bar{F} = -\frac{Q}{2\pi}[\ln(\bar{z}+a) + \ln(\bar{z}-a)]$$

and we find

$$V^2 = \frac{dF}{dz}\cdot\frac{d\bar{F}}{d\bar{z}} = \frac{Q^2}{4\pi^2}\left[\frac{1}{(z+a)} + \frac{1}{(z-a)}\right]\left[\frac{1}{(\bar{z}+a)} + \frac{1}{(\bar{z}+a)}\right]$$

which may be simplified to the following (since $z = x + iy$ and $\bar{z} = x - iy$):

$$V^2 = \frac{Q^2(x^2+y^2)}{\pi^2[(x^2+y^2)^2 - 2a^2(x^2-y^2) + a^4]}$$

and from Bernoulli's equation $p^2/\rho + \frac{1}{2}V^2 = p_0/\rho$ we find the pressure difference across the wall as $p - p_0 = -\frac{1}{2}V^2\rho$, and the total force per unit length of the wall (out of the paper) on the wall (positive to the right) is

$$\text{Force} = \int_{-\infty}^{+\infty} (p - p_0) \Big|_{x=0} \, dy = -\frac{1}{2}\rho \int_{-\infty}^{+\infty} V^2 \Big|_{x=0} \, dy = -\rho Q^2 \frac{\rho Q^2}{2\pi^2} \int_{-\infty}^{+\infty} \frac{y^2 \, dy}{(y^2 + a^2)^2} = -\frac{\rho Q^2}{4\pi a}$$

The negative result indicates that the net force is to the left and the wall is sucked toward the source.

6.3. Rework Problem 6.2 except that the source is replaced by a sink.

　　The complex potential is identical except that Q is now a negative number. However, in the solution for the force on the wall, Q appears squared so that the sign is irrelevant. Hence we conclude that the net force on the wall is the same for a source or a sink. The wall tends to be sucked toward the source or sink.

　　The result might seem surprising, but in each case the velocity increase near the wall causes a pressure defect and the result is the same.

6.4. Wind flows down a hill and over a rise, in the shape of a sector of a circle of radius b, and out onto the plains. Map the flow. Find ϕ and ψ explicitly. This flow is shown in Fig. 6-27.

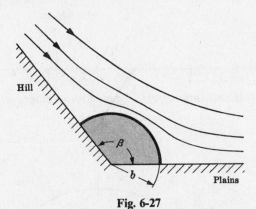

Fig. 6-27

　　We begin with flow over a circular cylinder, then map it into an angle with the transformation of equation (6.29).

　　In the ζ plane, referring to Fig. 6-28,

$$F(\zeta) = -U_0(\zeta + a^2/\zeta)$$

so that from $\zeta = z^\alpha$ we have

$$F(z) = -U_0(z^\alpha + a^2/z^\alpha)$$

as the complex potential in the physical z plane. α is given by $\pi/\alpha = \beta$. It is important to note that the free stream velocity V_0 in the z plane is not uniform but varies with r and θ. U_0 does not map directly from the ζ

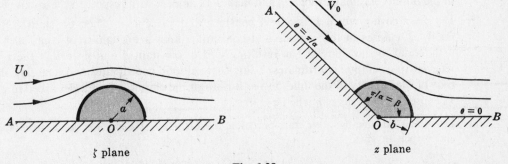

Fig. 6-28

to the z plane as it did for the Joukowski airfoil transformation. Furthermore, the radius a of the circle in the ζ plane is not the same as the radius b of the sector in the z plane.

We have
$$z^\alpha = r^\alpha e^{\alpha i\theta} = r^\alpha(\cos \alpha\theta + i \sin \alpha\theta)$$

and
$$F = -U_0[r^\alpha(\cos \alpha\theta + i \sin \alpha\theta) + (a^2/r^\alpha)(\cos \alpha\theta - i \sin \alpha\theta)]$$

so that
$$\phi = \operatorname{Re} F = -U_0 r^\alpha[\cos \alpha\theta + a^2 r^{-2\alpha} \cos \alpha\theta]$$

$$\psi = \operatorname{Im} F = -U_0 r^\alpha[\sin \alpha\theta - a^2 r^{-2\alpha} \sin \alpha\theta]$$

The behavior of the free stream velocity may be seen by finding the radial velocity v_r, given by $-\partial\phi/\partial r$.

$$v_r = \alpha U_0 r^{\alpha-1}[\cos \alpha\theta - a^2 r^{-2\alpha} \cos \alpha\theta]$$

so that for $\theta = \pi/\alpha$ (along the wall), we have

$$v_r\Big|_{\theta=\pi/2} = -\alpha U_0[r^{\alpha-1} - a^2 r^{-(1+\alpha)}]$$

For large r, and $\alpha > 1$, $v_r|_{\theta=\pi/2}$ becomes large and is approximately $-\alpha U_0 r^{(\alpha-1)}$. The negative sign indicates flow inward from infinity.

The radius b can be related to the radius a. The radius b is given by setting $v_r = 0$. Thus

$$1 - a^2 r^{-2\alpha}\Big|_{r=b} = 0 \qquad \text{from which} \qquad b = a^{1/\alpha}$$

6.5.   The transformation

$$z = C(\zeta + \lambda\zeta^{-1}), \qquad 0 \le \lambda \le 1, \qquad 0 \le C$$

transforms a circle with unit radius in the ζ plane to an ellipse in the z plane which is shown in Fig. 6-29.

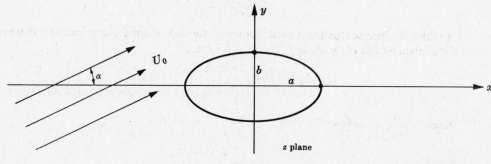

Fig. 6-29

(a)   Making use of this transformation, find the complex potential $F(z)$ for the steady flow past an elliptic cylinder with semi-axes a and b as shown in Fig. 6-29. The undisturbed velocity at the infinity has magnitude U_0 and makes an angle α with respect to the x axis.

(b)   Let $z = k \cosh \gamma$ where k is a real constant and $\gamma = \xi + i\eta$. Then constant ξ lines are a family of ellipses in the z plane and constant η lines are a family of confocal hyperbolas orthogonal to the ellipses. The resulting grid of constant ξ and η lines in the z plane are referred to as elliptic coordinates. Using these elliptic coordinates and denoting the magnitude of the velocity of the fluid on the surface of the elliptic cylinder described in part (a) by V, express V in terms of η when $\alpha = 0$.

(a)   Let us first examine the $\zeta - z$ transformation, $z = C(\zeta + \lambda\zeta^{-1})$. Let $\zeta = e^{i\theta}$ for a circle of unit radius and $z = x + iy$. Then

$$z = C(e^{i\theta} + \lambda e^{-i\theta}) = C(1 + \lambda) \cos \theta + iC(1 - \lambda) \sin \theta$$

and $x = C(1 + \lambda) \cos \theta$, $y = C(1 - \lambda) \sin \theta$ so that $\dfrac{x^2}{C^2(1 + \lambda)^2} + \dfrac{y^2}{C^2(1 + \lambda)^2} = 1$. Hence

$$a = C(1 + \lambda), \qquad b = C(1 - \lambda); \qquad C = (a + b)/2, \qquad 4c^2\lambda = a^2 - b^2 \qquad (1)$$

Now for flow over a unit circle in a γ plane shown in Fig. 6-30,

$$F(\gamma) = A(\gamma + \gamma^{-1})$$

Then $\zeta = \gamma e^{i\alpha}$ so that $\qquad\qquad F(\zeta) = A(\zeta e^{-i\alpha} + \zeta^{-1} e^{i\alpha})$

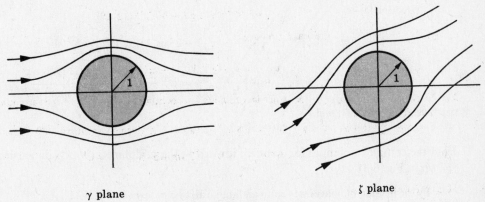

γ plane   ζ plane

Fig. 6-30

From the given transformation $z = C(\zeta + \lambda\zeta^{-1})$ we obtain

$$\zeta = \frac{1}{2C}\left(z \pm \sqrt{z^2 - 4C^2\lambda}\right)$$

We take $(+)$ here so that the domain outside of the circle in the ζ plane, Fig. 6-30, is transformed to the domain outside of the ellipse in the z plane. Then

$$F(z) = A\left[\frac{e^{-i\alpha}}{2C}\left(z + \sqrt{z^2 - 4C^2\lambda}\right) + \frac{2Ce^{i\alpha}}{\left(z + \sqrt{z^2 - 4C^2\lambda}\right)}\right] \qquad (2)$$

Now $\qquad \left.\dfrac{dF}{dz}\right|_{z\to\infty} = \dfrac{A}{C}e^{-i\alpha} \qquad$ but $\qquad -\left.\dfrac{dF}{dz}\right|_{z\to\infty} = (u - iv)\Big|_{z\to\infty} = U_0 e^{-i\alpha}; \qquad$ hence

$$A/C = -U_0 \qquad (3)$$

Using values of C, λ and A obtained by equations (1) and (3), equation (2) becomes

$$F(z) = -\tfrac{1}{2}U_0(a + b)\left\{\frac{[z + \sqrt{z^2 - (a^2 - b^2)}]e^{-i\alpha}}{a + b} + \frac{[z - \sqrt{z^2 - (a^2 - b^2)}]e^{i\alpha}}{a - b}\right\} \qquad (4)$$

(b)   We are given $z = k \cosh \gamma$, $\gamma = \xi + i\eta$ and we can write

$$x + iy = k \cosh(\xi + i\eta) = k(\cosh \xi \cos \eta + i \sinh \xi \sin \eta)$$

Hence $\qquad \dfrac{x^2}{k^2 \cosh^2 \xi} + \dfrac{y^2}{k^2 \sinh^2 \xi} = 1, \qquad \dfrac{x^2}{k^2 \cos^2 \eta} - \dfrac{y^2}{k^2 \sin^2 \eta} = 1$

Let $\xi = \xi_0$ represent the ellipse with semi-axes a and b in the z plane; then $k \cosh \xi_0 = a$ and $k \sinh \xi_0 = b$, or

$$\left.\begin{array}{c} k^2 = a^2 - b^2, \qquad \tanh \xi_0 = b/a \\[2mm] \dfrac{e^{2\xi_0}}{(a + b)^2} = \dfrac{e^{-2\xi_0}}{(a - b)^2} = \dfrac{1}{k^2} \end{array}\right\} \qquad (5)$$

Equation (4) becomes (with $\alpha = 0$)

$$F(\gamma) = -\tfrac{1}{2}U_0(a+b)\sqrt{a^2-b^2}\left(\frac{e^\gamma}{a+b} + \frac{e^{-\gamma}}{a-b}\right)$$

$$\bar{F}(\bar{\gamma}) = -\tfrac{1}{2}U_0(a+b)\sqrt{a^2-b^2}\left(\frac{e^{\bar{\gamma}}}{a+b} + \frac{e^{-\bar{\gamma}}}{a-b}\right)$$

$$(V_\gamma^2)_{\xi=\xi_0} = \frac{dF}{d\gamma}\cdot\frac{d\bar{F}}{d\bar{\gamma}} = \frac{U_0^2}{4}(a+b)^2(a^2-b^2)\left\{\frac{e^{2\xi}}{(a+b)^2} + \frac{e^{-2\xi}}{(a-b)^2} - \frac{e^{2i\eta}}{(a^2-b^2)} - \frac{e^{-2i\eta}}{(a^2-b^2)}\right\}$$

Equation (5) gives

$$V_\gamma^2\bigg|_{\xi=\xi_0} = \tfrac{1}{4}U_0^2(a+b)^2\{2 - 2\cos 2\eta\} = U^2(a+b)^2\sin^2\eta$$

$$V^2 = V_z^2\bigg|_{\xi=\xi_0} = \frac{V_\gamma^2|_{\xi=\xi_0}}{|dz/d\gamma|_{\xi=\xi_0}^2} = U_0^2\left(\frac{a+b}{a-b}\right)\left(\frac{\sin^2\eta}{\sin^2\eta + \sinh^2\xi_0}\right)$$

where $\xi_0 = \tanh^{-1} b/a$.

6.6. Find the complex potential for a source located in the midplane of a two-dimensional channel as shown in Fig. 6-31.

By the method of images we set up an infinite array as shown in Fig. 6-32.

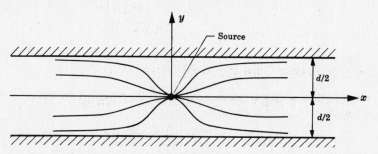

Fig. 6-31

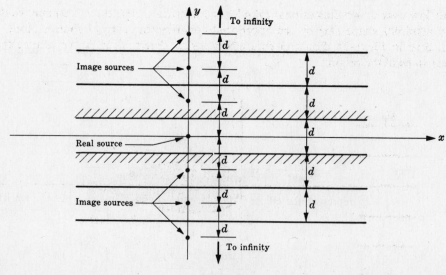

Fig. 6-32

The complex potential then is found by summing over all the images:

$$F = \sum_{n=0}^{\infty} -\frac{Q}{2\pi} \ln (z + ind) + \sum_{n=1}^{\infty} -\frac{Q}{2\pi} \ln (z - ind) = \sum_{n=-\infty}^{+\infty} -\frac{Q}{2\pi} \ln (z + ind) = -\frac{Q}{2\pi} \ln \left(\sinh \frac{\pi z}{d} \right)$$

By evaluating F for $-d/2 < y < d/2$, we have the flow between the walls at $\pm d/2$.

We can find ϕ and ψ as

$$\phi = -\frac{Q}{2\pi} \ln \left(\sinh \frac{\pi x}{d} \cos \frac{\pi y}{d} \right), \qquad \psi = -\frac{Q}{2\pi} \ln \left(\cosh \frac{\pi x}{d} \sin \frac{\pi y}{d} \right)$$

6.7. A cylinder of 1 in. diameter, shown in Fig. 6-33, rotates as indicated at 3600 rpm in standard air which is flowing over the cylinder at 100 ft/sec. Estimate the lift on the cylinder (per unit length).

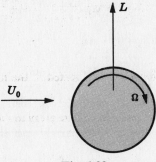

Fig. 6-33

Use $L = -\rho U_0 \Gamma$, where Γ here may be estimated by assuming

$$\Gamma = \oint \mathbf{V} \cdot d\mathbf{l} = \int_0^{2\pi} -(r\Omega)(r \, d\theta) = -2\pi r^2 \Omega$$

$$= -2\pi (1/24 \text{ ft})^2 (3600 \times 2\pi/60 \text{ rad/sec}) = -4.1 \text{ ft}^2/\text{sec}$$

Thus $L = -\rho U_0 \Gamma = -(0.0023 \text{ slug/ft}^3)(100 \text{ ft/sec})(-4.1 \text{ ft}^2/\text{sec}) = 0.94 \text{ lb/ft}$.

6.8. The flow over a two-dimensional cylinder (symmetrical about the x axis) may be determined for any arbitrary shape $f(x)$ by the appropriate distribution, $q(x)$, of sources along the x axis as indicated in Fig. 6-34. For a given shape $y = f(x)$, determine $q(x)$. We assume the nose of the body to be at the origin.

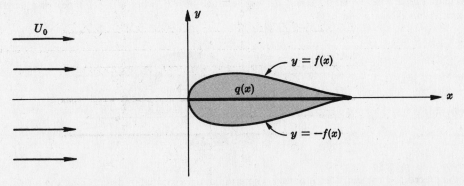

Fig. 6-34

The body surface $\pm f(x)$ is a streamline, and clearly the distribution $q(x)$ must change sign along the x axis and no fluid crosses the body boundary, $\pm f(x)$.

For any elemental source $q(x)\, dx$ the complex potential is

$$dF = -\frac{q(x')\, dx'}{2\pi} \ln (z - x')$$

where we use the primes on x to indicate a variable of integration as opposed to x in $z = x + iy$, the position where F is evaluated. Then

$$F = \int_0^L -\frac{q(x')}{2\pi} \ln (z - x')\, dx'$$

But the imaginary part of F is ψ, and $\psi(x, y) = $ arbitrary constant $= 0$ is the same as $y = f(x)$. Taking the imaginary part of F, we have

$$\text{Im } F = \psi = \int_0^L -\frac{q(x')}{2\pi} \tan^{-1} \frac{y}{x - x'}\, dx'$$

and on the surface, $y = f(x)$ and $\psi = 0$ so that

$$0 = \int_0^L -\frac{q(x')}{2\pi} \tan^{-1} \frac{f(x)}{x - x'}\, dx'$$

which is an integral equation to be solved for $q(x)$. In general, the solution is not simple. The flow over the Rankine oval (Fig. 6-10) was a special, simple example of this method.

This technique is useful for flow calculations over airships and ship hulls.

Supplementary Problems

6.9. Can a vortex tube terminate in the fluid or must it form a ring?

6.10. In a tornado, can you define a vortex filament or only a tube? Suppose the core is of uniform cross section.

6.11. In a viscous shear flow, can a vortex filament be defined? Can a vortex tube be defined? In Poiseuille flow, what do the vortex lines look like?

6.12. Why does the air over the top of a wing flow inward toward the center of the wing and outward toward the wing tips on the bottom of the wing?

6.13. An arctic quonset hut experiences a lift due to wind blowing over it. Its shape is approximately semi-circular. For a wind of 80 mph and air temperature of 0°F, what is the lift per unit length on a 10 ft diameter hut shown in Fig. 6-35?

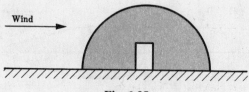

Fig. 6-35

6.14. In the preceding problem, does it matter where the door is located in determining the lift when the door is left open? What determines the pressure inside the hut?

6.15.   Discuss the flow $z = C \cosh F$. Show that

$$x = C \cosh \phi \cos \psi, \qquad y = C \sinh \phi \sin \psi$$

and the streamlines (ψ = constant) are confocal hyperbolas and this pattern might represent flow through an aperture.

6.16.   Calculate the force on a wall due to a dipole of strength m located a distance a from the wall. Assume the wall is parallel to the x axis and the dipole is aligned parallel to the x axis.

6.17.   In the preceding problem, what is the force if the dipole is inclined at an angle α to the wall?

6.18.   What happens in Problems 6.16 and 6.17 if we take the dipole strength as $-m$? Does the force on the wall have a different sign?

6.19.   Discuss the motion $F = z^2 - 1$.

6.20.   Find F, ϕ, and ψ and plot the streamlines for flow due to a source in the wall of a channel of width d shown in Fig. 6-36. (*Hint*. Set up an infinite array of images and sum up.)

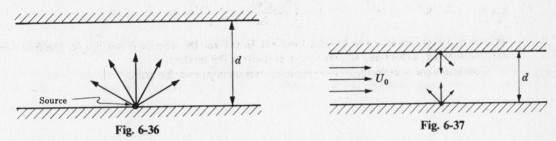

Fig. 6-36   Fig. 6-37

6.21.   Discuss the flow $F = Az^2$.

6.22.   Discuss the flow $z^2 = F^3$.

6.23.   Find F and draw the streamlines for flow through a channel with two sources located opposite each other as shown in Fig. 6-37.

6.24.   A potential vortex is located a distance d from a wall. Discuss the flow. See Fig. 6-38.

6.25.   Discuss the flow about a circular cylinder near a wall, Fig. 6-39, assuming $d \gg D$.

6.26.   In Problem 6.25 the cylinder rotates about its axis with angular speed Ω. Discuss the flow.

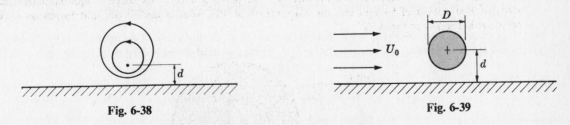

Fig. 6-38   Fig. 6-39

6.27.   Discuss the force on the cylinder in Problem 6.26.

6.28.   A light piece of cardboard with a pin stuck through it and inserted up through a spool, is shown in Fig. 6-40. If one blows air lightly down through the spool the cardboard disk is held up and does not fall. Why? Actually the air is deflected radially and the flow between the cardboard and spool is similar to a potential

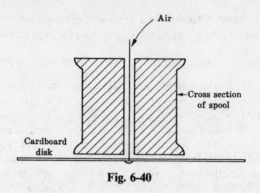

Fig. 6-40

source flow. What would the pressure distribution be along the surface of the disk? Does this explain the lift?

If one blows very hard, the disk falls. Why? What about the momentum of the air rushing down through the spool? The air must change its direction if it begins to flow radially. Does this help you explain it?

6.29. A source is located at position z_1 and a sink at position z_2. For general strengths, what is F, ϕ and ψ?

6.30. For a source and sink of equal strength lying along a line $z = e^{i\alpha}$ and equidistant, a, from the origin, and submerged in a uniform flow along the x axis, plot the streamlines.

6.31. A source is located as shown in Fig. 6-41 between a right angle. What is the complex potential? Find the complex velocity.

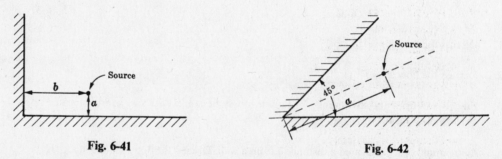

Fig. 6-41   Fig. 6-42

6.32. A source is located midway between a 45° angle wall, at a distance a from the origin. Find F and describe the flow. See Fig. 6-42.

6.33. A tornado of circulation Γ has a solid core of diameter a. Calculate the pressure distribution in the tornado. Do you think this low pressure might contribute to the devastating effect of a tornado?

6.34. An ocean whirlpool is shown in Fig. 6-43. The pressure along any line A-A decreases from the hydrostatic value $\rho g h$ at large radii to zero (gage) at point P. From potential vortex theory estimate the equation of the

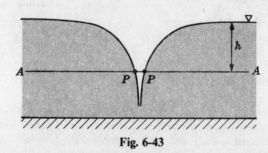

Fig. 6-43

free surface of the whirlpool. What happens in the center at sufficient depths? Can the whirlpool just stop, or must it go all the way to the ocean floor? Can you explain why objects caught in a whirlpool tend to get sucked down?

6.35.   The motion of a water sprinkler and the reaction torques were discussed in Chapter 3, Problem 3.14. The sprinkler spins because of the recoil as the water leaves the nozzle. Consider the following problem. The same sprinkler is completely submerged in a large swimming pool or lake. Water is sucked out of the pool through the nozzle and back through the hose that feeds the nozzle. Will the sprinkler rotate in the opposite direction when water is pumped out of the nozzles? Will it rotate at all? *Hint*: Operating as a sprinkler, the water leaves the nozzles in a jet. But as the water is sucked into the nozzles does it come in as jets or do the nozzles act like sinks in potential flow? This problem was mentioned by the late physicist Richard Feynman in his book *Surely You're Joking, Mr. Feynman!* (Norton, 1985). His anecdotal solution is discussed in his book and by John Wheeler in the journal *Physics Today*, February 1989, p. 24.

NOMENCLATURE FOR CHAPTER 6

F = complex potential

Im = denotes imaginary part

L = lift

M = moment

p = pressure

p_0 = stagnation pressure

p_∞ = free stream pressure, where $V = U_0$

Q = volumetric flow rate

Re = denotes real part

U_0 = free stream velocity

u = x component of velocity

V = velocity vector

v = y component of velocity

v_r = r component of velocity

v_θ = θ component of velocity

w = z component of velocity

X = x component of force

Y = y component of force

z = $(x + iy)$ = complex variable

Γ = circulation

ρ = density

ϕ = velocity potential

ψ = stream function, gravitational potential

Ω = angular velocity vector

ω = vorticity or rotation vector

$(\bar{\ })$ = denotes complex conjugate

Chapter 7

One-Dimensional Compressible Flow

7.1 INTRODUCTION

For many physical flows the assumptions of "frictionless" and "incompressible" lead to a reasonably accurate model representation (Chapter 6). In other flows the viscosity was seen to be important (Chapter 5). In this chapter we consider one-dimensional flows where the density variations are of major importance in determining the character of the flow. Although the restriction of one-dimensional flow may seem rather severe, this physical model is found to be a good approximation to many actual flows.

This chapter is restricted exclusively to internal flows. In applying the one-dimensional assumption, it is assumed that the quantities (pressure, temperature, velocity, etc.) are uniform over any cross section of the channel.[1]

Internal flows (flow in pipes, channels, and ducts) may involve property changes along the channel resulting from area change, heating and friction. And, indeed, all of these do take place in real flows. However, we will investigate these effects by looking at them one at a time. This is done for two reasons: first, there are many flows where only one effect is actually important; second, this leads to a better understanding of the role of these effects in more general flows.

Ideal Gas Approximation

The ideal gas approximation is valid for low and moderate density gases. This approximation yields property relations which are convenient in making up the total mathematical model. For example, if we assume the fluid is an ideal gas, we have

$$pv = RT \tag{7.1}$$

where p is absolute pressure, v is specific volume, R is the gas constant, and T the absolute temperature. In addition, if it is assumed that the specific heats at constant volume and constant pressure do not vary with temperature, we have between two states 1 and 2 the following additional property relationships

$$u_2 - u_1 = c_v(T_2 - T_1) \tag{7.2}$$

$$h_2 - h_1 = c_p(T_2 - T_1) \tag{7.3}$$

where u is the specific internal energy, h is specific enthalpy, and c_v and c_p are the specific heats at constant volume and constant pressure respectively. Further, if the flow is assumed to be frictionless and adiabatic (and thus isentropic) and follows the ideal gas equation of state, then we have the following equation describing the process

$$p(1/\rho)^k = pv^k = \text{constant} \tag{7.4}$$

where k is the ratio of specific heats, c_p/c_v.

Propagation of an Infinitesimal Disturbance

A disturbance in a fluid will propagate through the fluid at a well-defined velocity which depends on the properties of the fluid. The velocity of propagation depends on the magnitude of the disturbance.

---

[1] Strictly speaking, one-dimensional flow is defined as a flow where only one spatial coordinate is required for description. This could be realized by fully developed pipe flow. However, we are concerned only with the flows of uniform conditions at each cross section.

If the disturbance is very small, then the velocity of propagation of the disturbance is called the velocity of sound or acoustic velocity. This velocity is itself a property of the fluid—a very important property in compressible flow.

The velocity of sound may be determined by considering a fluid in a long tube as shown in Fig. 7-1. An infinitesimal disturbance has occurred and the wave front is moving at velocity a. We will fix our coordinate system to the wave so that the undisturbed fluid is moving relative to the wave with velocity a.

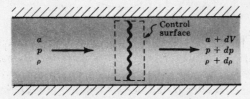

Fig. 7-1. Propagation of a sound wave.

The momentum equation for the control volume of cross sectional area A shown is

$$A[p - (p + dp)] = \rho A a[(a + dV) - a]$$

which gives

$$-dp = \rho a \, dV$$

where ρ is the density and a is the speed of the sound wave and is known as the sonic speed. The continuity equation for the control volume is

$$\rho A a = (\rho + d\rho)(a + dV)A$$

Simplifying and neglecting higher order terms we have

$$\frac{d\rho}{\rho} = -\frac{dV}{a}$$

Combining the momentum and continuity equations,

$$dp/d\rho = a^2$$

This equation is often written

$$(\partial p/\partial \rho)_s = a^2 \tag{7.5}$$

since the disturbance is infinitesimal and thus the process is reversible and adiabatic and hence isentropic as the subscript s on the derivative indicates.

For an ideal gas we may use equation (7.4) to show that

$$a^2 = kRT \tag{7.6}$$

The Mach Cone

A small disturbance at a point in a stationary fluid will be propagated radially in all directions, with the wave fronts for different times forming concentric spheres as shown in Fig. 7-2. Next, if we allow the disturbance source to move at velocity V, which is less than a, we have again spheres depicting the wave fronts for different times, but these are no longer concentric. This is shown in Fig. 7-3 below. For the case where $V > a$, i.e., the velocity of the disturbance is greater than the acoustic velocity, a conical surface is formed where the flow is undisturbed on one side and has felt the effect of the wave on the other. The half angle of the cone is $\alpha = \sin^{-1}(a/V) = \sin^{-1}(1/M)$ as shown in Fig. 7-4. Here M is the Mach number V/a.

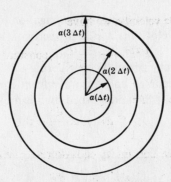

Fig. 7-2. Stationary infinitesimal disturbance.

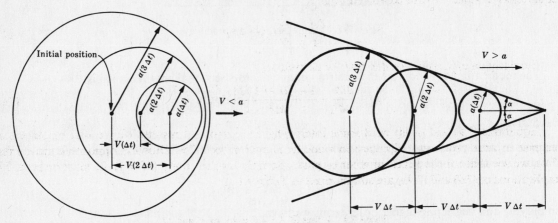

Fig. 7-3. Infinitesimal disturbance having $V < a$.

Fig. 7-4. Infinitesimal disturbance having $V > a$. $\alpha = \sin^{-1}(1/M)$.

We can imagine the disturbances of Fig. 7-3 and Fig. 7-4 as being stationary and the fluid having velocity V from right to left and obtain the same pictures. These figures demonstrate the basic difference in subsonic and supersonic flow. In subsonic flow, $M < 1$, an infinitesimal disturbance will be felt throughout the entire flow. In supersonic flow, $M > 1$, a disturbance is felt only over a portion of the flow. This leads to some interesting and important differences in the behavior of subsonic and supersonic flows.

7.2 ISENTROPIC FLOW

Many flows may be described with reasonable accuracy by assuming that they are isentropic. This assumption implies they are frictionless and adiabatic with no discontinuities in the flow properties. Examples of such flows are (1) external flows in regions of small velocity and temperature gradients and (2) internal flows such as in nozzles and diffusers where the area change is the predominant cause of change of flow conditions.

Effect of Area Change

The energy equation for steady, one-dimensional, adiabatic flow of an ideal gas with no shaft work is

$$V^2/2 + c_p T = \text{constant}$$

Using equation (7.6) for the acoustic velocity, we have

$$V^2/2 + \frac{k}{k-1}\frac{p}{\rho} = \text{constant}$$

Differentiation gives

$$V\,dV + a^2\frac{d\rho}{\rho} = 0 \tag{7.7}$$

The one-dimensional steady flow continuity equation may be differentiated to give

$$\frac{d\rho}{\rho} + \frac{dA}{A} + \frac{dV}{V} = 0 \tag{7.8}$$

Equations (7.7) and (7.8) are combined to give

$$\frac{dA}{A} = \frac{dV}{V}(M^2 - 1) \tag{7.9}$$

Similarly an expression for dM may be derived as

$$\frac{dM}{M} = \frac{1 + [(k-1)/2]M^2}{M^2 - 1}\,dA \tag{7.9a}$$

Equations (7.9) and (7.9a) yield some interesting and important results. We see that for $M < 1$ an increase in area produces a decrease in velocity. However, for $M > 1$ just the opposite is true. From (7.9a) we see that for the Mach number to pass smoothly through unity ($M = 1$) dA must be zero. The implications of (7.9) and (7.9a) are summarized in Table 7.1.

Table 7.1   Effect of Area Change on V and M

	$M < 1$	$M > 1$
$dA < 0$	V increases M increases	V decreases M decreases
$dA > 0$	V decreases M decreases	V increases M increases

Converging Nozzle Flow

Let us investigate the flow in a converging passage such as shown in Fig. 7-5. We will assume that the fluid is an ideal gas, the flow is one-dimensional, steady, adiabatic and frictionless. The continuity equation is

$$\dot{m} = A_2 V_2/v_2 \tag{7.10}$$

where $\dot{m}$ is the mass rate of flow and v is the specific volume, $1/\rho$. The energy equation in terms of enthalpy h is

$$\tfrac{1}{2}(V_2^2 - V_1^2) = h_1 - h_2 \tag{7.11}$$

If we assume that $V_1 \ll V_2$, the equation (7.11) may be written as (using isentropic and property relationships)

$$V_2 = \left\{\frac{2k}{k-1}\,p_1 v_1[1 - (p_2/p_1)^{(k-1)/k}]\right\}^{1/2} \tag{7.12}$$

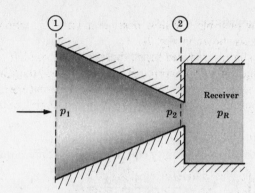

Fig. 7-5.   Converging nozzle.

Equations (*7.10*) and (*7.12*), along with the isentropic relationship $p_1 v_1^k = p_2 v_2^k$, are combined to give

$$\dot{m}/A_2 = \left\{ \frac{2k}{k-1} \frac{p_1}{v_1} [(p_2/p_1)^{2/k} - (p_2/p_1)^{(k+1)/k}] \right\}^{1/2} \qquad (7.13)$$

If we take the inlet conditions as fixed, then the mass flow change takes place as a result of a change in p_2 only. This result is shown in Fig. 7-6 as plotted from equation (*7.13*). Also, the actual mass flow results are shown in a plot of $\dot{m}$ versus p_R/p_1 where p_R is the receiver pressure.

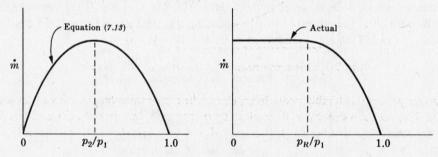

Fig. 7-6.   Mass flow in a converging nozzle.

There is obviously some discrepancy between the actual and predicted results. The actual results and those predicted by the equation are in good agreement from the point where $p_R/p_1 = 1.0$ down to the value of receiver pressure where the mass flow reaches a maximum. We see that a further reduction in receiver pressure yields no change in the mass flow rate. We also observe that experimentally the throat pressure p_2 is never less than the value for maximum mass flow. This minimum throat pressure is called the critical pressure p_c and is found by differentiating equation (*7.13*) and setting the result equal to zero. For this we obtain

$$(p_2/p_1)_{\text{max flow}} = p_c/p_1 = [2/(k+1)]^{k/(k-1)} \qquad (7.14)$$

By combining equations (*7.12*) and (*7.14*), the Mach number is found to be equal to unity where the pressure is critical. These results are not surprising, then, in light of the results of the previous section. We see that in order to increase the Mach number above unity, a diverging section extension would need to be added.

Converging-Diverging Nozzle

We saw that in a converging nozzle the maximum Mach number was unity. And from equation (*7.9*) we saw that in order to increase the Mach number above unity, the area must increase. Thus in order to make supersonic flow possible, the flow passage must be a section of decreasing area followed by a section of increasing area as shown in Fig. 7-7.

If the receiver pressure is slightly reduced (a, b), there is flow from left to right. The flow is subsonic throughout, with the convering portion functioning as a nozzle and the diverging portion as a diffuser. If the receiver pressure is further reduced (c) the pressure at the throat reaches a minimum (critical) and the velocity there will equal the sonic velocity. The flow in the diverging section is subsonic.

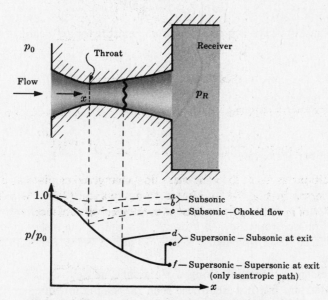

Fig. 7-7. Flow regimes of a converging-diverging nozzle.

If the receiver pressure is further reduced (d, e), then the flow following the throat for a distance will be supersonic. This is followed by a discontinuity in pressure (normal shock) and the flow will be subsonic for the remaining distance to the exit. Condition (e) is just that where the shock has moved to the nozzle exit.

There is only one receiver pressure (f) where the flow can be isentropic for supersonic flow. If the pressure of the receiver is between e and f, then the nozzle is said to be overexpanded. Here shock patterns occur outside the nozzle where the pressure adjusts from a lower to a higher value. If the receiver pressure is below f, then the nozzle is said to be underexpanded. For this case a series of expansion waves and oblique shock waves occur outside the nozzle, with the pressure going from a higher to a lower value.

The important assumptions of this section are one-dimensional, steady adiabatic flow of an ideal gas. Now, in most real flows these assumptions are not fully satisfied. However, this model gives predicted results (say, of exit velocity or thrust) which are within a few percent of the experimental results for many flows.

In some problems the designer would like not to be even a few percent off in the predictions. Then the fact that the flow is not actually one-dimensional, that there is a boundary layer, that there is some heat transfer, and that the gas is not an ideal gas must be taken into account.

Isentropic Equations

If the flow is one-dimensional, steady, and adiabatic and the fluid is an ideal gas, the energy equation between two positions 1 and 2 is

$$c_p T_1 + \frac{V_1^2}{2} = c_p T_2 + \frac{V_2^2}{2} = c_p T_0 = h_0 \qquad (7.15)$$

Here T_0 is the stagnation or reservoir temperature which the fluid would achieve if brought to rest adiabatically. h_0 is the stagnation specific enthalpy. It should be pointed out that the stagnation temperature T_0 is constant along a streamline if the flow is *adiabatic*. It is not necessary that the flow be isentropic. Writing the velocity in terms of Mach number and using $c_p - c_v = R$, the energy equation becomes

$$\frac{T_2}{T_1} = \frac{1 + \frac{1}{2}(k-1)M_1^2}{1 + \frac{1}{2}(k-1)M_2^2} \qquad (7.16)$$

If the isentropic relationship between pressure and temperature for an ideal gas is used, we have

$$\frac{p_2}{p_1} = \left[\frac{1 + \frac{1}{2}(k-1)M_1^2}{1 + \frac{1}{2}(k-1)M_2^2}\right]^{k/(k-1)} \qquad (7.17)$$

The mass flow can be determined from

$$\dot{m} = \rho A V$$

or in terms of Mach number, we have

$$\dot{m}/A = \sqrt{k/RT}\, p M \qquad (7.18)$$

The stagnation pressure p_0 is the pressure that would be achieved if the fluid were brought to rest in an isentropic manner. For isentropic flow in a nozzle it is simply the reservoir pressure. For any fluid state one can always think of an imaginary isentropic nozzle being fed from a reservoir at p_0 that delivers flow to that state.

The stagnation pressure is constant along a streamline in isentropic flow so that $p_{01} = p_{02}$. If, however, the flow is adiabatic but not isentropic then the stagnation temperature T_0 is still constant along the flow, but p_0 is not.

Some further results for isentropic flow of an ideal gas can be obtained by using the expression $a^2 = kRT$ with the energy equation for adiabatic flow (but not necessarily isentropic).

$$\frac{V^2}{2} + \frac{a^2}{k-1} = \frac{a_0^2}{k-1} \qquad (7.19)$$

$$\frac{a_0^2}{a^2} = \frac{T_0}{T} = 1 + \frac{k-1}{2}M^2 \qquad (7.20)$$

Then for isentropic flow equation (7.4) may be used to obtain

$$\frac{p_0}{p} = \left(1 + \frac{k-1}{2}M^2\right)^{k/(k-1)} \qquad (7.21)$$

$$\frac{\rho_0}{\rho} = \left(1 + \frac{k-1}{2}M^2\right)^{1/(k-1)} \qquad (7.22)$$

In a duct, particularly a nozzle, the area A is an important parameter. We use a starred quantity ( )* to denote the condition or property where the Mach number is unity (just sonic). By continuity

$$\dot{m} = \rho A V = \rho^* A^* a^* \qquad (7.23)$$

A^* is a fictitious throat area which would be required to achieve sonic conditions by an *isentropic* path. A^* may or may not actually exist in the duct depending on whether sonic conditions are achieved or not. If sonic conditions are achieved, then $A_t = A^*$ where A_t is the actual throat area. The sonic speed a varies through the flow, of course, and a^* is the value just where $M = 1$.

Making use of the nozzle relationships, the ratio A/A^* may also be found in terms of the local Mach number M (where the area is A) as

$$\left(\frac{A}{A^*}\right)^2 = \frac{1}{M^2}\left[\frac{2}{k+1}\left(1 + \frac{k-1}{2}M^2\right)\right]^{(k+1)/(k-1)} \tag{7.24}$$

Note that the critical pressure p_c given in equation (7.14) is simply p^*.

For a given isentropic flow with a constant p_0 and T_0, equations (7.20), (7.21), (7.22), and (7.24) give T, p, ρ, and A/A^* in terms of the local Mach number M. Since the ratios T/T_0, p/p_0, ρ/ρ_0, and A/A^* are functions of M only, these ratios have been numerically tabulated and are useful for calculations. A brief table is given in the Appendix.

7.3 NORMAL SHOCKS

We saw that there was a discontinuity in pressure (density and temperature) in a converging-diverging nozzle when the exit pressure was within a certain range of values. This discontinuity is called a normal shock. In this section we are going to (1) show that such a discontinuity may occur and (2) develop equations describing the change in conditions across a shock.

Let us assume that the normal shock illustrated in Fig. 7-8 can be represented by the following model:

1. Area is constant through the shock. (The channel area may vary but not appreciably through the shock thickness.)

2. Ideal gas

3. Steady flow

4. One-dimensional flow

5. Adiabatic

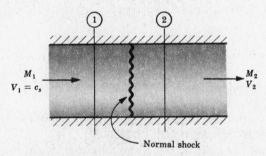

Fig. 7-8.   Flow model for normal shock.

The control volume between lines 1 and 2 is very thin and at rest with respect to the shock (which may or may not be actually moving with respect to some fixed observer). It is important to realize that the flow through a shock is irreversible and hence the isentropic equations cannot be used. The control volume forms of energy, momentum and continuity become (see previous section)

Energy:
$$\frac{T_2}{T_1} = \frac{1 + \frac{1}{2}(k-1)M_1^2}{1 + \frac{1}{2}(k-1)M_2^2} \tag{7.16}$$

Continuity:
$$p_2/p_1 = \sqrt{\frac{T_2}{T_1}}\frac{M_1}{M_2} \tag{7.25}$$

Momentum:
$$(p_1 - p_2)A = \dot{m}(V_2 - V_1)$$

which becomes
$$\frac{p_2}{p_1} = \frac{1 + kM_1^2}{1 + kM_2^2} \tag{7.26}$$

The equation of state combined with the thermodynamic expression for entropy change gives

$$s_2 - s_1 = c_p \ln (T_2/T_1) - R \ln (p_2/p_1) \tag{7.27}$$

Equations (7.16), (7.25) and (7.26) involve three unknowns (T_2, p_2, M_2) and may be combined to give

$$M_2 = \left[\frac{M_1^2(k - 1) + 2}{2kM_1^2 - k + 1} \right]^{1/2} \tag{7.28}$$

Equations (7.16), (7.26) and (7.28) completely describe the conditions behind a shock in terms of the upstream conditions.[1]

A bit of algebraic manipulation yields several relationships between the upstream and downstream properties, although shock tables are generally used for numerical calculations. The Rankine-Hugoniot relationships are

$$\frac{\rho_2}{\rho_1} = \frac{1 + \dfrac{k + 1}{k - 1} \dfrac{p_2}{p_1}}{\dfrac{k + 1}{k - 1} + \dfrac{p_2}{p_1}} = \frac{u_1}{u_2} \tag{7.29}$$

$$\frac{T_2}{T_1} = \frac{p_2}{p_1} \frac{\dfrac{k + 1}{k - 1} + \dfrac{p_2}{p_1}}{1 + \dfrac{k + 1}{k - 1} \dfrac{p_2}{p_1}} \tag{7.30}$$

Some additional useful relations are

$$\frac{p_2}{p_1} = 1 + \frac{2k}{k + 1} (M_1^2 - 1) \tag{7.31}$$

$$\frac{\rho_2}{\rho_1} = \frac{u_1}{u_2} = \frac{(k + 1)M_1^2}{(k - 1)M_1^2 + 2} \tag{7.32}$$

$$\frac{T_2}{T_1} = \frac{p_2}{p_1} \frac{\rho_1}{\rho_2} \tag{7.33}$$

The ratios of downstream to upstream properties are functions of M_1 only, and the shock tables tabulate these ratios as a function of the independent variable M_1.

The question we have not answered is: when can a normal shock occur? Let us assume that all of the conditions of the model are satisfied except the momentum equation. Then equations (7.16), (7.25) and (7.27) (i.e., energy, continuity and equation of state) define a curve on the enthalpy-entropy diagram which passes through the initial state 1, as shown in Fig. 7-9. This curve is called the Fanno line.

Now if all of the conditions of the model are satisfied except the energy equation, we have another curve on the enthalpy-entropy diagram. This curve is called the Rayleigh line.

In order that all the conditions of the model be satisfied, the equilibrium states of the Fanno and Rayleigh lines must be satisfied simultaneously. Since there are two intersections of the curves, two states will satisfy the conditions and all other states are not allowed. We note also that the allowed states have finite differences in properties.

---

[1] The property variations across a normal shock have been numerically tabulated in convenient form, and their use eliminates the necessity of making calculations using the above equations. Shock tables may be found in the references and in abbreviated form in the Appendix.

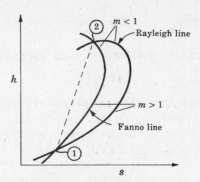

Fig. 7-9. Fanno and Rayleigh lines.

Thus if the flow has conditions of state 1, it would be possible for a discontinuity to occur and the flow to take on conditions of state 2. Would it be possible for the flow conditions to go from state 2 to state 1? The answer is no because this would involve an entropy decrease and would be a violation of the second law of thermodynamics. The flow for the intersection on the lower portion of the curves is supersonic, while the intersection on the upper portion is subsonic flow. Thus we have shown that a discontinuity may occur if the flow is supersonic and cannot occur if the flow is subsonic.

7.4 ADIABATIC CONSTANT AREA FLOW (FANNO LINE)

Consider adiabatic, one-dimensional, steady flow of an ideal gas in a constant area duct. The equations describing this flow are (where the subscript 1 denotes a reference upstream state):

Energy: $$h_1 + V_1^2/2 = h + V^2/2 = h_0 \tag{7.34}$$

where h_0 is the stagnation enthalpy defined by the above equation.

Momentum: $$(p_1 - p)A - \tau_0 LC = A(\rho V^2 - \rho_1 V_1^2) \tag{7.35}$$

where τ_0 = wall shear stress, L = length and C = circumferential distance or wetted perimeter.

Continuity: $$V_1/v_1 = V/v \quad \text{or} \quad V_1\rho_1 = V\rho \tag{7.36}$$

Equation of state: $$p_1 v_1/T_1 = pv/T \tag{7.37}$$

Property relation: $$s - s_1 = c_v \ln (h/h_1) + R \ln (v/v_1) \tag{7.38}$$

Combining the energy and continuity equations gives

$$h = h_0 - \tfrac{1}{2}(V_1/v_1)^2 v^2 \tag{7.39}$$

Further, by combining this equation with the property relation, we obtain the Fanno line curve of Fig. 7-9. This curve represents the possible conditions for the fixed upstream conditions as designated by 1. We observe that there is a maximum entropy for this curve. This has important significance. For example, if the upstream conditions 1 are on the lower portion of the curve as shown in Fig. 7-9, then there will be an increasing value of enthalpy as shown. Note that the entropy must increase according to the second law of thermodynamics. This means that if the flow conditions reach those corresponding to the maximum entropy, then the flow conditions cannot change further. The condition of maximum entropy represents the flow at the exit of the constant area channel for "choked" flow.

Combining the $T\,ds$ equation (i.e., $T\,ds = dh - v\,dp$, a properties relationship) with the differential form of the energy and continuity equations gives

$$\frac{ds}{dh} = \frac{1}{T}\left(1 - \frac{1}{V^2}\frac{dp}{d\rho}\right) \tag{7.40}$$

If we set this equation equal to zero, we get the conditions at the exit of the pipe for choked flow, obtaining

$$V^2 = dp/d\rho = (\partial p/\partial \rho)_s = a^2 \tag{7.41}$$

or

$$M = 1.0$$

Thus we see that the frictional effects cause the fluid to tend toward a Mach number of unity for both initially subsonic (upper curve) and supersonic (lower curve) conditions.

We can determine the property changes for one-dimensional adiabatic flow of a gas in a constant area duct from the equations. Summarizing the results, we have

Property	Subsonic Flow	Supersonic Flow
s	increases	increases
h	decreases	increases
T	decreases	increases
M	increases	decreases
V	increases	decreases
p	decreases	increases
p_0	decreases	increases
ρ	decreases	increases

The flow in a duct is said to be choked when the Mach number at the exit is unity and a further decrease in the receiver or back pressure (the pressure of the environment to which the pipe is exhausted) will not increase the flow rate or change the properties in the pipe. Generally a flow may be choked by either lowering the back pressure sufficiently or increasing the length of the duct (or pipe) sufficiently. A subsonic flow ($M < 1$) can never become supersonic in the duct and may only reach sonic conditions ($M = 1$) at the exit. Similarly a supersonic flow ($M > 1$) can never become subsonic by passing through $M = 1$ in a smooth fashion; only at the exit may $M = 1$.

However, shock waves may form in ducts with supersonic flow if certain conditions occur. There, of course, supersonic flow becomes subsonic but not in a smooth continuous manner. If there were no friction the flow would be uniform in properties and Mach number along the duct, and it is due to the wall friction that the flow changes along the duct and shock waves may form.

Working formulas for adiabatic flow with friction may be obtained as follows: Equation (7.35) may be put into differential form as

$$\frac{1}{p}\frac{dp}{dx} + \frac{2kfM^2}{D} + \frac{kM^2}{2V^2}\frac{d}{dx}(V^2) = 0 \tag{7.42}$$

where f is the coefficient of friction defined as $f = \tau_0/(\rho V^2/2)$ (which is different by a factor of 4 from the Darcy friction factor f used in hydraulics and defined in Chapters 3, 4 and 5). The f, which is generally used in compressible flow, is called the Fanning friction factor and is the same as the friction coefficient used in boundary layer theory. D is the hydraulic diameter of the duct defined as

$$D = \frac{4A}{C}$$

and the Mach number has been introduced by definition

$$M^2 = \frac{V^2}{a^2} = \frac{V^2}{kRT} = \frac{\rho V^2}{kp}$$

The other equations (7.34) to (7.39) may be put into differential form and combined to give property variations in terms of M and position x. Of particular interest is the expression for M in terms of position x along the duct

$$\frac{dM^2}{M^2} = \frac{kM^2[1 + M^2(k-1)/2]}{1 - M^2}\frac{4f}{D}dx \tag{7.43}$$

CHAP. 7]   ONE-DIMENSIONAL COMPRESSIBLE FLOW   189

This equation may be integrated between an initial Mach number M_1 at $x = 0$ and the value of M at an arbitrary position x along the duct to give M as a function of initial Mach number and downstream position. By integrating between M_1 at $x = 0$ and $M_2 = 1$ a value of x where the Mach number becomes unity can be found. This value of x is the critical length L_{max} where choking occurs. The result is

$$4\bar{f}\frac{L_{max}}{D} = \frac{1 - M_1^2}{kM_1^2} + \frac{k+1}{2k}\ln\left(\frac{(k+1)M_1^2}{2\{1 + [(k-1)/2]M_1^2\}}\right) \tag{7.44}$$

where $\bar{f}$ is an average friction coefficient over the length. By using values of $4\bar{f}L_{max}/D$ as a function of M the length L of duct required to pass from an initial Mach number M_1 to another Mach number M_2 may be found as

$$4\bar{f}\frac{L}{D} = \left(\frac{4\bar{f}L_{max}}{D}\right)_{M_1} - \left(4\bar{f}\frac{L_{max}}{D}\right)_{M_2} \tag{7.45}$$

Values of $4\bar{f}L/D$ as a function of M are tabulated and provide a quick means of calculation. Appropriate tables are given in several of the references.

For the purpose of property calculations it is convenient to denote properties at the place where $M = 1$ as ( )*. In an adiabatic duct with friction this can only occur at the exit and under choking conditions and for an actual duct may represent an imaginary position in a longer duct. The ratio of properties at any Mach number M to the value at $M = 1$ (starred value), such as p/p^*, depends only on the Mach number. Hence property ratios between two positions in the duct (p_2/p_1, for example) may be found from the ratio

$$\frac{p_2}{p_1} = \frac{(p/p^*)_{M_2}}{(p/p^*)_{M_1}} \tag{7.46}$$

The property ratios ( )/( )* have been tabulated and may be found in references 1, 4, 6, and 8. We list below the analytical expressions for these property ratios

$$\frac{p}{p^*} = \frac{1}{M}\frac{k+1}{2\{1 + [(k-1)/2]M^2\}} \tag{7.47}$$

$$\frac{\rho^*}{\rho} = \frac{V}{V^*} = M\frac{k+1}{2\{1 + [(k-1)/2]M^2\}} \tag{7.48}$$

$$\frac{T}{T^*} = \frac{k+1}{2\{1 + [(k-1)/2]M^2\}} \tag{7.49}$$

$$\frac{p_0}{p_0^*} = \frac{1}{M}\left(\frac{2\{1 + [(k-1)/2]M^2\}}{k+1}\right)^{(k+1)/2(k-1)} \tag{7.50}$$

Choking Effects

We have mentioned that under some conditions the flow chokes. There are two ways in which a duct may be fed, with either a converging nozzle or a converging-diverging nozzle. With a converging nozzle feed the flow must remain subsonic throughout the duct (Fig. 7-10). Along the duct the Mach number increases and the pressure decreases. For a given back pressure, as the length of the duct is increased the exit Mach number increases until it becomes unity and choking occurs. If the back pressure is then further reduced the exit pressure remains constant, but external expansion occurs outside the duct to accommodate the back pressure and the flow in the duct is unchanged. On the other hand (referring to Fig. 7-10), for a shorter duct with a given back pressure without choking (P_a), a sufficient reduction in back pressure to P_b will cause the duct to eventually choke at that length. Further decreases in back pressure (P_c) will cause external expansions to occur without influence inside the duct.

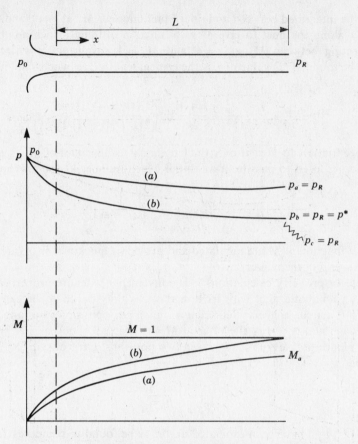

Fig. 7-10.   Subsonic duct fed with a converging nozzle. (a) shows
an unchoked and (b) a choked subsonic flow.

If the duct is fed with a converging-diverging nozzle the flow entering the duct may be subsonic or supersonic and shock waves may stand in the duct, depending on the back pressure and duct length. The effect of varying the back pressure is complex and will only be sketched here. The reader is referred to the references, particularly Shapiro (reference 6), for a more complete discussion.

For a given reservoir state, and supersonic flow in the duct, the Mach number at the throat is unity and unique at the duct inlet (M_1) as shown in Fig. 7-11. The Mach number decreases and the pressure increases downstream. For a given length there must be a unique back pressure for smooth flow. As the length is increased (assuming the back pressure is continuously adjusted to ensure smooth flow) the Mach number at the exit decreases until the value of L_{max} is reached, where the Mach number becomes unity (Fig. 7-11a). A further increase in duct length causes a shock to form in the duct. The shock moves upstream as the duct length is increased until it is finally swallowed by the nozzle and the flow becomes subsonic throughout the duct. If L is equal to L_{max} (for supersonic flow) and the back pressure p_R is decreased from its matched value (p^*) the flow remains unchanged in the duct but expansion waves occur outside the duct to accommodate the pressure. If the back pressure is increased above p^* a weak shock forms at the duct outlet and moves upstream as the back pressure is increased. Eventually the shock moves through the nozzle and is swallowed, giving rise to subsonic flow throughout the duct. Further increases in back pressure reduce the subsonic flow.

For any value of $L < L_{max}$ (Fig. 7-11b) there is a unique value of M at the exit and unique back pressure (p_a, less than p^*) for smooth supersonic flow. If the back pressure is decreased (p_b) an external expansion occurs but the duct flow is unchanged. If the back pressure is increased shocks appear external to the duct (p_c). A further increase in back pressure will cause the shock to just enter the duct

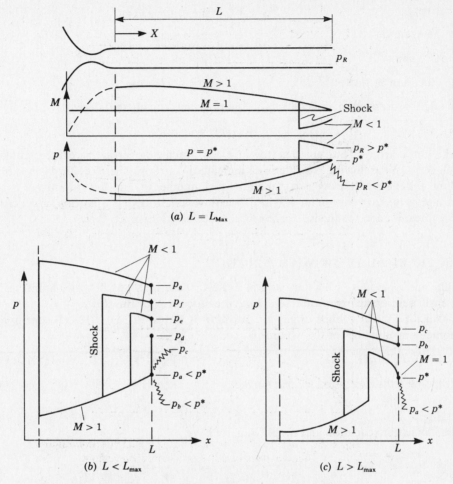

Fig. 7-11. Supersonic flow with a shock in a duct fed with a
converging-diverging nozzle.

(p_d) and move upstream as the back pressure is increased (p_e, p_f). Eventually the back pressure reaches a value where the shock stands in the nozzle (p_g). A further increase in back pressure causes the shock to move to the throat of the nozzle and then disappear as the flow becomes subsonic throughout the duct. A shock that passes back through the nozzle throat is said to be "swallowed".

Referring to Fig. 7-11c for $L > L_{max}$, if the back pressure p_a is lower than a critical value p^* a shock forms at a unique position in the duct which gives a value of $M = 1$ at the exit. The flow is subsonic after the shock. Free expansion occurs external to the duct to accommodate the back pressure (p_a). As the back pressure is raised, the shock remains in the same position but the free expansion becomes weaker until the back pressure just equals the unique pressure (p^*) corresponding to the $M = 1$ condition at the exit. A further increase in the back pressure (p_b) will cause the shock to move upstream, and the matching of the exit pressure (p_b) to the back pressure and to the appropriate value of the exit Mach number (which is subsonic) locates the position of the shock. Eventually with a further increase in back pressure (p_c) the shock stands in the nozzle and is eventually swallowed as the back pressure is increased and flow becomes subsonic throughout the duct.

7.5 FRICTIONLESS CONSTANT AREA FLOW WITH HEATING AND COOLING

The equations for frictionless, one-dimensional, steady flow in a constant area duct with heating and cooling are (where q is the heat transferred to unit mass of fluid flowing):

Energy: $\qquad\qquad\qquad h_1 + V_1^2/2 + q = h + V^2/2$

Momentum: $\qquad\qquad\quad p_1 - p = \rho V^2 - \rho_1 V_1^2$

Continuity: $\qquad\qquad\qquad\quad \rho_1 V_1 = \rho V$

Equation of state: $\qquad\qquad p_1/\rho_1 T_1 = p/\rho T$

Combining the momentum and continuity equations and writing in differential form gives

$$(\partial p/\partial \rho)_s = a^2 = V^2$$

for maximum entropy. Thus, for this flow too, the Mach number at the exit for choked conditions is unity. Heating of the flowing gas causes an increase in entropy. Therefore for both subsonic and supersonic flow, the effect of heating is to cause the Mach number to tend toward unity.

An interesting observation for Rayleigh line flow is that there is a portion of the curve which shows that when heat is added to the gas there will be a decrease in temperature.

7.6 ISOTHERMAL FLOW WITH FRICTION

There are some flows which are approximately isothermal, e.g., flow of natural gas in a long pipeline. We will develop an expression describing pressure drop for this flow.

Consider a control volume of gas for this flow as shown in Fig. 7-12. The one-dimensional steady flow momentum equation for this volume is

$$\sum F_x = \dot{m}(V + dV - V) \qquad [p - (p + dp)](\tfrac{1}{4}\pi D^2) - \tau_0 \pi D\ dx = (\tfrac{1}{4}\pi D^2)\rho V\ dV$$

Then combining the expression for friction factor

$$f \equiv \tau_0/(\tfrac{1}{2}\rho V^2)$$

with the momentum equation and simplifying, we have

$$\frac{v}{V^2}\,dp + \frac{2f}{D}\,dx + \frac{dV}{V} = 0$$

If the continuity equation is combined with the ideal gas equation of state for constant temperature, we have

$$\frac{v}{V^2} = \frac{v_1}{V_1^2 p_1}\,p$$

This can be combined with the momentum equation to give

$$\frac{v_1}{V_1^2 p_1}\,p\,dp + \frac{2f}{D}\,dx + \frac{dV}{V} = 0$$

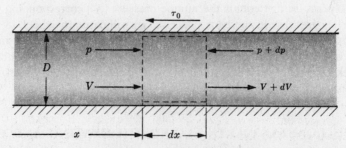

Fig. 7-12. Volume element for isothermal flow in a constant area duct.

We will assume that the friction factor depends only on the Reynolds number (for a given pipe). Then the Reynolds number is a constant for isothermal flow and the friction factor is also constant. (Note $\rho V = $ constant and $\mu = f(T) = $ constant.)

Now the momentum equation can be integrated, giving

$$(p/p_1)^2 = 1 - kM_1^2(2 \ln V/V_1 + 4fL/D) \tag{7.51}$$

where L is the length between stations as designated by the subscript 1 and those quantities having no subscript. We can rewrite equation (7.51) as

$$fL/D = \frac{1}{4kM_1^2} [1 - (p/p_1)^2] + \tfrac{1}{2} \ln p/p_1 \tag{7.52}$$

Thus the pressure is a nonlinear function of length, whereas in incompressible flow it is a linear function of length.

We can combine the momentum and continuity equations to show that the pipe exit Mach number for choked flow is $M^* = 1/\sqrt{k}$. The corresponding pressure is $p^*/p_1 = M_1\sqrt{k}$.

7.7 INCOMPRESSIBLE FLOW FOR LOW MACH NUMBERS

Starting with the energy equation

$$\tfrac{1}{2}(V_1^2 - V^2) = h - h_1$$

and writing in terms of an ideal gas for an isentropic process, we have

$$\tfrac{1}{2}(V_1^2 - V^2) = \frac{kRT}{k-1} [1 - (p_1/p)^{(k-1)/k}]$$

Letting $V_1 \to 0$ we obtain the expression for the stagnation pressure,

$$p_0/p = [1 + \tfrac{1}{2}(k-1)M^2]^{k/(k-1)}$$

If this equation is written in terms of a binomial expansion for the Mach number, we have

$$p_0 = p + \tfrac{1}{2}\rho V^2[1 + \tfrac{1}{4}M^2 + \tfrac{1}{24}(2-k)M^4 + \cdots] \tag{7.53}$$

For small values of Mach numbers the equation becomes

$$p_0 = p + \tfrac{1}{2}\rho V^2 \tag{7.54}$$

which is the same as the expression for stagnation pressure for incompressible flow.

The importance of this result is that adiabatic flow of gases around objects and in ducts may be considered as incompressible as long as the Mach number is small (say, for $M < 0.3$), and as we have seen there is a substantial simplification in the model for constant density.

7.8 THE SHOCK TUBE

The shock tube is a device used to produce a high velocity, high temperature gas over a short interval of time. The tube is sealed, with a diaphragm separating two portions of the tube, one of which is at a high pressure, the other at a low pressure. The diaphragm is suddenly burst and a shock wave propagates into the still gas on the low pressure side, and an expansion wave propagates into the still gas on the high pressure side. The disturbed gas (between the shock and expansion waves) travels at a high velocity which may be supersonic (relative to the undisturbed gas and tube). Shock tubes are used for short duration wind tunnels much like blow-down tunnels.

The Piston Generated Shock

Before looking at the shock tube itself, let us examine the types of waves generated in the tube. Consider a shock wave generated in a tube by a moving piston as shown in Fig. 7-13. The shock is a normal shock for which we have already calculated the jump conditions. The gas between the piston and the shock has the same velocity as the shocks.

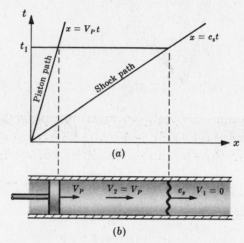

Fig. 7-13.   A shock wave generated by a moving piston. (a) The x-t plot, (b) the tube at time t_1. V is the gas absolute velocity.

The x-t plot shows the development or motion of the piston and shock wave. The slopes of the lines are the reciprocals of the speeds.

Centered Expansion Waves

If the piston described above is withdrawn from the tube instead of driven in, an expansion wave develops and propagates away from the piston. This expansion wave is isentropic.

The sudden withdrawal of the piston creates a step function change in particle velocity (the particles immediately adjacent to the piston move with it) and pressure. This step function flattens as the wave begins to propagate; and locally, within the wave, the disturbances travel at the local sonic speed. Since the pressure (and temperature) vary through the wave (which began as a step in these variables) the local speed of disturbance propagation varies through the wave. On the edge of the wave adjacent to the undisturbed gas (region 4) the temperature and sonic speed are greatest. On the opposite side (region 3) the temperature is the smallest and hence the sonic speed is also. As a consequence, the expansion wave expands or "fans" out as it propagates as shown in Fig. 7-14. The x-t plot is a fan of constant sonic speed lines which show the development of the wave. These lines are called "characteristics" and follow the path of local isentropic disturbances. Clearly the absolute speed of the disturbances is the sum of the local sonic speed and the local absolute gas speed. The terminating characteristic on the right (condition 4) has slope $dx/dt = a_4$, and the one on the left has slope $dx/dt = a_3 + V_3$. V_3 is defined positive to the right and is negative here (since V_3 must be equal to $-V_P$). Hence dx/dt may be positive or negative depending on whether $a_3 > V_3$ or $a_3 < V_3$ respectively.

We may explicitly evaluate a_3 and the structure of the fan as follows. In terms of a_4 the local sonic speed in the fan may be written (using the isentropic relationships)

$$a = a_4(\rho/\rho_4)^{(k-1)/2}$$

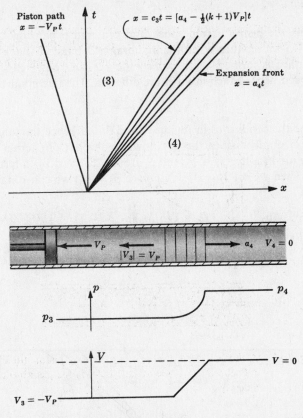

Fig. 7-14. The centered expansion fan generated by a piston being withdrawn from a tube.

and the particle velocity in the wave may be written locally

$$dV = a \frac{d\rho}{\rho}$$

and may be integrated by using the previous expression to yield

$$a = a_4 + \tfrac{1}{2}(k - 1)V$$

The absolute wave speed c (of the portion of the wave behind the front) then is the sum of the local sonic speed and fluid velocity,

$$c = a + V = a_4 + \tfrac{1}{2}(k + 1)V$$

Now, V is negative, so c decreases through the wave as we go from region 4 to 3. The terminating characteristic on the left has slope

$$c_3 = a_4 + \tfrac{1}{2}(k + 1)V_3 = a_4 - \tfrac{1}{2}(k + 1)V_P \tag{7.55}$$

(which may be positive or negative and may slope to the right or left). From the isentropic relationships the density and pressure change across the fan are given by

$$\rho_3/\rho_4 = [1 - \tfrac{1}{2}(k - 1)(V_P/a_4)]^{2/(k-1)} \tag{7.56}$$

$$p_3/p_4 = [1 - \tfrac{1}{2}(k - 1)(V_P/a_4)]^{2k/(k-1)} \tag{7.57}$$

where V_P is a positive number.

The Shock Tube Flow

Now we can combine the piston driven shock and centered expansion fan to describe the shock tube. Referring to Fig. 7-15 we see that when the diaphragm is burst, a shock and expansion wave are generated. A contact surface is formed across which the pressure and velocity are constant, but the temperature and density and hence Mach number are different. The two conditions

$$p_2 = p_3, \qquad V_2 = V_3$$

are sufficient to determine the conditions in regions 2 and 3 and hence the jumps across the shock and fan. ($V_2 = V_3$ is a negative number since the gas flows to the left.) In terms of known conditions in regions 1 and 4, we solve for V_3 in terms of p_3 and p_4 from (7.57) and V_2 in terms of p_1 and p_2 from the normal shock relations. Then we set $V_2 = V_3$ (and since $p_2 = p_3$) we can obtain a relationship for p_2. The result is

$$\frac{p_4}{p_1} = \frac{p_2}{p_1}\left[1 - \frac{(k-1)(a_1/a_4)(p_2/p_1 - 1)}{\sqrt{2k}\sqrt{2k + (k+1)(p_2/p_1 - 1)}}\right]^{-2k/(k-1)} \tag{7.58}$$

which gives p_2/p_1 implicitly. Once p_2 is known, all the other parameters of the problem can easily be found.

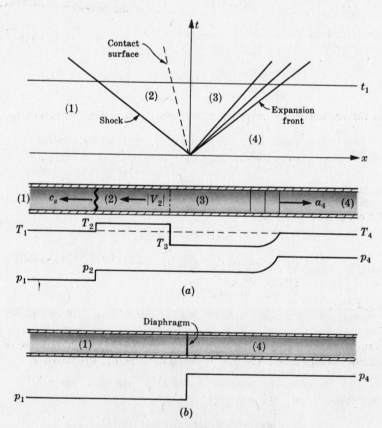

Fig. 7-15. The shock tube. (a) Time $t = t_1$, (b) time $t = 0$.

References

1. John, J. E. A., *Gas Dynamics*, 2nd ed., Allyn and Bacon, 1984.
2. Liepmann, H. W., and Roshko, A., *Elements of Gasdynamics*, John Wiley, 1957.
3. Owczarek, J. A., *Fundamentals of Gas Dynamics*, International Textbook, 1964.

4. Saad, M. A., *Compressible Fluid Flow*, Prentice-Hall, 1985.

5. Shames, I. H., *Mechanics of Fluids*, McGraw-Hill, 1962.

6. Shapiro, A. H., *The Dynamics and Thermodynamics of Compressible Fluid Flow*, Ronald Press, 1953.

7. Thompson, P. A., *Compressible-Fluid Dynamics*, McGraw-Hill, 1972.

8. Zucrow, M. J., and Hoffman, J. D., *Gas Dynamics*, Vol. 1, John Wiley, 1976.

Solved Problems

7.1. An airplane is flying at 10,000 ft where the air temperature is 20°F and the pressure is 21.0 in. of mercury. At a certain point on the wing (Fig. 7-16) the static pressure is 980 lb/ft^2 and the local Mach number is 0.95. Assuming frictionless adiabatic flow, what is the velocity of the airplane relative to the undisturbed air?

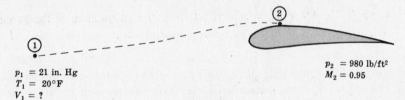

$p_1 = 21$ in. Hg
$T_1 = 20°F$
$V_1 = ?$

$p_2 = 980$ lb/ft^2
$M_2 = 0.95$

Fig. 7-16

Consider the airplane as stationary and the air moving as shown. For isentropic flow of an ideal gas, we have

$$T_2 = T_1(p_2/p_1)^{(k-1)/k} = (460 + 20)\left[\frac{980}{21(0.491)(144)}\right]^{0.286} = 427°R$$

and

$$V_2 = M_2 a_2 = M_2\sqrt{kRT_2} = 0.95\sqrt{1.4(32.2)(53.3)(427)} = 955 \text{ ft/sec}$$

Then from the energy equation,

$$V_1^2/2 + h_1 = V_2^2/2 + h_2$$

and

$$V_1^2 = V_2^2 + 2c_p(T_2 - T_1) = (955)^2 + 2(32.2)(778)(0.24)(427 - 480)$$

$$= 275,000 \text{ ft}^2/\text{sec}^2$$

so that $V_1 = 524$ ft/sec.

7.2. Air is flowing in a converging channel shown in Fig. 7-17 with conditions as indicated. Determine the pressure, temperature, velocity and Mach number at the throat 2.

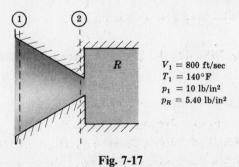

$V_1 = 800$ ft/sec
$T_1 = 140°F$
$p_1 = 10$ lb/in^2
$p_R = 5.40$ lb/in^2

Fig. 7-17

Assume frictionless adiabatic flow. Next, determine the conditions upstream where the velocity is negligible. From the energy equation we have

$$T_0 = T_1 + \frac{V_1^2}{2c_p} = (140 + 460) + \frac{(800)^2}{2(32.2)(778)(0.24)} = 653°R$$

and

$$p_0 = p_1(T_0/T_1)^{k/(k-1)} = 10[653/(140 + 460)]^{3.5} = 13.52 \text{ lb/in}^2$$

so that

$$p_R/p_0 = 5.40/13.52 = 0.407$$

where T_0 and p_0 are the stagnation temperature and pressure respectively.

The critical pressure ratio is $p_c/p_0 = [2/(k + 1)]^{k/(k-1)} = 0.528$. Therefore the receiver pressure is below critical pressure and the nozzle is operating under choked conditions. Thus $p_2 = p_c$ where p_c is the critical pressure, $0.528 \, p_0$.

Hence we find

$$p_2 = 0.528(13.52) = 7.14 \text{ lb/in}^2$$

$$T_2 = T_1(p_2/p_1)^{(k-1)/k} = (140 + 460)(7.14/10)^{0.286} = 546°R$$

$$M_2 = 1.0 \text{ (choked flow)}$$

$$V_2 = a_2 = \sqrt{kRT_2} = \sqrt{1.4(32.2)(53.3)(546)} = 1140 \text{ ft/sec}$$

7.3. For air flowing in a converging-diverging nozzle, the pressure and temperature at the inlet are 14.7 psi and 70°F. At some location in the diverging portion of the nozzle the pressure is 1.37 psi. Determine the area at this location. The area at the throat is 1.0 in².

Assume frictionless adiabatic flow. We note that the pressure in the diverging portion of the nozzle is below critical pressure. Thus the nozzle is operating under choked conditions and the pressure at the throat, p_t, is

$$p_t = p_c = 14.7[2/(k + 1)]^{k/(k-1)} = 14.7(0.528) = 7.79 \text{ lb/in}^2$$

so that

$$T_t = T_0(p_t/p_0)^{(k-1)/k} = (460 + 70)(1.37/14.7)^{0.286} = 269°R$$

and

$$V_t = \sqrt{kRT_t} = \sqrt{1.4(32.2)(53.3)(269)} = 1030 \text{ ft/sec}$$

The energy equation gives

$$V_3^2/2 + h_3 = h_0$$

so that

$$V_3 = \sqrt{2c_p(T_0 - T_3)} = \sqrt{2(32.2)(778)(0.24)(530 - 269)} = 1770 \text{ ft/sec}$$

The continuity equation is

$$\rho_t A_t V_t = \rho_3 A_3 V_3$$

and hence

$$A_3 = A_t \frac{\rho_t}{\rho_3} \frac{V_t}{V_3} = A_t \frac{p_t}{p_3} \frac{T_3}{T_t} \frac{V_t}{V_3} = 1.0 \frac{7.79}{1.37} \frac{269}{442} \frac{1030}{1770} = 2.00 \text{ in}^2$$

7.4. Air is flowing in an insulated constant diameter horizontal duct shown in Fig. 7-18. The inlet conditions are $p_1 = 100 \text{ lb/in}^2$, $T_1 = 100°F$ and $V_1 = 500 \text{ ft/sec}$. The diameter of the pipe is 6.0 in. and the flow is "choked". Determine the net force of the fluid on the pipe.

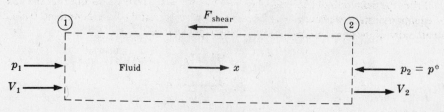

Fig. 7-18

The control volume momentum equation can be applied to the control volume enclosing the fluid between the entrance and exit as shown.

$$\sum F_x = \dot{m}(V_2 - V_1)$$

which is

$$(p_1 - p_2)(\pi D^2/4) - F_{shear} = \rho_1 A_1 V_1 (V_2 - V_1)$$

Now we need to determine the conditions at the exit. We have

$$M_2 = M^* = 1.0 \text{ (choked flow)}$$

so that

$$V_2 = \sqrt{kRT_2}$$

From the energy equation,

$$V_2^2 = V_1^2 + 2c_p(T_1 - T_2) = V_1^2 + 2c_p(T_1 - V_2^2/kR)$$

Then

$$V_2^2 = \frac{V_1^2 + 2c_p T_1}{1 + 2c_p/kR} = \frac{V_1^2 + 2c_p T_1}{1 + 2/(k-1)}$$

$$V_2 = \left[\frac{(500)^2 + 2(32.2)(778)(0.24)(460 + 100)}{1 + 2/(1.4 - 1)} \right]^{1/2} = 1030 \text{ ft/sec}$$

The temperature at the exit is

$$T_2 = \frac{V_2^2}{kR} = \frac{(1030)^2}{1.4(32.2)(53.3)} = 442°R$$

From the continuity equation we have

$$\rho_2 V_2 = \rho_1 V_1$$

Hence

$$(p_2/T_2)V_2 = (p_1/T_1)V_1$$

and

$$p_2 = p_1 \frac{T_2 V_1}{T_1 V_2} = 100\left(\frac{442}{460 + 100} \right)\left(\frac{500}{1030} \right) = 38.2 \text{ lb/in}^2$$

Then

$$(p_1 - p_2)(\pi D^2/4) = (100 - 38.2)(36\pi/4) = 1750 \text{ lb}_f$$

and

$$\rho_1 A_1 V_1 (V_2 - V_1) = \frac{100}{53.3(32.2)(460 + 100)} (36\pi/4)(500)(1030 - 500) = 780 \text{ lb}_f$$

so that finally the force of the pipe on the fluid is

$$F_{shear} = 1750 - 780 = 970 \text{ lb}_f$$

and the direction is as shown in Fig. 7-18.

7.5. Show that for a compressible fluid in isothermal flow with no external work,

$$\frac{dM^2}{M^2} = 2 \frac{dV}{V}$$

From the definition of Mach number we have $M^2 = V^2/a^2$. If we assume that the fluid follows the ideal gas equation of state, we have $a = \sqrt{kRT}$. Thus for isothermal flow a is a constant. By taking the logarithm of both sides of the equation defining Mach number, we have

$$\ln M^2 = \ln V^2 - \ln a^2$$

or

$$\ln M^2 = 2 \ln V - \ln a^2$$

Then differentiation gives

$$\frac{dM^2}{M^2} = 2 \frac{dV}{V}$$

7.6. Show that for a perfect gas flowing through a constant area duct at constant temperature conditions,

$$\frac{dp}{p} = -\frac{dV}{V} = -\frac{1}{2}\frac{dM^2}{M^2}$$

Assuming one-dimensional and steady flow, the continuity equation is $\rho V A =$ constant; and since the area is constant, we have $\rho V =$ constant. Using the ideal gas equation of state the density may be eliminated from the above equation to give $pV =$ constant. Differentiating this equation gives

$$p\,dV + V\,dp = 0$$

and dividing by pV gives

$$\frac{dp}{p} = -\frac{dV}{V}$$

Then using the results of Problem 7.5 we have

$$\frac{dp}{p} = -\frac{dM^2}{M^2}$$

7.7 Show that for adiabatic flow of a perfect gas,

$$T/T_0 = 1 - \tfrac{1}{2}(k-1)(V/a_0)^2$$

where T_0 is the stagnation temperature and a_0 is the velocity of sound for stagnation conditions $(a_0 = \sqrt{kRT_0})$.

From the one-dimentional steady flow energy equation for zero work and adiabatic flow we have

$$h - h_0 = -V^2/2$$

For a perfect gas and constant c_p we have

$$c_p(T - T_0) = -V^2/2$$

Dividing by T_0 and c_p gives

$$T/T_0 = 1 - V^2/2c_p T_0$$

But $c_p = Rk/(k-1)$ and $T_0 = a_0^2/Rk$, which substituted into the above equation gives

$$T/T_0 = 1 - \tfrac{1}{2}(k-1)(V/a_0)^2$$

7.8. A converging-diverging nozzle with a flow of air has an exit area ratio of $A^*/A_e = 0.2362$ where A_e is the nozzle exit area and A^* is the throat area. If a shock occurs at the location where $A^*/A = 0.5926$ find the back pressure and exit Mach number M_e and important properties upstream and downstream of the shock in terms of the reservoir pressure p_0.

The ratio A^*/A changes across the shock since the flow is not isentropic. Although A does not change across the shock, A^* does change since the gas states have different stagnation pressures across the shock. Using the isentropic table we find that the exit Mach number would be 3 if no shock were present. At position 1, just upstream of the shock where $A^*/A = 0.5926$, we find $M_1 = 2$.

Using the shock tables $M_1 = 2$ gives $M_2 = 0.5773$ downstream of the shock. Further, from the isentropic tables

$$\frac{p_1}{p_{01}} \text{ (for } M_1 = 2) = 0.1228$$

From the shock tables

$$\frac{p_{02}}{p_{01}} = 0.7209 \qquad \text{and} \qquad \frac{p_2}{p_1} = 4.5$$

so that

$$p_2 = 4.5\ p_1 = 4.5 \times 0.1278\ p_{01} = 0.5751\ p_{01}$$

The flow downstream of the shock is isentropic, proceeding from Mach number $M_2 = 0.5773$ and $p_2 = 0.57571\ p_{01}$ to the exit valves.

Now we know the actual exit area A_e in terms of the actual throat area A_1^*. Across the shock $A_1 = A_2 = A_1^*/0.5926$.

Using the isentropic tables we find for $M_2 = 0.5773$ that $A_2^*/A_2 = 0.82$. We wish to find A_2^*/A_e, from which we find the actual M_e.

We can write

$$\frac{A_2^*}{A_e} = \frac{A_1^*}{A_e} \cdot \frac{A_1}{A_1^*} \cdot \frac{A_2^*}{A_2} = (0.2362)\left(\frac{1}{0.5926}\right)(0.82) = 0.327$$

From the isentropic tables this value of A^*/A_e corresponds to $M_e = 0.20$. Further the tables give $p_e/p_{02} = 0.973$. Hence $p_e/p_{01} = p_e/p_{02} \cdot p_{02}/p_{01} = (0.973)(0.7209) = 0.70$. Therefore, if the back pressure is set at 0.70 p_{01} the shock will occur at the specified location and the exit Mach number will be $M_e = 0.20$.

7.9. A long adiabatic pipe with friction is fed air by a converging nozzle from a tank where the stagnation pressure and temperature are 520°R and 300 psia. If the pipe is 1 inch in diameter and has a friction factor (average) 0.002 find L_{max} if the back pressure is maintained at 100 psia. What is the flow rate under these conditions if the pipe length is L_{max}?

The solution requires a trial and error approach. The pressure and Mach number at the throat (station 1 or inlet to the duct) are unknown. The procedure is as follows: We assume a value of M_1 and find the corresponding value of p_1 from the isentropic tables (for flow in the nozzle). Using M_1 we obtain L_{max} from the value of $4\bar{f}(L_{max}/D)$ from the Fanno flow tables. p_1/p^* is also found (which depends only on M_1) from the Fanno tables. p^* is exactly the exit pressure of 100 psia. The exit Mach number is of course unity. Then p_1 at the duct inlet obtained from the Fanno tables is compared to the value of p_1 obtained from the isentropic nozzle flow. The correct value of M_1 has been chosen when these two methods of finding p_1 give the same result.

After several trials we find the following

$$M_1 = 0.4$$

$$\frac{4\bar{f}L_{max}}{D} = 2.31$$

using $\bar{f} = 0.002$ and $D = 1$ inch we find

$$L_{max} = 289 \text{ inches}$$

From the isentropic tables, $p_1/p_0 = 0.896$, which gives $p_1 = 0.896 \times 300 = 269$ psia. Using $M_1 = 0.4$ in the Fanno tables gives $p_1/p^* = 2.696$ and since $p^* = 100$ psia, $p_1 = (2.696)(100) = 267$ psia. The check is acceptably close, and we stop here.

The other properties in the duct may be found in a similar manner once M_1 is known. The mass rate of flow in may be found by evaluating the properties and velocity at any place. The inlet (where $M_1 = 0.4$) or the exit (where $M_3 = 1$) are both convenient and may both be calculated as a check since they both should give the same answer. Using the duct inlet values which are obtainable from the isentropic tables we find

$$\rho_1 = 0.9243\ \rho_0$$

$$a_1 = 0.9844\ a_0$$

$$V_1 = M_1 a_1$$

and using

$$k = 1.4$$

$$T_0 = 520°R$$

$$R = 1718 \frac{ft^2}{°R\ sec^2}$$

$$p_0 = 300 \text{ psia}$$

we calculate

$$\rho_0 = \frac{p_0}{RT_0} = \frac{300 \times 144}{1718 \times 520} = 0.048 \text{ slug/ft}^3 = 0.000028 \text{ slug/in.}^3$$

$$a_0 = \sqrt{kRT_0} = \sqrt{1.4 \times 1718 \times 520} = 1122 \text{ ft/sec} = 13{,}460 \text{ in./sec}$$

$$\dot{m} = \rho_1 A_1 V_1 = (0.9243)(0.000028)\left(\frac{\pi(1)^2}{4}\right)(0.4)(0.9844)(13460)$$

$$= 0.108 \text{ slug/sec}$$

It should be remembered that the unit of mass is slugs so that ρ must be in slugs/ft^3 or slugs/in.^{3}. If lb$_m$ were used then a factor of g_c would have to be used in the momentum equations such as that to obtain the sonic speed. Further, the units of R must be consistent with those used for ρ and the equation of state. A value of $R = 53.3$ ft/°R is often seen in engineering calculations, but there the mass is given in lb$_m$ and a g_c factor must be used in the equation for sonic speed. We use the unit of slug or kg for mass throughout this book and hence never use the g_c conversion factor.

Supplementary Problems

Distinguish between the true and false statements of Problems 7.10–7.27.

7.10. Sound waves and shock waves

(a) have no similarities.

(b) are the same except for the magnitude of the property changes across the waves.

(c) are both irreversible processes.

(d) are both reversible processes.

(e) travel through still air at different velocities.

7.11. For isentropic flow of an ideal gas

(a) the stagnation temperature is constant.

(b) the stagnation pressure is constant.

(c) the entropy is constant.

(d) the maximum Mach number is one.

(e) the temperature decreases as the pressure increases.

7.12. In case of flow of an ideal gas in a converging channel

(a) the temperature decreases in the direction of flow.

(b) there is a maximum possible flow for fixed inlet conditions and throat diameter.

(c) it may be possible for a normal shock to occur.

(d) the velocity will always increase in the direction of flow.

(e) the Mach number may be greater than one.

7.13. For an ideal gas flowing in a converging channel

    (a)   there is a maximum rate of flow for fixed exit area and inlet conditions.

    (b)   the Mach number at the throat is always one.

    (c)   the maximum temperature will be at the throat.

    (d)   the velocity of the gas at the throat is equal to the velocity of sound at the entrance for choked conditions.

    (e)   a normal shock will occur at the throat for critical conditions.

7.14. A supersonic nozzle with isentropic flow

    (a)   may have a normal shock occurring in the converging section.

    (b)   may have a normal shock occurring in the diverging section.

    (c)   will be "shockless".

7.15. For a converging-diverging flow channel

    (a)   when the Mach number at the exit is greater than unity no shock has occurred.

    (b)   when the critical pressure ratio is exceeded, the Mach number at the throat is greater than unity.

    (c)   for sonic velocity at the throat only one value of pressure is possible at a given downstream position.

    (d)   the Mach number at the throat is always unity.

7.16. For flow of a compressible fluid in a converging-diverging flow channel operating under choked conditions,

    (a)   the flow is supersonic in all of the diverging section.

    (b)   a normal shock may occur in the converging section.

    (c)   a normal shock may occur in the diverging section.

    (d)   the Mach number is always equal to one at the throat.

    (e)   the ratio of the throat pressure to the inlet pressure does not depend upon the flow geometry.

7.17. In a normal shock in one-dimensional flow

    (a)   velocity, pressure, and density increase.

    (b)   pressure, density, and temperature increase.

    (c)   velocity, temperature, and density increase.

    (d)   the stagnation enthalpy is unchanged.

7.18. A normal shock

    (a)   may occur only when the flow is supersonic.

    (b)   always involves a temperature increase.

    (c)   always involves a pressure decrease.

    (d)   always moves (relative to the fluid) with a velocity equal to the velocity of sound.

    (e)   may occur for subsonic flow.

7.19. A normal shock

    (a)   will form ahead of a blunt object moving at supersonic velocity through a compressible fluid.

    (b)   will occur only when the flow is supersonic.

    (c)   will cause the entropy of the fluid to increase.

    (d)   involves a frictionless adiabatic process.

    (e)   is impossible.

7.20. For adiabatic subsonic flow with friction in a constant area duct

(a) the temperature will increase in the direction of flow.

(b) the stagnation temperature decreases.

(c) the stagnation pressure decreases.

(d) the velocity increases.

(e) the limiting Mach number is one.

7.21. The Fanno Line

(a) represents a series of equilibrium states on the h-s diagram for adiabatic flow.

(b) represents a series of equilibrium states on the h-s diagram for flow with friction.

(c) was developed from energy, continuity and equations of state.

(d) represents only subsonic flow.

(e) represents isothermal flow in a constant area duct.

7.22. For adiabatic supersonic flow with friction in a constant area duct

(a) the temperature will increase in the direction of flow.

(b) the stagnation temperature decreases.

(c) the stagnation pressure decreases.

(d) the velocity increases.

(e) the limiting Mach number is one.

7.23. For isothermal flow of an ideal gas in a constant area duct

(a) the pressure decreases in the direction of flow.

(b) heat is removed from the fluid.

(c) the fluid particles are accelerated.

(d) the fluid density is decreased.

(e) the maximum Mach number is one.

7.24. For flow of an ideal gas in an insulated constant area duct having an initial Mach number less than one

(a) the velocity decreases in the direction of flow.

(b) the temperature decreases in the direction of flow.

(c) the Mach number can increase to a value greater than one.

(d) the entropy will always increase in the direction of flow.

(e) the pressure is constant.

7.25. Some of the results of heating and cooling of an ideal fluid flowing in a constant area frictionless duct are

(a) heating will always cause an increase in entropy.

(b) heating will always cause an increase in velocity.

(c) heating will always cause an increase in temperature.

(d) heating will always cause a decrease in velocity.

(e) the limiting Mach number is one for subsonic flow.

7.26. For adiabatic flow of an in ideal gas in a frictionless constant area duct

(a) the stagnation temperature increases as the entropy increases.

(b) the maximum temperature occurs at $M = 1$.

(c) the plot of equilibrium states on the *h-s* diagram is called the Rayleigh line.

(d) the isentropic equations apply.

(e) the limiting Mach number is one for supersonic flow.

7.27. For adiabatic flow of an ideal gas in a frictionless constant area duct

(a) the pressure must decrease in the direction of flow.

(b) the temperature will remain constant.

(c) the velocity will remain constant.

(d) the entropy will increase.

(e) the Mach number will remain constant.

7.28. What is the velocity of sound in carbon dioxide at 300°F? What is the acoustic velocity in water at 60°F?

7.29. At point *A* in the undisturbed region of an airstream which flows past a body, the density is 0.002378 slugs per cubic foot, the pressure is 14.7 psi and the velocity is 350 fps. The pressure at a point *B* on the body is 7.35 psi. What is the Mach number at each point for frictionless adiabatic flow?

7.30. An airplane is flying through standard air at 600 fps. Calculate the local Mach number at a point on the plane where the air velocity relative to the plane is 200 fps.

7.31. A body moves through standard air at 400 fps. What is the pressure at a point on the body where the velocity of the air relative to the body is zero?

7.32. A blunt projectile moves through stationary air at 14.7 psia and 60°F at 1900 fps. Find the stagnation pressure, the stagnation enthalpy and the stagnation temperature.

7.33. A nozzle is to be designed to expand air isentropically from 50 psia and 200°F to 18 psia at a rate of 0.30 lb/sec. Calculate (a) the Mach number at the nozzle exit and (b) the exit area (in ft^2).

7.34. Air flows isentropically through a converging nozzle having a throat area of 0.1 ft^2. Air which enters with negligible velocity is at 60 psia, 200°F. The receiver pressure is 15 psia. Calculate the mass rate of flow in lb/sec.

7.35. Air is to be expanded through a converging-diverging nozzle by a frictionless adiabatic process from a pressure of 160 psia and a temperature of 24°F to a pressure of 20 psia. Determine (a) the throat and exit areas for a well designed shockless nozzle if the rate of flow is 4.5 lb/sec and (b) the Mach number at the nozzle outlet.

7.36. Air, initially at 50 psia and 250°F, flows isentropically through a converging-diverging nozzle with a throat area of 0.4 in^2. The downstream pressure is 14.7 psia and the temperature is 40°F. Calculate (a) the rate of flow in lb/sec and (b) the velocity at the nozzle outlet.

7.37. Air at 100 psia and 140°F flows through a converging tube into a receiver in which the pressure is 20 psia. Assuming frictionless adiabatic flow and negligible upstream velocity, calculate the velocity at the throat.

7.38. Air at 100 psia and 140°F flows through a converging tube into a receiver in which the pressure is 20 psia. Assuming frictionless adiabatic flow and negligible upstream velocity, calculate the velocity at the throat.

7.39. Sketch the *h-s* and *T-s* diagrams for an ideal gas flowing adiabatically through a (1) long pipe line, (2) nozzle, (3) diffuser.

7.40. A gas (molecular weight 18, $k = 1.3$) is to be pumped through a 36 inch I.D. pipe connecting two compressor stations 40 miles apart. At the upstream station the pressure is not to exceed 90 psig and the downstream station is to be at least 10 psig. Calculate the maximum allowable rate of flow (cubic feet per day at 70°F and 1 atm), assuming that there is sufficient heat transfer through the pipe to maintain the gas at 70°F. The kinematic viscosity is 4×10^{-4} ft²/sec.

7.41. Air enters a 6 inch diameter pipe at 200 psia and 60°F with a velocity of 190 fps. The friction coefficient f is 0.016. What is the Mach number and distance from entrance at the section where the pressure is 75 psia for isothermal flow?

7.42. Air enters a horizontal pipe at 600 psia, temperature of 200°F, and a velocity of 300 fps. What is the limiting pressure for isothermal flow? What is the limiting pressure for adiabatic flow?

7.43. Air at a pressure of 120 psia and a temperature of 80°F enters a clean steel pipe 950 ft long and 5 inches in diameter. The inlet velocity is 80 fps. Assuming isothermal flow, find (a) the pressure drop (psi) if the fluid is considered compressible, (b) the pressure drop (psi) if the fluid is considered incompressible.

7.44. Hydrogen at an initial pressure of 60 psia and a temperature of 140°F flows through a horizontal insulated pipe 900 feet long and 0.2 feet in diameter. The friction factor is 0.015. What is the maximum possible rate of flow and what pressure drop would be required to maintain this flow?

7.45. A large tank in the laboratory can be filled with air at 200 psia and at room temperature. Design a supersonic nozzle which can be screwed into the wall of this tank and discharge air to the atmosphere at supersonic velocities. (If a small airfoil or test body is placed in the stream of this nozzle, we have in effect a blowdown supersonic windtunnel.) Assume that the air in the tank remains at 200 psia for the short period of time that we are interested in the performance. The throat should have an area of 0.5 in².

7.46. Discuss what happens if the back pressure is raised or lowered in Problem 7.9.

7.47. A long adiabatic duct of diameter 0.01 m is fed from a reservoir at a pressure of 2 MPa through a converging-diverging nozzle with a throat area ratio of $A^*/A_1 = 0.2362$. A_1 is the duct cross sectional area. The average friction factor is 0.001. Find L_{max}. What back pressure gives shockless (smooth) flow throughout the duct? If this back pressure is maintained but the length is increased to $2 L_{max}$ locate the shock in the duct. If the duct were reduced to $1/2 L_{max}$ in length (with the same back pressure) where would the shock be located? In each case above discuss in detail what happens as the back pressure is increased or lowered.

NOMENCLATURE FOR CHAPTER 7

A = area

a = speed of sound

C = wetted perimeter

c_p = specific heat at constant pressure

c_s = shock speed

c_v = specific heat at constant volume

D = diameter or hydraulic diameter of a duct

f = Fanning friction factor

h = enthalpy per unit mass

k = specific heat ratio, c_p/c_v

K = degree Kelvin

L = length of duct

M = Mach number, V/a

$\dot{m}$ = mass flow rate

p = static pressure

$°R$ = degree Rankine

R = gas constant

s = entropy per unit mass

T = absolute temperature

t = time

u = internal energy per unit mass

V = velocity

v = specific volume = $1/\rho$

x = axial coordinate

ρ = density

τ_0 = wall shear stress

$(\ )_c$ = denotes critical condition

$(\ )_R$ = denotes receiver condition (back pressure)

$(\ )_0$ = denotes stagnation condition

$(\ )_t$ = denotes throat condition

$(\ )^*$ = denotes condition where flow is choked or where $M = 1$

Chapter 8

Two-Dimensional Compressible Flow Gasdynamics

8.1 EQUATIONS OF FRICTIONLESS COMPRESSIBLE FLOW

In Chapter 6 we discussed two-dimensional incompressible potential flow and in Chapter 7 we considered one-dimensional compressible flow. We can now extend both these approaches to develop the more general treatment of compressible frictionless (ideal) flow. Since the flow is ideal we can again show that it is irrotational throughout (if the free stream flow is irrotational) and hence potential. We are concerned then with general three-dimensional compressible potential flow, although for explicit calculations we will usually restrict ourselves to two dimensions.

As before, we will discuss the flow outside the boundary layer (where the flow is irrotational) and will be concerned mainly with aerodynamic problems—flow over wings and aircraft shapes.

At low speeds (Mach number M less than about 0.3) the fluid behaves nearly incompressibly and the analysis of incompressible potential flow (Chapter 6) is adequate. As the flow speed increases, the compressibility effects become more and more important; and as the Mach number passes unity, shock waves can form on the body about which we are studying the flow. If the Mach number of the free stream is greater than unity ($M > 1$), the flow is supersonic and is very different from flow which is subsonic ($M < 1$). Transonic flow over a body occurs when the flow over one part of the body is subsonic and over another part is supersonic. In steady transonic flow over a body one can find regions over the body where $M > 1$, where $M = 1$, and where $M < 1$. Mathematically, transonic flow is rather more complicated than either purely subsonic or supersonic flow. When an aircraft accelerates from subsonic to supersonic flow it must pass through a transonic condition.

In this chapter we will study the full range of Mach numbers, beginning in the subsonic region and extending to very large Mach numbers. For M greater than about 6 the flow is considered hypersonic and many assumptions made for ordinary supersonic flow are not valid and the analysis becomes more complicated. As M increases into the hypersonic region the boundary layer friction increases and the stagnation temperature becomes so great (for M greater than about 6) that the temperature in the boundary layer rises and the aircraft surface may become so hot that special refractory or ablative material must be used to prevent structural damage. Re-entry vehicles pass through a range of such high Mach numbers. In Chapter 10 we will discuss the structure of the hypersonic boundary layer. Now, let us concentrate on the potential flow solution outside the boundary layer. The laws governing the flow are basically the same for subsonic through hypersonic flow, but various terms in the fundamental equations become more or less important as M changes and, indeed, the entire character of the differential equations changes as M is greater or less than unity. As a result the flow pattern is drastically different for subsonic and supersonic flow.

In most compressible aerodynamic flow problems it is accurate to assume that the flow is frictionless and irrotational and isentropic throughout. However, in supersonic flow, whenever shock waves may occur, the flow is *not isentropic* through the shock waves.

Let us now briefly review the basic equations of frictionless, isentropic flow which may be used except in shock wave calculations. These equations, in general, will be nonlinear and various approximate solutions must be developed. In some instances exact solutions may be constructed even when shock waves occur.

The basic equations are the same as those developed in Chapter 5 except that compressibility effects must be taken into account and consequently the energy equation or an equivalent must be also considered. The basic equations are continuity, motion and energy. If the energy equation is written in terms of temperature, an equation of state is also necessary. If the flow is isentropic throughout, the isentropic relationship between the pressure and density may be used, obviating the necessity of using the energy equation.

As we showed in Chapter 5, adiabatic frictionless flow is isentropic along a streamline if it is irrotational. Further if the flow has a uniform value of total (or stagnation) enthalpy, $h_0 = h + V^2/2$ throughout, in addition to being irrotational, then the flow is isentropic throughout. Such a flow is sometimes called *homentropic*.[1]

For homentropic flow (in which body forces are negligible) for a perfect gas in two dimensions, the basic relationships are

Continuity:
$$\frac{\partial p}{\partial t} + \frac{\partial}{\partial x}(\rho u) + \frac{\partial}{\partial y}(\rho v) = 0 \qquad (8.1)$$

Momentum:
$$\rho \frac{\partial u}{\partial t} + u \frac{\partial u}{\partial x} + v \frac{\partial u}{\partial y} = -\frac{\partial p}{\partial x} \qquad (8.2)$$

$$\rho \frac{\partial v}{\partial t} + u \frac{\partial v}{\partial x} + v \frac{\partial v}{\partial y} = -\frac{\partial p}{\partial y}$$

Isentropic relationship:
$$p/p_0 = (\rho/\rho_0)^k \qquad (8.3)$$

where $k = c_p/c_v$ is the ratio of specific heats, and p_0 and ρ_0 are the pressure and density at arbitrary reference states, usually taken as free stream values or stagnation values.

Equations (8.1) through (8.3) are independent but not unique. In place of (8.3) we could use the complete energy equation along with the equation of state. If the flow is not adiabatic such a precedure would be necessary.

Since the flow is irrotational the velocity potential, defined by $\mathbf{V} = -\nabla\phi$, may be introduced. Then the term ∇p may be written as

$$\nabla p = (\partial p/\partial \rho)_s \nabla \rho = a^2 \nabla \rho \qquad (8.4)$$

where a is the sonic speed. Then the scalar product of $\mathbf{V}$ and the vector equation of motion may be formed and (8.4) and (8.1) used to eliminate p and ρ. The result is (for two-dimensional flow)

$$(u^2 - a^2)\frac{\partial u}{\partial x} + (v^2 - a^2)\frac{\partial v}{\partial y} + uv\left(\frac{\partial u}{\partial y} + \frac{\partial v}{\partial x}\right) = 0 \qquad (8.5)$$

and the condition $\dfrac{\partial v}{\partial x} - \dfrac{\partial u}{\partial y} = 0$, or in terms of ϕ

$$\frac{1}{a^2}\left[\left(\frac{\partial\phi}{\partial x}\right)^2 \frac{\partial^2\phi}{\partial x^2} + \left(\frac{\partial\phi}{\partial y}\right)^2 \frac{\partial^2\phi}{\partial y^2} + \frac{\partial\phi}{\partial x}\frac{\partial\phi}{\partial y}\left(2\frac{\partial^2\phi}{\partial x\,\partial y}\right)\right] = \frac{\partial^2\phi}{\partial x^2} + \frac{\partial^2\phi}{\partial y^2} \qquad (8.6)$$

which reduces to $\nabla^2\phi = 0$ as $a \to \infty$ which is the appropriate limit for incompressible flow. (8.6) is a single equation for ϕ and describes the flow, but the sonic speed a must be related to the velocity components through the energy equation and complicates the situation. In Section 8.3, when we discuss small perturbation theory, we will carry out this calculation explicitly. The method of small perturbations allows a linearization of (8.6) (and resultant approximate solutions) for thin bodies such as airfoils and slender bodies of revolution.

Exact shock-expansion solutions will be discussed in Section 8.2, and in Section 8.4 methods for the solution of (8.5) will be discussed.

---

[1] This statement follows from Crocco's theorem which states that

$$T\nabla s + \mathbf{V} \times (\nabla \times \mathbf{V}) = \nabla h_0 + \partial\mathbf{V}/\partial t$$

where s is specific entropy. In steady flow $\nabla s = 0$ if $\nabla \times \mathbf{V} = 0$ and $\nabla h_0 = 0$. We will not prove this theorem here but it follows from the equation of motion and the thermodynamic relationship $T\,ds = dh - (1/\rho)\,dp$. In most aerodynamic flows the fluid originates in the same "reservoir," the free stream, and hence h_0 is constant.

8.2 SHOCK-EXPANSION THEORY

Exact solutions for flow in certain simple geometries may be obtained if the flow consists entirely of shock waves and/or expansion waves. We will discuss the oblique shock wave and simple expansion wave, then show how these solutions may be combined to describe the flow over certain simple shapes.

The Oblique Shock

Fig. 8-1 shows an oblique shock wave. By conservation of momentum the tangential component of velocity V_t is continuous across the shock so that $V_{t1} = V_{t2}$. Then V_{n1} and V_{n2} are related by the normal shock relations of Chapter 7. Hence in terms of the normal velocities an oblique shock is identical to a normal shock and the Mach numbers (in terms of V_n, $M_1 = V_{n1}/a_1$), pressure, density, etc., are related in the same way as in the normal shock. Since $V_{t1} = V_{t2}$ but $V_{n1} \neq V_{n2}$, the actual velocity vector V is rotated and changed in magnitude as it passes through the shock as shown in Fig. 8-2. V_{n1} is greater than the sonic speed a_1 and V_{n2} must be less than the sonic speed a_2. However, although $V_1 > a_1$, V_2 may also be greater than a_2. In an oblique shock the magnitude of the velocity does not necessarily drop to a subsonic value through the shock, although the normal component does.

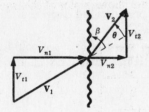

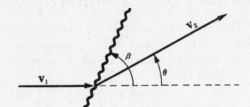

Fig. 8-1.   The oblique shock.

Fig. 8-2.   Rotation of the velocity vector through the shock.

We define M_1 as V_1/a_1 and $V_{n1} = V_1 \sin \beta$ so that $V_{n1}/a_1 = M_1 \sin \beta$. The normal shock relationships of Chapter 7 can be used with $M_1 \sin \beta$ replacing the M_1 there.[1] For reference we list the following.

$$\frac{V_{n1}}{V_{n2}} = \frac{\rho_2}{\rho_1} = \frac{(k+1)(M_1^2 \sin^2 \beta)}{(k-1)M_1^2 \sin^2 \beta + 2}$$

$$\frac{p_2 - p_1}{p_1} = \frac{2k}{k+1}(M_1^2 \sin^2 \beta - 1)$$

$$\frac{T_2}{T_1} = \frac{a_2^2}{a_1^2} = 1 + \frac{2(k-1)}{(k+1)^2} \frac{M_1^2 \sin^2 \beta - 1}{M_1^2 \sin^2 \beta}(kM_1^2 \sin^2 \beta + 1) \qquad (8.7)$$

$$\frac{s_2 - s_1}{R} = \ln\left\{\left[1 + \frac{2k}{k+1}(M_1^2 \sin^2 \beta - 1)\right]^{1/(k-1)} \cdot \left[\frac{(k+1)M_1^2 \sin^2 \beta}{(k-1)M_1^2 \sin^2 \beta + 2}\right]^{-k/(k-1)}\right\}$$

$$M_2^2 \sin^2(\beta - \theta) = \frac{1 + \frac{1}{2}(k-1)M_1^2 \sin^2 \beta}{kM_1^2 \sin^2 \beta - \frac{1}{2}(k-1)}$$

It may be noted that $0 < \beta < \pi/2$ and $M_1 \sin \beta = V_{n1}/a_1 \geq 1$, so that β has a minimum value for a given M_1. Hence

$$\sin^{-1} 1/M_1 \leq \beta \leq \pi/2 \qquad (8.8)$$

For a normal shock, $\beta = \pi/2$. For a weak shock, as the shock becomes a sonic wave, $\sin^{-1} 1/M = \beta$.

---

[1] Normal shock tables may be used with $M_1 \sin \beta$ considered as M_1 in the tables.

The value of M_2 may be found by noting that $M_2 = V_2/a_2$ and $V_{n2}/a_2 = M_2 \sin(\beta - \theta)$. From the normal shock relation

$$M_2^2 = \frac{1 + \frac{1}{2}(k-1)M_1^2}{kM_1^2 - \frac{1}{2}(k-1)}$$

we get by substituting $M_1 \sin \beta$ and $M_2 \sin(\beta - \theta)$ for M_1 and M_2 respectively,

$$M_2^2 \sin^2(\beta - \theta) = \frac{1 + \frac{1}{2}(k-1)M_1^2 \sin^2 \beta}{kM_1^2 \sin^2 \beta - \frac{1}{2}(k-1)} \tag{8.9}$$

We can relate β and θ by using the first of (8.7) and relating V_{2n} to V_{1n} by geometry, ($\tan \beta = V_{n1}/V_{t1}$, $\tan(\beta - \theta) = V_{n2}/V_{t1}$). Hence

$$\frac{\tan(\beta - \theta)}{\tan \beta} = \frac{(k-1)M_1^2 \sin^2 \beta + 2}{(k+1)M_1^2 \sin^2 \beta}$$

which can be written $$\tan \theta = 2 \cot \beta \frac{M_1^2 \sin^2 \beta - 1}{M_1^2(k + \cos 2\beta) + 2} \tag{8.10}$$

For a given M_1, there are two values of β for each value of θ. Fig. 8-3 shows a plot of this equation.

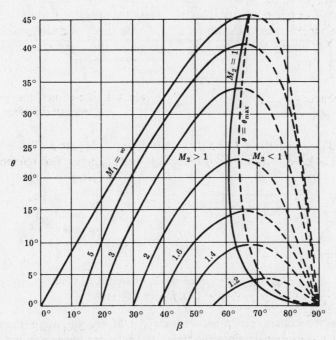

Fig. 8-3. β versus θ for an oblique shock.

For any value of θ (less than θ_{max}) there are two solutions. In practice the weaker solution (β smaller) usually occurs (if $\theta < 45°$). These correspond to M_2 remaining > 1, except for a small region between the lines $M_2 = 1$ and $\theta = \theta_{max}$.

In practice, what determines β and θ? In other words, how are oblique shocks generated? A simple example is flow through a corner as shown in Fig. 8-4(a). A symmetrical wedge would generate the same flow pattern symmetrical about the horizontal centerline [see Fig. 8-4(b)]. The vector $\mathbf{V}_2$ is parallel to the surface of the wedge (or corner) which sets θ. Then if M_1 is specified, β must follow.

For any given Mach number there is a maximum value of θ that is possible. What happens if the actual wedge angle θ is larger than this maximum? The answer is that the shock then becomes detached

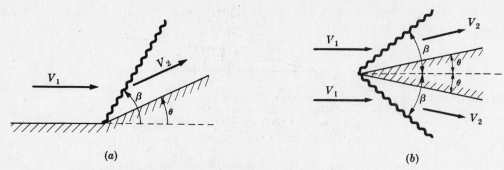

Fig. 8-4. Oblique shock generated by flow (*a*) through a corner and (*b*) over a symmetrical wedge.

from the body and stands in front. The detached shock is curved around the body and is known as a "bow shock." The location of the bow shock is determined by the downstream flow which will always have a region which is subsonic. (The bow shock will always appear as a normal shock just at the apex with $M_2 < 1$ and the jump conditions continuously change along the curved shock.) The complete description of the bow shock is complicated by the fact that its location cannot be found explicitly (because of the subsonic influence upstream) and the entire flow field must be considered together. There is an absolute limit of θ just greater than $45°$ where the shock will always detach no matter how large the Mach number M_1. Hence, if the object is blunt the shock will always detach (since $\theta = 90°$ at the nose). The dotted line where $M = 1$ (in Fig. 8-5) is called the sonic line and divides the regions of supersonic and subsonic flow.

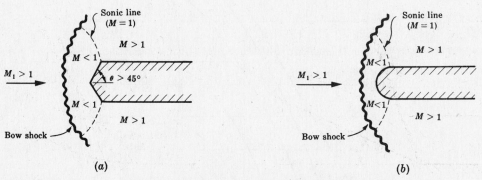

Fig. 8-5. A detached shock. (*a*) wedge, (*b*) blunt.

As the wedge angle θ becomes zero, β becomes $\sin^{-1} 1/M$ and the wave is called a *Mach line*. It is merely the locus of sonic disturbances since the shock strength has become infinitesimal and any infinitesimal or sonic disturbance would propagate out along this line. The angle $\sin^{-1} 1/M$, denoted as μ, is the *Mach angle*. In three dimensions the Mach lines would form a cone (from a point disturbance as discussed in Chapter 7).

In two dimensions, at any point in the flow, there are two Mach lines (at $\pm\beta = \sin^{-1} 1/M$ to the streamlines). These lines are called *characteristics* and must point downstream since there can be no upstream influence in supersonic flow (Fig. 8-6).

In three dimensions the locus of Mach lines form a cone, known as the Mach cone. The Mach cone formed on the front of a plane or missile delineates the zone of influence as discussed in Section 7.1. The Mach line to the right of the velocity vector is known as the right-running characteristic (or Mach line), and the one to the left as the left-running characteristic. Sometimes these lines are denoted by a ($+$) or ($-$), respectively, or vice versa. This is arbitrary.

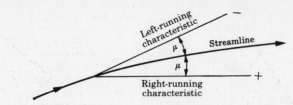

Fig. 8-6. Characteristics in supersonic flow. $(\theta \to 0$ and $\beta \to \mu = \sin^{-1} 1/M$). One set runs to the right of the streamlines and one set to the left.

As the angle θ becomes small but finite the shock becomes weak, and the shock relationships may be simplified. It may be shown that across the shock $\Delta p \sim \theta$ and $\Delta s \sim \theta^3$ so that for very weak shocks the flow becomes nearly isentropic. The change in speed ΔV across the weak shock is approximately (for flow deflection $\Delta\theta$)

$$\frac{\Delta V}{V} = -\frac{\Delta\theta}{\sqrt{M_1^2 - 1}} \qquad (8.11)$$

which is exact for isentropic flow. If compression takes place through a continuous bend (Fig. 8-7), then a continuous set of Mach lines is formed and (8.11) may be written in differential form

$$\frac{dV}{V} = -\frac{d\theta}{\sqrt{M_1^2 - 1}} \qquad (8.12)$$

The Mach lines coalesce to form a shock wave, but near the body the flow is approximately isentropic in the compression fan of the Mach lines.

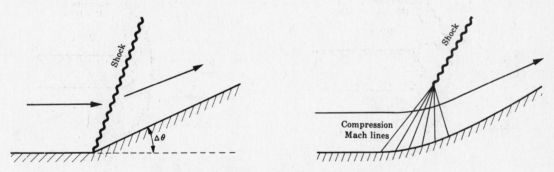

Fig. 8-7. Compression of supersonic flow by turning.

An oblique shock will be reflected from a wall such that the flow downstream of the reflected shock is parallel to the wall. Fig. 8-8 shows such a reflection. Generally the interaction of shocks by intersection (*b*), etc., can be described by using the condition that the flow direction must accommodate appropriate solid boundaries and satisfy continuity.

If the reflected shocks require an angle of turn greater than possible for the upstream Mach number, detachment from the wall will occur (*c*).

If the shock hits the wall at a corner and the wall downstream is aligned with the downstream flow no reflection occurs and the shock is "canceled" (*d*).

Supersonic Expansion and the Prandtl-Meyer Function

In flow over a convex corner or convex curved surface, no shock can form and the flow expands isentropically (Fig. 8-9). A single wave cannot accommodate the change (it would lead to a decrease in

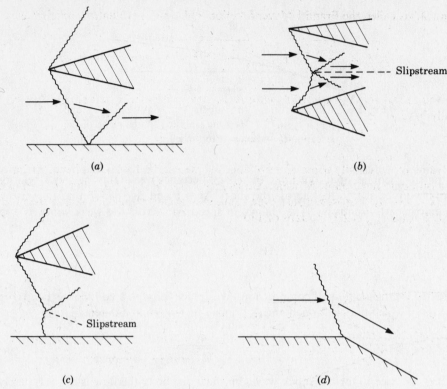

Fig. 8-8. Shock interactions. (*a*) reflection from a wall, (*b*) intersection of two shocks, (*c*) detachment from wall, (*d*) cancellation.

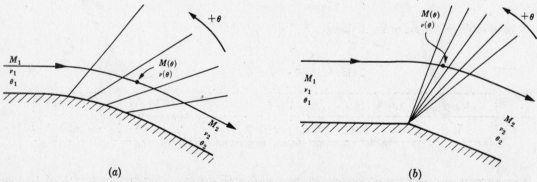

Fig. 8-9. Supersonic expansion on a convex surface. (*a*) continuous expansion around a curve and (*b*) an expansion fan centered at a sharp corner.

entropy) and a fan of waves must exist. The lines shown in Fig. 8-9 are actually Mach lines or characteristics. At any point in the flow they are inclined at an angle $\sin^{-1} 1/M$ to the streamlines. Along a Mach line all fluid properties are constant, as in the velocity vector.

Equation (8.12) holds throughout the expansion and the angle θ is the inclination of the streamline with respect to an arbitrary datum. Usually we take $\theta_1 = 0$ and measure θ positive counterclockwise (see Fig. 8-9, where θ_2 is negative if we take $\theta_1 = 0$). Equation (8.12) may be integrated:

$$-\theta + \text{constant} = \int_{V_1}^{V} \sqrt{M^2 - 1}\, \frac{dV}{V} = v(M) \qquad (8.13)$$

The function $v(M)$ is called the Prandtl-Meyer function and may be evaluated explicitly as

$$v(M) = \sqrt{\frac{k+1}{k-1}} \tan^{-1} \sqrt{\frac{k-1}{k+1}(M^2-1)} - \tan^{-1}\sqrt{M^2-1} \qquad (8.14)$$

where v is given in radians. We arbitrarily choose the constant in (8.13) so that $v = 0$ when $M = 1$, and thus we finally have

$$-(\theta - \theta_1) = v(M) - v_1(M_1) \qquad (8.15)$$

For a given value of θ_1 (usually taken as zero) and M_1, we can find $v(M)$ and hence M for any value of θ. $v(M)$ is a monotonic function of M which varies from 0 for $M = 1$ to v_{max} for $M \to \infty$, where $v_{max} = \frac{1}{2}\pi(\sqrt{(k+1)/(k-1)} - 1)$. For supersonic expansion, θ is negative; and for compression, θ is positive (see Fig. 8-10). For air v_{max} is 2.77 radians or 130°.

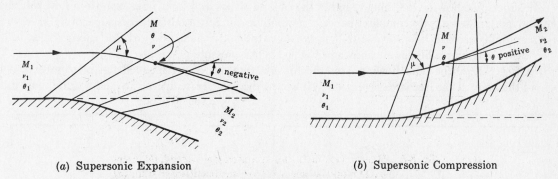

(a) Supersonic Expansion   (b) Supersonic Compression

Fig. 8-10.   Supersonic expansion and compression. Locally the angle between the Mach line and the streamline is μ.

A brief table of $v(M)$ is given in degrees in Table 8.1.[1]

Table 8.1.   The Prandtl-Meyer Function

v (degrees)	0.0	26.5	49.8	65.7	77.0	85.0	91.0	95.6	99.3
M	1.00	2.00	3.00	4.00	5.00	6.00	7.00	8.00	9.00

For example, consider the flow over a convex surface (Fig. 8-9a) through a total angle of 10 degrees, with $M_1 = 2$. From Table 8.1, $v_1 = 26.5°$ and $\theta_2 = -10°$ so that from (8.15) $v_2 = 26.5° + 10° = 36.5°$ and hence M_2 is approximately 2.39. Note that the rate of curvature is irrelevant in determining the final state. A sharp corner and a smooth curve give the same final result for expansion waves. For compression, only a smooth curve is isentropic and a sharp concave corner would lead to a shock wave.

What happens when expansion occurs around a corner where $|\theta|$ is greater than v_{max}? Then the streamlines behave as though the flow occurred over an expansion of v_{max}, and a region of stagnant fluid lies between the "slipstream" or "tangential discontinuity" and the body. Across the slipstream the pressure is continuous but the velocity is discontinuous. See Fig. 8-11. (In practice, as the vortex sheet forms, the slipstream and the fluid below it may be turbulent and not stagnant.)

---

[1] A more complete table may be found in the references, and in the Appendix.

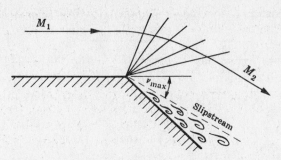

Fig. 8-11. Expansion around a corner of $|\theta| > v_{max}$, with the formation of a slipstream.

Combined Oblique Shocks and Expansions

In supersonic flow, by patching together the oblique shock and isentropic expansion flow solutions, the flow pattern over fairly complex shapes can be obtained. Because the flow is supersonic, there is no upstream influence and it is possible to "march" over the body calculating the flow along the way. The flows over some simple bodies are shown in Fig. 8-12. The pressure plot is also shown.

In Fig. 8-12(b) the fluid flowing over the top and bottom of the inclined flat plate must end up behind the body at the free stream pressure. However, the velocity, temperature (and density) may differ

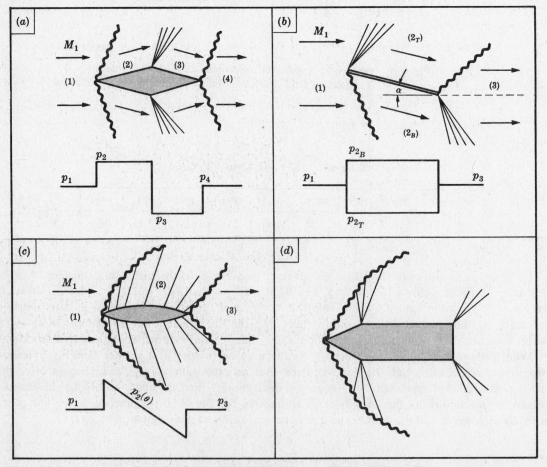

Fig. 8-12. Shock-expansion flow over two-dimensional bodies.

slightly across a "contact surface". In this figure, the pressure on the bottom is increased through the shock wave, and decreased on top through the expansion wave, giving rise to a lift on the plate. Unlike a subsonic lifting airfoil, the supersonic airfoil has a component of this pressure force in the direction of flow. This force is a drag force, known as "wave drag".

In practice, supersonic aircraft have wings which in some respects behave as subsonic wings, thus reducing the wave drag. Subsonic wings are possible on supersonic airplanes because the sweepback on the wings gives rise to a component of velocity, normal to the wing leading edge, which is subsonic.

Simple and Nonsimple Regions

We have been discussing simple oblique shock waves and isentropic waves of expansion or compression. In regions where these waves interact and become curved (the characteristics are no longer straight lines) these simple solutions are no longer valid and the complete equations must be solved. In general these equations are nonlinear, but in some simple cases, as in the flow around thin bodies, the equations may be linearized. These solutions will be discussed in Section 8.3. An example of nonsimple flow in a nozzle is shown in Fig. 8-13.

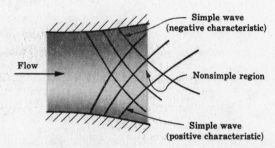

Fig. 8-13. Nonsimple flow in a nozzle. The expansion waves interact and the characteristics are no longer straight lines.

Thin Airfoil Theory

Flow over thin airfoils at small angles of attack may be considered isentropic if the shocks are weak enough, and shock-expansion theory may be simplified considerably. (The same results will be obtained later from linearized perturbation theory.) For weak shocks or expansion waves the pressure change is, from (8.7) and (8.9) approximately

$$\frac{\Delta p}{p} \approx \frac{kM^2}{\sqrt{M^2 - 1}} \Delta\theta$$

and assuming M is close to M_1 and p close to p_1,

$$\frac{p - p_1}{p_1} \approx \frac{kM_1^2}{\sqrt{M_1^2 - 1}} \theta \tag{8.16}$$

The pressure coefficient C_p is defined as

$$C_p = \frac{p - p_1}{\frac{1}{2}\rho_1 V_1^2} = \frac{p - p_1}{\frac{1}{2}kp_1 M_1^2} = \frac{2}{kM_1^2}(p/p_1 - 1) \tag{8.17}$$

where the subscript 1 refers to free stream values. Hence we obtain from (8.16) and (8.17),

$$C_p = \frac{2\theta}{\sqrt{M_1^2 - 1}} \tag{8.18}$$

For a flat plate at incidence α, Fig. 8-12(b), C_p is simply $C_{pB} = +2\alpha/\sqrt{M_1^2 - 1}$ for the bottom and $C_{pT} = -2\alpha/\sqrt{M_1^2 - 1}$ for the top. The lift and drag coefficients are then

$$C_L = \frac{\text{lift per unit length}}{\frac{1}{2}\rho_1 V_1^2 c} = \frac{(p_B - p_T)c \cos \alpha}{\frac{1}{2}\rho_1 V_1^2 c} = (C_{pB} - C_{pT}) \cos \alpha$$

$$C_D = \frac{\text{drag per unit length}}{\frac{1}{2}\rho_1 V_1^2 c} = \frac{(p_B - p_T)c \sin \alpha}{\frac{1}{2}\rho_1 V_1^2 c} = (C_{pB} - C_{pT}) \sin \alpha$$

where c is the chord length (width of the airfoil). Since α is small, $\cos \alpha \approx 1$, $\sin \alpha \approx \alpha$, and we finally have

$$C_L = \frac{4\alpha}{\sqrt{M_1^2 - 1}}, \qquad C_D = \frac{4\alpha^2}{\sqrt{M_1^2 - 1}} \qquad (8.19)$$

for the flat plate at a small angle of attack, α.

In general it can be shown that the pressure coefficients for thin airfoils may be expressed as follows:

$$C_{pT} = \frac{2}{\sqrt{M_1^2 - 1}}\left(\frac{df_T}{dx}\right), \qquad C_{pB} = \frac{2}{\sqrt{M_1^2 - 1}}\left(-\frac{df_B}{dx}\right) \qquad (8.20)$$

where $f_T(x)$ and $f_B(x)$ are the equations of the top and bottom surfaces, respectively, of the airfoil. (See Fig. 8-14.) From (8.20) the expressions for C_L and C_D can be found as

$$C_L = \frac{4\bar{\alpha}}{\sqrt{M_1^2 - 1}}, \qquad C_D = \frac{4}{\sqrt{M_1^2 - 1}}\left[\overline{\left(\frac{dh}{dx}\right)^2} + (\bar{\alpha})^2 + \overline{\alpha_c^2(x)}\right] \qquad (8.21)$$

where $h(x)$ is the half thickness of the airfoil, $\bar{\alpha}$ is the mean angle of attack and $\alpha_c(x)$ is the local angle of attack of the camber line with respect to the mean angle of attack line inclined at angle $\bar{\alpha}$, as indicated in Fig. 8-14.

These same results may be obtained from the perturbation theory of the next section.

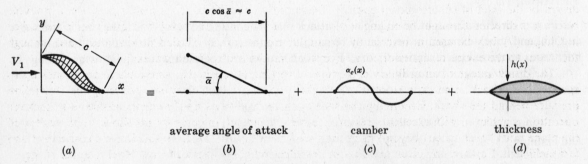

Fig. 8-14. Thin airfoil at a small angle of attack. The airfoil in
(a) may be decomposed into (b), (c) and (d).

The Plug Nozzle

The converging-diverging nozzle was described in Chapter 7. Subsequently in Section 8.4 we will discuss two-dimensional aspects of this nozzle.

Another type of nozzle which has wide application in supersonic jet engine design is the plug nozzle. The flow in this nozzle is external and the expansion is achieved through a Prandtl-Meyer expansion fan. Fig. 8-15 shows half of the cross section of a two-dimensional plug nozzle. The center

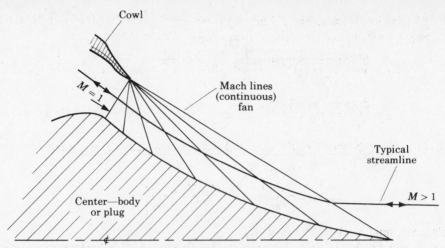

Fig. 8-15. The plug nozzle. A well-designed isentropic nozzle is reversible in flow direction for accelerating or diffusing the flow. The cross section (upper half) is shown here. A similar design may be effected for a body of revolution (axisymmetric flow).

body generates continuous expansion (or compression if the flow is reversed). The characteristics (Mach lines) may be focused by proper design to a single point. Along each Mach line the velocity vector is parallel and all properties are constant. The shape of the surface of the center body may be easily found from the properties of the Prandtl-Meyer function and continuity.

Jet Engine Inlets at Supersonic Speeds

Nozzles are used to generate supersonic flow as in a rocket nozzle and in reverse to diffuse a supersonic flow to a subsonic higher pressure flow. The latter application is of particular interest for the design of the inlet of supersonic jet engines. There the air is diffused to as high a pressure as possible before it is directed through the engine compressor. An ordinary converging-diverging nozzle (discussed in Chapter 7) may be used in reverse, the flow entering the nozzle exit and flowing backward through the throat to the engine compressor, ideally approaching its isentropic stagnation pressure.

The disadvantage of an ordinary converging-diverging nozzle is that as the plane accelerates up to speed a bow shock appears in front of the engine, creating a large drag and reducing the stagnation pressure behind the shock. Eventually the shock can be swallowed by the engine nozzle and shockless operation is achieved with excellent pressure recovery. However, in order for the shock to be swallowed the plane must "over-speed" beyond its cruise speed, then throttle back. This action is inconvenient and fuel inefficient. Further, any given nozzle operates effectively only over a narrow Mach number range.

A more effective means for diffusing the inlet air without significant loss of stagnation pressure is by means of the plug nozzle discussed in the previous section. This type of nozzle has the advantage of not requiring over-speed and is smoother over a wider Mach number range.

A cylindrical version of the plug nozzle was used in the SR-71 "Blackbird" reconnaissance airplane where high altitude sustained flight was required. For planes which must operate over a wider range of speeds a variable (in flight) nozzle is preferable. This is achievable in practice in only a two-dimensional configuration where one side of a rectangular duct is adjustable. Two or three short hinged plates or ramps are controllable to approximate the curved shape of the plug nozzle. Instead of a continuous isentropic compression, a set of discrete but weak shocks are generated. Although the weak shocks are not as efficient as a smooth curved surface in recovering the stagnation pressure, the improvement over a single normal shock is significant. Figure 8-16 shows such an inlet in cross section.

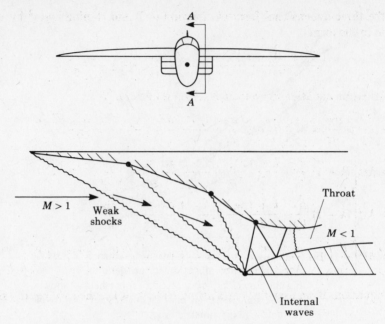

Fig. 8-16.   Supersonic adjustable ramped inlet in a modern fighter jet. Section A-A is shown.

8.3   SMALL PERTURBATIONS AND THE LINEARIZED THEORY

Perturbation Theory

If we consider homentropic flow (isentropic throughout), that is, irrotational and frictionless and such that h_0 is the same throughout, then we can use equation (8.6) as the basic governing equation. Shocks are not permitted, so we must confine ourselves to weak, virtually isentropic shocks and expansions. Physically this means that the bodies over which the flow occurs must be thin two-dimensional airfoils, or thin cylindrical bodies which limit the shock strength. Mathematically we can express this condition by saying that the velocity is everywhere close to the free stream velocity, V_0, or V_1 as we have called it in the previous section. Referring to Fig. 8-17, we express the velocity at any point in the flow as the free stream velocity V_0 plus a small perturbation velocity v'. Hence

$$
\begin{aligned}
u &= V_0 + u' & u'/V_0 &\ll 1 \\
v &= v' & v'/V_0 &\ll 1 \\
w &= w' & w'/V_0 &\ll 1
\end{aligned}
\qquad (8.22)
$$

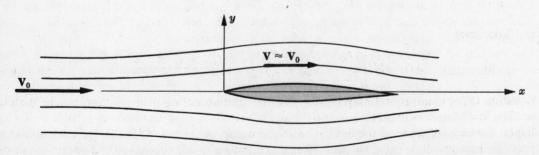

Fig. 8-17.   Thin body perturbing a uniform flow, V_0.

Substituting into the three-dimensional form of equation (8.5) and eliminating a^2 by using the perfect gas energy equation in the form

$$\frac{(V_0 + u')^2 + v'^2 + w'^2}{2} + \frac{a^2}{k-1} = \frac{V_0^2}{2} + \frac{a_0^2}{k-1} \tag{8.23}$$

we obtain the perturbation equation for velocities which may be written

$$(1 - M_0^2)\frac{\partial u'}{\partial x} + \frac{\partial v'}{\partial y} + \frac{\partial w'}{\partial z} = M_0^2\left[(k+1)\frac{u'}{V_0} + \frac{k+1}{2}\frac{u'^2}{V_0^2} + \frac{k-1}{2}\frac{v'^2 + w'^2}{V_0^2}\right]\frac{\partial u'}{\partial x}$$

$$+ M_0^2\left[(k-1)\frac{u'}{V_0} + \frac{k+1}{2}\frac{v'^2}{V_0^2} + \frac{k-1}{2}\frac{w'^2 + u'^2}{V_0^2}\right]\frac{\partial v'}{\partial y}$$

$$+ M_0^2\left[(k-1)\frac{u'}{V_0} + \frac{k+1}{2}\frac{w'^2}{V_0^2} + \frac{k-1}{2}\frac{u'^2 + v'^2}{V_0^2}\right]\frac{\partial w'}{\partial z}$$

$$+ M_0^2\left[\frac{v'}{V_0}\left(1 + \frac{u'}{V_0}\right)\left(\frac{\partial u'}{\partial y} + \frac{\partial v'}{\partial x}\right) + \frac{w'}{V_0}\left(1 + \frac{u'}{V_0}\right)\left(\frac{\partial u'}{\partial z} + \frac{\partial w'}{\partial x}\right) + \frac{v'w'}{V_0^2}\left(\frac{\partial w'}{\partial y} + \frac{\partial v'}{\partial z}\right)\right] \tag{8.24}$$

which is an exact equation. If we assume that u', v' and w' are $\ll V_0$, then we get the simplified second order equation

$$(1 - M_0^2)\frac{\partial u'}{\partial x} + \frac{\partial v'}{\partial y} + \frac{\partial w'}{\partial z} = M_0^2(k+1)\frac{u'}{V_0}\frac{\partial u'}{\partial x} + M_0^2(k-1)\frac{u'}{V_0}\left(\frac{\partial v'}{\partial y} + \frac{\partial w'}{\partial z}\right)$$

$$+ M_0^2\frac{v'}{V_0}\left(\frac{\partial u'}{\partial y} + \frac{\partial v'}{\partial x}\right) + M_0^2\frac{w'}{V_0}\left(\frac{\partial u'}{\partial z} + \frac{\partial w'}{\partial x}\right) \tag{8.25}$$

which is good for the full range of Mach numbers from subsonic through transonic to supersonic and hypersonic flow. However, a simpler linear form, good for subsonic and supersonic flow, but not for transonic flow where $M \approx 1$ nor hypersonic flow where $M > 6$, may be obtained by retaining only first order terms. It is

$$(1 - M_0^2)\frac{\partial u'}{\partial x} + \frac{\partial v'}{\partial y} + \frac{\partial w'}{\partial z} = 0 \tag{8.26}$$

Equation (8.25) may be simplified for transonic flow to

$$(1 - M_0^2)\frac{\partial u'}{\partial x} + \frac{\partial v'}{\partial y} + \frac{\partial w'}{\partial z} = M_0^2(k+1)\frac{u'}{V_0}\frac{\partial u'}{\partial x} \tag{8.27}$$

which is still good, actually better than (8.26), for subsonic and supersonic flow, but it is nonlinear. It is often convenient to express these perturbation equations in terms of a perturbation velocity potential defined as $\mathbf{v}' = -\nabla\phi$. (8.26) becomes

$$(1 - M_0^2)\frac{\partial^2 \phi}{\partial x^2} + \frac{\partial^2 \phi}{\partial y^2} + \frac{\partial^2 \phi}{\partial z^2} = 0 \tag{8.28}$$

and (8.27) becomes

$$(1 - M_0^2)\frac{\partial^2 \phi}{\partial x^2} + \frac{\partial^2 \phi}{\partial y^2} + \frac{\partial^2 \phi}{\partial z^2} = -\frac{M_0^2(k+1)}{V_0}\frac{\partial \phi}{\partial x}\frac{\partial^2 \phi}{\partial y^2} \tag{8.29}$$

Equation (8.28) is linear and may be easily solved for thin airfoils in the subsonic and supersonic regime. For low Mach numbers, $(1 - M_0^2) \approx 1$ and we have $\nabla^2\phi = 0$ which is exactly the equation used in Chapter 6 where we assumed the flow to be incompressible (and hence $M \ll 1$). We now can discuss compressible subsonic flow using equation (8.28). The form of (8.28) changes from elliptic for subsonic flow in which $M < 1$ (where the influence of the body is felt throughout the flow field) to hyperbolic for supersonic flow where $M > 1$ (where the body can exert no influence upstream and the solutions are wavelike disturbances propagating along characteristics).

The Pressure Coefficient in the Linearized Theory

The pressure coefficient defined by equation (8.17) and the pressure p may be written in terms of the components of $\mathbf{V}$ (using the energy equation) as

$$C_p = \frac{2}{kM_1^2} \{[1 + \tfrac{1}{2}(k-1)M_1^2(1 - V^2/V_0^2)]^{k/(k-1)} - 1\} \tag{8.30}$$

Inserting the perturbation velocities, expanding and retaining second order terms, we obtain

$$C_p = -\left[\frac{2u'}{V_0} + (1 - M_0^2)\frac{u'^2}{V_0^2} + \frac{v'^2 + w'^2}{V_0^2}\right] \tag{8.31}$$

For two-dimensional and planar flows it is adequate to retain only first order terms, and

$$C_p = -\frac{2u'}{V_0} \tag{8.32}$$

For flow over thin cylindrical bodies the second order term must be retained (but is $< v'$), and

$$C_p = -\frac{2u'}{V_0} - \frac{v'^2 + w'^2}{V_0^2} \tag{8.33}$$

Boundary Conditions

Physically, the boundary condition is that the velocity vector must be tangential to the body at the surface of the body. If the surface of the body is given by the equation

$$f(x, y, z) = 0$$

then this boundary condition is $\qquad\qquad \mathbf{V} \cdot \nabla f = 0 \tag{8.34}$

Explicitly writing out (8.34) in terms of u', v' and w' leads to the condition, accurate to a consistent order, for two-dimensional flow in the xy plane (with small y dimension)

$$v'(x, y = 0) = V_0(dy/dx)_{\text{body}} \tag{8.35}$$

For "planar" flow (for thin, essentially flat three-dimensional bodies like wings) the flow is quasi-two-dimensional, and the boundary condition on the surface becomes

$$v'(x, 0, z) = V_0(\partial y/\partial x)_{\text{body}} \tag{8.36}$$

where the body lies in the xz plane and is almost two-dimensional in the xy plane.

For flow about thin cylindrical objects, the situation is somewhat more complicated and will not be discussed here. The reader is referred to the references.

It is interesting to note that the velocity $v'(x)$ is evaluated at $y = 0$ and not actually on the surface of the body. This approximation is consistent with the thin body approximation.

Supersonic Thin Airfoil Theory

The perturbation theory just discussed may be applied to two-dimensional thin airfoils. The results are identical to the weak shock solutions discussed previously in Section 8.2.

8.4 THE METHOD OF CHARACTERISTICS

In complex flows where simple shock-expansion theory or the linearized theory is not adequate, the original nonlinear equations must be solved by numerical techniques. The method of characteristics allows a simplification of these equations in the supersonic range and provides a well-established mathematical technique readily adaptable to numerical analysis.

Elliptic and Hyperbolic Equations

The equations describing subsonic flow are elliptic and those describing supersonic flow are hyperbolic, and entirely different methods of numerical analyses must be used for the two. In subsonic flow, any disturbance can be felt throughout the entire flow field and the method of relaxation must be used to solve the elliptic equations. Here the boundary conditions must be prescribed along a boundary completely enclosing the domain of interest and the entire flow field must be viewed simultaneously. However, in supersonic flow (with hyperbolic equations), disturbances and information can propagate only downstream, and boundary conditions must be specified only along a curve located upstream. Then one can "march" downstream, calculating values along the way, without the necessity of iterating back over the flow field in order to take account of a downstream condition as in subsonic flow.

A hyperbolic equation is one that gives rise to characteristics, but we will not give an exact mathematical definition here. Some differences between elliptic and hyperbolic differential equations are tabulated below, considering ϕ as the dependent variable.

Elliptic equation	Hyperbolic equation
(1) Either ϕ or $\partial\phi/\partial n$ must be prescribed on a closed boundary to avoid singularities.	(1) ϕ or $\partial\phi/\partial n$ may be specified on an open boundary. Singularities may result if boundary conditions are prescribed on a closed boundary.
(2) A change in boundary conditions affects the whole region of flow.	(2) A change in boundary condition affects only a limited domain of the flow.
(3) The solution must be analytic.	(3) Solution need not be analytic (shocks are singularities).

Characteristics

We will not discuss the general mathematical theory of characteristic and hyperbolic differential equations here. Rather, we will present a simple physical discussion of characteristics as they appear in two-dimensional gasdynamics and briefly outline the numerical method of calculation.

We begin by considering steady two-dimensional homentropic flow. We cannot describe the structure of shock waves with the equations of such flow since shock waves involve irreversible processes. However, shocks may be admissible solutions (if the boundary conditions are of a certain type) since shocks correspond to singularities in the solution to the hyperbolic system. The flow is then homentropic within domains which do not contain shocks or which are bounded by shocks. If shocks occur, their location may necessitate some trial and error calculations. For example, in a converging-diverging nozzle under critical operation, if the inlet pressure and outlet pressure are both specified, a shock wave may occur downstream of the throat. In this section downstream of the throat the flow is supersonic and the method characteristics may be used to map the flow. However, if we were to simply march downstream to the exit, the final pressure may not be that which was specified. Only by admitting a shock in the nozzle could the pressure at the outlet be made to correspond to the prescribed value. Hence, whenever the boundary conditions are overspecified in a hyperbolic system singularities (shocks) may result. (This overspecification may be a physical one, of course, and not merely a mathematical one.) Usually the method of characteristics is most useful where shocks do not occur, and only the boundary conditions on a curve upstream are specified. Then one can simply "march" downstream, calculating along the way, without any iteration or trial and error.

Equation (8.28) in terms of ϕ,

$$(1 - M_0^2) \frac{\partial^2 \phi}{\partial x^2} + \frac{\partial^2 \phi}{\partial y^2} + \frac{\partial^2 \phi}{\partial z^2} = 0$$

is elliptic if $M_0 < 1$ and hyperbolic if $M_0 > 1$. However, we are generally not interested in applying the method of characteristics to this simple linearized equation. Rather, the complete nonlinear equations are of concern. Consider the general equations (8.5):

$$(u^2 - a^2) \frac{\partial u}{\partial x} + (v^2 - a^2) \frac{\partial v}{\partial y} + uv \left(\frac{\partial u}{\partial y} + \frac{\partial v}{\partial x} \right) = 0$$

$$\frac{\partial v}{\partial y} - \frac{\partial u}{\partial x} = 0$$

We ask the question: under what condition are these equations reducible to a form which has a one-dimensional variation (at any point in space)? That is, the variation is only normal to a characteristic, and not along it. We look for a condition such that a combination of the above equations may be put into the form

$$A \frac{\partial u}{\partial \alpha} + B \frac{\partial v}{\partial \alpha} = 0 \qquad (8.37)$$

where $\frac{\partial}{\partial \alpha} = \cos \chi \frac{\partial}{\partial x} + \sin \chi \frac{\partial}{\partial y}$. Let $\tan \chi = \zeta$. By multiplying the first of equation (8.5) by λ_1, and the second by λ_2, and adding we can obtain the form of (8.37) if certain conditions are satisfied by λ_1 and λ_2. These are

$$\frac{\lambda_2 - \lambda_1 uv}{\lambda_1 (a^2 - u^2)} = \zeta, \qquad \frac{\lambda_1 (v^2 - a^2)}{(\lambda_2 - \lambda_1 uv)} = \zeta$$

Eliminating λ_1 and λ_2 we obtain

$$(a^2 - u^2) \zeta^2 - 2uv\zeta + (a^2 - v^2) = 0 \qquad (8.38)$$

The direction α is defined by ζ which exists only if

$$(u^2 + v^2) > a^2; \qquad M > 1$$

that is, the flow is supersonic with $M > 1$. The directions α defined by two roots of (8.38) are the $+$ and $-$ characteristics, which we denote as ζ_+ and ζ_-. It can be shown that

$$\zeta_+ \zeta_- = (a^2 - v^2)/(a^2 - u^2)$$

Then by finding λ_1 and λ_2 explicitly, it is possible to show that if we take $dy = \zeta_+ \, dx$ then $du = -\zeta_- \, dv$, and if we take $dy = -\zeta_- \, dx$ then $du = -\zeta_+ \, dv$. Hence we arrive at the relationships, equivalent to the original differential equation (8.5), that:

$$\text{on } dy = \zeta_+ \, dx: \qquad du = -\zeta_- \, dv$$

$$\text{on } dy = -\zeta_- \, dx: \qquad du = -\zeta_+ \, dv \qquad (8.39)$$

However, it is more convenient to work in polar coordinates. We take $u = V \cos \theta$ and $v = V \sin \theta$, Fig. 8-18, so that

$$du = \cos \theta \, dV - V \sin \theta \, d\theta \qquad \text{and} \qquad dv = \sin \theta \, dv = \sin \theta \, dV + V \cos \theta \, d\theta$$

Substituting into (8.39) gives on the $(+)$ and $(-)$ characteristics respectively

$$\frac{\cot \mu}{V} \, dV = +d\theta, \qquad \frac{\cot \mu}{V} \, dV = -d\theta \qquad (8.40)$$

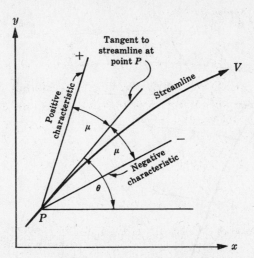

Fig. 8-18.   Polar coordinate system and characteristics.

and $\zeta_+ = \tan(\theta + \mu)$, $\zeta_- = \tan(\theta - \mu)$. Then the energy equation may be used to express V in terms of M and

$$\pm d\theta = \frac{\sqrt{M^2 - 1}\ dM}{M[1 + \frac{1}{2}(k - 1)M^2]} = dv(M)$$

which may be integrated to give $v(M)$. Hence since $\pm d\theta - dv = 0$, $(v - \theta)$ is constant on a (+) characteristic, and $(v + \theta)$ is constant on a (−) characteristic. Then if v and θ are both specified on a line L-M, we can construct a grid as we work downstream determining v and θ at points of intersection. Once v and θ are known, the Mach number, flow direction and velocity may be found. Referring to Fig. 8-19(a), let v and θ be specified along line segment L-M. Then the direction of the characteristics are known and at a set of arbitrary points $A_1, \ldots, A_n$. Let the (+) and (−) characteristics be drawn to the first set of intersections $B_1, \ldots, B_{n-1}$. At each of these points, $(v - \theta)$ and $(v + \theta)$ are known and hence v and θ can be found explicitly and then μ, M, V and the new direction of the $\pm$ characteristic may be drawn. This new set of lines intersect at $C_1, \ldots, C_{n-2}$. The process is continued throughout the range of influence which is bounded by the characteristics through A_1 and A_n. In general, in nonsimple flow regions the characteristics will be curved but the segments between intersections are drawn as straight lines. The finer the mesh, the more exact the solution.

To illustrate in more detail, consider an enlarged diagram, Fig. 8-19(b), of a section of Fig. 8-19(a). At point B_1,

$$v_3 = \tfrac{1}{2}(v_1 + v_2) + \tfrac{1}{2}(\theta_1 - \theta_2)$$

and

$$\theta_3 = \tfrac{1}{2}(v_1 - v_2) + \tfrac{1}{2}(\theta_1 + \theta_2)$$

Between mesh points the characteristic lines are assumed to be straight.

The complete theory of characteristics for compressible flow is beyond the scope of this discussion and the reader is referred to the references.

The Method of Weak Waves or Region to Region Method

In two-dimensional flow a simple computational method may be used which is essentially equivalent to the method of characteristics but is somewhat simpler and more easily understood from a physical point of view. A characteristic may be drawn at any point in a supersonic flow, and there is a continuum or characteristics in the flow. On the other hand, the method of weak waves is based on actual expansion or compression waves. For flow over a solid wall or surface, characteristics may be drawn from any point. However, finite waves may be generated at discrete points where straight seg-

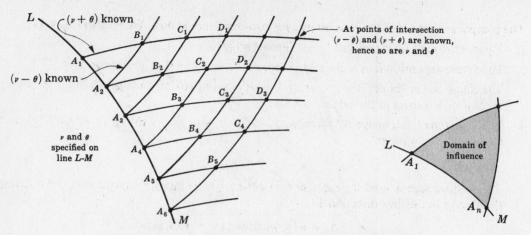

Fig. 8-19(a).   Characteristics construction in polar coordinates.

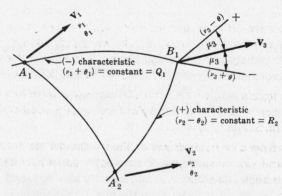

Fig. 8-19(b).   Details of the computational method.

ments form an apex. If the wall is convex and is approximated by a series of straight segments, expansion fans form at each apex. If the wall is concave (with small angles), weak shocks are formed at each internal corner (Fig. 8-20).

In the method of weak waves each expansion or weak shock is replaced by a single isentropic wave which is equivalent (in terms of velocity vector and property changes) to the fan or weak shock.

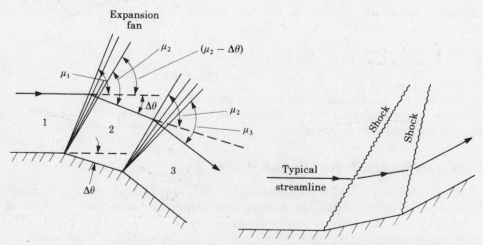

Fig. 8-20.   The formation of expansion waves and weak shocks
at corners.

The computational technique is as follows for two-dimensional flow. We refer to Fig. 8-21.

1. A curved surface (a) is replaced by straight segments (b).

2. The actual expansion fans and shocks of (b) are replaced by the single waves (c).

3. The single waves "bisect" the fan which is delineated by the Mach lines which are inclined at an angle μ with respect to the velocity vectors.

4. The flow direction change $\Delta\theta$ through a wave is related to the change in Prandtl-Meyer function across the wave as

$$\Delta v = \pm |\Delta\theta|$$

The positive sign is used for expansion and the negative sign for compression. $\Delta\theta$ is defined as the change in the flow direction, i.e.,

$$\Delta v = v \text{ (downstream)} - v \text{ (upstream)}$$

A wave emanating from a convex corner is always expansive and from a convex concave, compressive.

5. The strength of the wave $\Delta\theta$ is unchanged along a given wave even when the wave intersects and passes through another wave. However, the wave may change direction (its orientation). Therefore, $\Delta\theta$ and Δv remain constant along a wave. The actual values of v on either side of the wave may change across intersections, of course, but Δv and $\Delta\theta$ are constant along a wave.

6. A wave that reflects from a wall changes from an expansion wave to a compression wave (of the same strength) and vice versa. Hence the angle of incidence α_i must equal the reflection angle α_r, referring to Fig. 8-21a. Similarly $\beta_i = \beta_r$.

7. A wave that reflects from a contact surface or slip stream changes from an expansion wave to a compression wave and vice versa. Again, the strength remains the same and the angle of incidence must spiral the angle of reflection.

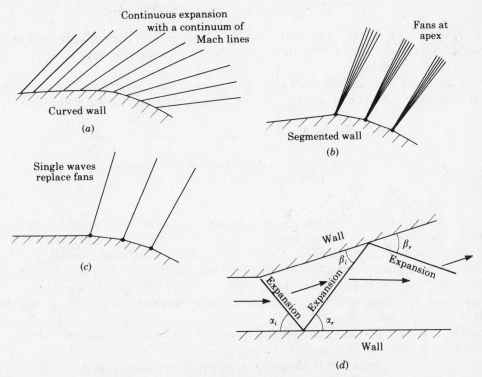

Fig. 8-21. The weak waves approximation for curved surfaces.

If the point of incidence of the wave on the wall is an apex which accommodates the flow turn of $\Delta\theta$ it is possible for the wave to cancel and not reflect since the flow is already parallel to the wall (Fig. 8-8d).

As an illustration of this method we will design a simple two-dimensional supersonic nozzle. In Chapter 7 we discussed nozzle flow, but the only physical parameter of importance was the area ratio without regard to length. We will see now that the actual shape of the nozzle is crucial in order to provide a smooth uniform flow exiting the nozzle and the nozzle must indeed be designed in two dimensions. We confine ourselves to two-dimensional flow here as in the flow between two parallel curved surfaces, not axisymmetric flow. Flow in circular nozzles may be treated by similar methods but is beyond our scope here. The two-dimensional aspect of nozzle design is vital in the design of rocket nozzles for maximum thrust and in the design of wind tunnels for smooth flow.

In general a nozzle should be designed with smooth curved walls. We approximate the curved wall by short straight segments. Further, the nozzle must not be so long that the boundary layer grows to significant thickness, and the walls must not diverge so fast that the boundary layer separates. Hence there is a range of acceptable nozzle designs of various lengths, for a given exit Mach number M_e and corresponding exit area to throat area ratio (A_e/A^*). The more straight segments used (and the smaller the $\Delta\theta$ for each step) the more accurate the method.

The simplest nozzle is a single straight diverging wall, Fig. 8-22. Only one set of intersecting (expansion) waves is needed. For $M_e = 1.435$, we see that $v_e = 10°$ and at the inlet $v_1 = 0$. The $\Delta\theta$ across each wave is the same and the fluid crosses two waves. Hence, $|\Delta\theta|$ across each wave is $5°$, $\mu_1 = 90°$, $v_2 = 5°$, $M_2 = 1.256$, $\mu_2 = 52.738°$, $v_3 = 10°$, $M_3 = 1.435$, $\mu_3 = 51.642°$. An inspection of Fig. 8-22 and a little geometry allows complete specification of the nozzle using either a graphical or simple analytical calculation. The ratio A_e/A^* may be found this way and compared to the value from the supersonic tables of 0.881. The smaller the angle $\Delta\theta$, the more accurate the results.

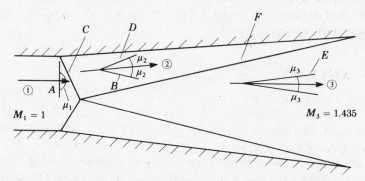

Fig. 8-22. A simple two-dimensional nozzle illustrating the method of weak waves. The orientation of C is midway between A and B. The orientation of F is midway between D and E.

A better design for larger Mach numbers requires the use of an arbitrary number of straight segments as shown in Fig. 8-23. We design for an exit Mach number M_e with a corresponding total $\Delta\theta$. $\Delta\theta$ is simply $v_3 - v_1$ (v_1 is zero since $M_1 = 1$). For the example shown there is an initial expansion region about two short segments of A and B of chosen arbitrary length, then final expansion through waves of the opposite family. Note in this example design we chose two segments A and B, and there are six expansions of $\Delta\theta/6$ each, with the flow finally realigned parallel to the incoming flow. Note that the lengths of the segments C, D, and E and their orientations are not arbitrary but are determined once A, B, and $\Delta\theta$ are chosen. The waves are canceled by proper orientation of segments C, D, and E, which are each rotated by $\Delta\theta/6$ with respect to the previous segment.

In general for a given exit Mach number M_e there corresponds a total $\Delta\theta$ which may be accomplished by expansion through n waves, which must be an even number if all waves are canceled by the

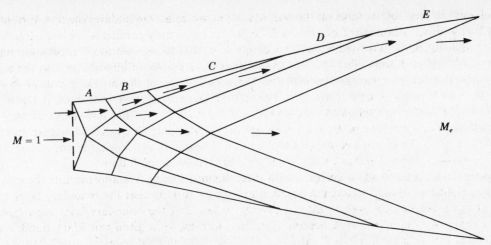

Fig. 8-23. A nozzle design using the weak wave technique.

wall. The number of initial wall segments to be chosen is $(n/2 - 1)$ and each may be successively rotated $\Delta\theta/n$. Their length is arbitrary but must be short enough to avoid reflection of the opposite family of waves. In the limit of zero length the design reduces to that shown in Fig. 8-22.

It is also possible to lengthen the nozzle by allowing reflections from the wall (by reducing the rotation angle of the segments). Then the waves reflect, interact in another non-simple region, and finally cancel at the wall farther downstream.

Each trapezoidal region bounded by the waves has different fluid properties. The waves change direction through the intersections, and their locations are determined by using the rule that the wave bisects the Mach lines on each side of the wave.

8.5 SUPERSONIC AIRCRAFT

In actual practice, the wave drag on a wing flying supersonically, with shock-expansion wave formation, is much greater than the drag on a subsonic type wing. Consequently, supersonic aircraft are made so that the wings actually operate subsonically. This type of operation is accomplished by making the sweepback angle greater than the Mach angle. Then the component of velocity normal to the leading edge is subsonic and the wing behaves essentially as a subsonic airfoil. However, the fuselage generates a nose shock but the wave drag is minimized by tapering the nose to a needle point to weaken

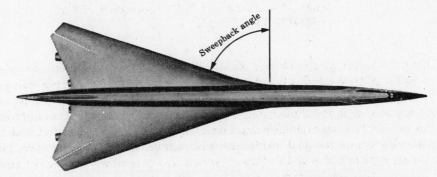

Fig. 8-24. A modern supersonic airplane. The sweepback angle is greater than the Mach angle so that the normal component of velocity on the leading edge of the wing is subsonic.

the shock, and by properly varying the cross sectional area of the fuselage (so that it looks something like a "coke" bottle) by the so called "Area Rule."

The wings are three-dimensional bodies and must be treated as such in the computation of flow over them. The details are beyond the scope of this discussion, but it is important to remember that the sweepback causes the wing to behave essentially as a subsonic airfoil in terms of lifting characteristics. One effect of increasing the sweepback is to stabilize the plane in a manner similar to the effect of a positive dihedral angle. In fact in highly sweptback wings the dihedral may be negative (giving a drooping appearance).

References

1. Anderson, John D., Jr., *Fundamentals of Aerodynamics*, McGraw-Hill, 1984.

2. Anderson, John D., Jr., *Introduction to Flight*, 2nd ed., McGraw-Hill, 1985.

3. Anderson, John D., Jr., *Modern Compressible Flow: With Historical Perspective*, McGraw-Hill, 1982.

4. Howarth, L. (Editor), *Modern Developments in Fluid Dynamics, High Speed Flow*, Vols. 1 and 2, Oxford University Press, 1956.

5. Jones, R. T., and Cohen, D., *High Speed Wing Theory*, Princeton University Press, 1960.

6. Kuethe, A. M., and Chow, C., *Foundations of Aerodynamics*, 4th ed., John Wiley, 1986.

7. Liepmann, H. W., and Roshko, A., *Elements of Gasdynamics*, John Wiley, 1957.

8. Owczarek, J. A., *Fundamentals of Gas Dynamics*, International Textbook, 1964.

9. Shapiro, A. H., *The Dynamics and Thermodynamics of Compressible Fluid Flow*, Vols. 1 and 2, Ronald Press, 1952.

10. Zucrow, M. J., and Hoffman, J. D., *Gas Dynamics*, Vol. I and II, John Wiley, 1976.

Solved Problems

8.1.   Air at 520°R and atmospheric pressure passes through an oblique shock wave inclined at 50° to the flow as shown in Fig. 8-25. The initial Mach number is 2. Determine the conditions downstream and the angle through which the flow is turned.

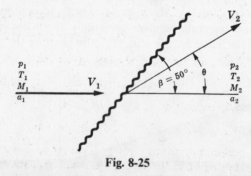

Fig. 8-25

From Fig. 8-3, we find that θ is approximately 18°. From equations (8.7), we can find downstream conditions (at point 2) in terms of the upstream conditions (at point 1). Alternatively, we could use normal shock tables with the numerical value of M_1 in the table interpreted as our value of $M_1 \sin \beta$, and the M_2 in the table as $M_2 \sin (\beta - \theta)$. We find then:

$p_2/p_1 = 2.57;$ $\qquad\qquad\qquad\qquad$ $p_2 = 37.8$ psia

$T_2/T_1 = 1.34;$ $\qquad\qquad\qquad\qquad$ $T_2 = 697$°R

$M_2 \sin (\beta - \theta) = 0.69;$ $\qquad\qquad\quad$ $M_2 = 1.3$

8.2.   A thin flat plate airfoil is shown in Fig. 8-26 inclined at an angle of attack of $10°$. Find the fluid properties in all regions of the flow. $M_1 = 3$ and the free stream air is at $T_1 = 500°R$ and $p_1 = 15$ psia.

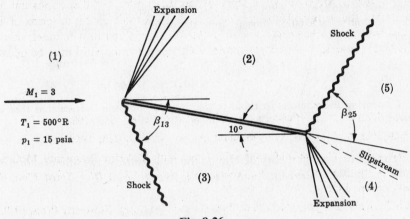

Fig. 8-26

Using equations (*8.7*) and Fig. 8-3, we can find conditions in region 3. We find $\beta_{13} = 27.5°$. Using $k = c_p/c_v = 1.4$ for air, we obtain

$$p_3/p_1 = 2.07, \quad p_3 = 31.0 \text{ psia}; \qquad M_3 = 2.49; \qquad T_3 = 1.25T_1 = 625°R$$

To find conditions in region 4 we use

$$v_4 = v_3 + |\theta_4 - \theta_3| = v_3 + 10° = 39.1° + 10° = 49.1°$$

and $M_4 = 2.97$ from Table 8.1. Since the expansion is isentropic, we write

$$p_4/p_3 = \left[\frac{2 + (k - 1)M_3^2}{2 + (k - 1)M_4^2}\right]^{k/(k-1)}$$

from which we find $p_4/p_3 = 0.48$ and $p_4 = 14.9$ psia. T_4 is found from

$$T_4/T_3 = (p_4/p_3)^{(k-1)/k}$$

and hence $T_4 = 0.81(T_3) = 0.81(1.25T_1) = 1.01T_1$.

Now proceeding along the upper surface of the plate, we find conditions in region 2.

$$v = v_1 + 10° = 49.8° + 10° = 59.8°$$

and hence $M_2 = 3.57$. Using the isentropic relationships we find

$$p_2 = 0.44p_1 = 6.60 \text{ psia}, \qquad T_2 = 0.79T_1$$

Then through the shock to region 5 we find

$$\beta_{25} = 24°, \qquad M_5 = 2.98, \qquad p_5 = 2.29p_2 = 1.01p_1, \qquad T_5 = 1.29T_2 = 1.02T_1$$

We see that the pressures p_5 and p_4 are approximately but not exactly equal, and the temperatures T_5 and T_4 are approximately equal. In actuality a slipstream or contact surface must extend from the rear of the airfoil separating regions of unlike temperature and density but the pressures p_4 and p_5 must be equal. Further, across this slipstream the velocities are different and a shear layer exists. It is a trial and error process to locate the exact inclination of the slipstream in order to make $p_4 = p_5 = p_1$.

The lift and drag per unit wing area are

$$L = (p_3 - p_2) \cos 10° = 24.0 \text{ psi}, \qquad D = (p_3 - p_2) \sin 10° = 4.23 \text{ psi}$$

Note that since p_5 and p_1 are almost equal the contact surface comes off the airfoil nearly parallel to the free stream. In general, in order to insure that $p_5 = p_1$, as it must in reality, the contact surface may be inclined at a slight angle to the free stream.

8.3.   Compute the lift and drag on the airfoil in the above problem, using thin wing theory and compare to the exact shock-expansion solution.

From equation (8.19), we find C_L and C_D. $C_L = 0.247$ and $C_D = 0.043$. (Remember here, $M_1 = 3$ and α is measured in radians; $10° = 0.1745$ radians.) The lift per unit area is $\frac{1}{2}\rho_1 V_1^2 C_L$ or $\frac{1}{2}kM_1^2 p_1 C_L$. Knowing T_1 and p_1 we find ρ_1 as 0.00253 slugs/ft^3 from the perfect gas law, and $V_1 = a_1 M_1 = (1090 \text{ ft/sec})(3) = 3270$ ft/sec.

$$L = \tfrac{1}{2}\rho_1 V_1^2 C_L = \tfrac{1}{2}(0.00253)(3270)^2(0.247) = 3340 \text{ psf} = 23.2 \text{ psi}$$

and
$$D = \tfrac{1}{2}\rho_1 V_1^2 C_D = 4.04 \text{ psi}$$

Comparing these results to those of the previous problem, we see that they are off by less than 5 percent.

8.4.   Using shock expansion theory determine the flow over the two-dimensional double wedge profile shown in Fig. 8-27. The Mach number of the free stream is 2.

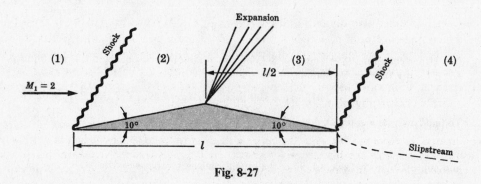

Fig. 8-27

A contact surface will extend from the trailing edge. The exact angle this contact surface makes with the free stream must be determined by imposing the condition that the pressure at 4 is the same as the pressure at 1. This would mean a slight turning of the free stream flow on the bottom, at the trailing edge. We neglect this effect here for the moment and assume the bottom flow to be undisturbed.

For flow through the leading shock, 1-2, $\theta = 10°$, and from Fig. 8-3, $\beta = 39.5°$. We find from shock tables or equations (8.7),

$$p_2/p_1 = 1.72, \qquad T_2/T_1 = 1.17, \qquad M_2 = 1.63$$

Then through the expansion fan 2-3 we find

$$v_2 = 15.9°, \qquad v_3 = v_2 + 20° = 35.9°, \qquad M_3 = 2.37$$

$$p_3/p_2 = \left[\frac{2 + (k-1)M_2^2}{2 + (k-1)M_3^2}\right]^{k/(k-1)} = 0.32, \qquad T_3/T_2 = (p_3/p_2)^{(k-1)/k} = 0.72$$

Through the shock 3-4, $\theta = 10°$ and $\beta = 34°$, and

$$p_4/p_3 = 1.88, \qquad T_4/T_3 = 1.21, \qquad M_4 = 1.90$$

Clearly then, $p_4 = 1.03p_1$, and the flow cannot leave the airfoil parallel to the free stream flow since p_4 would not then be the same as p_1. Hence the shock 3-4 must take place through slightly less than a $10°$ deflection angle and the flow on the bottom is slightly compressed through a shock. The determination of the exact angle of the contact surface may be done by trial and error, such that $p_4 = p_1$. The details are left as an exercise for the reader.

8.5.   A shock wave impinging on a wall shown in Fig. 8-28 will be reflected so that the final flow is parallel to the wall. Consider a shock reflecting as shown.

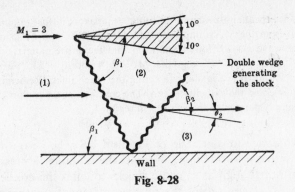

Fig. 8-28

For $\theta = 10°$ and $M_1 = 3$, we find $\beta_1 = 28°$ and $M_2 = 2.5$ from (8.7) or appropriate tables. Then for $\theta_2 = 10°$ and $M_2 = 2.5$ we find $\beta_2 = 32°$. Remember, if we use standard normal shock tables we insert $M_1 \sin \beta_1$ and $M_2 \sin (\beta_1 - \theta)$ for M_1 and M_2 in the tables respectively.

8.6. A well-designed two-dimensional nozzle for an exit Mach number $M_e = 2$ must be exhausted to a back pressure p_b of $p_b = p_e = 0.1278\, p_0$ (from the supersonic tables) to achieve smooth shock-less flow. p_0 is the stagnation pressure which is the pressure in the reservoir feeding the nozzle. The exit pressure p_e is then equal to the back pressure p_b. If the back pressure is lowered below p_e expansion waves occur outside the nozzle. Show a diagram of these expansion waves and indicate critical angles for a back pressure of $0.0550\, p_0$.

Referring to Fig. 8-29, as the flow leaves the nozzle it turns through an expansion fan to accommodate the back pressure p_b. We show the fan as a single wave and use the method of weak waves. Expansion waves reflect from the slip stream (or contact surface) as a compression wave and vice versa. The flow proceeds through an expansion wave to pressure p_b (region 1), through another expansion wave that realigns the flow but expands to a pressure less than p_b (region 2), through a compression wave back to pressure p_b (region 3), then is realigned and compressed again to pressure p_e (region 4). The entire pattern is repeated so that downstream of the nozzle a diamond pattern is established. The repetition continues until the waves diffuse and finally dissipate due to friction which may occur only after many repetitions.

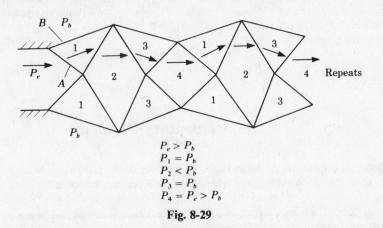

$$P_e > P_b$$
$$P_1 = P_b$$
$$P_2 < P_b$$
$$P_3 = P_b$$
$$P_4 = P_e > P_b$$

Fig. 8-29

Critical angles are shown on the figure using the method of weak waves. The value of ν_e at the exit is $26.5°$. For expansion down to $0.0550\, p_0$ the Mach number in region 1 is 2.54. Then for $M_1 = 2.54$, the corresponding ν_1 is $40°$. Hence the flow turn angle at the exit corner is $(40° - 26.5°) = 13.5°$. In region 2 an additional expansion turn angle of $13.5°$ is required which gives $\nu_2 = 53.5°$ and $M_2 = 3.202$ and from the isentropic tables $p_2/p_0 = 0.02$. Then compression through $13.5°$ occurs from region 2 to region 3. There $\nu_3 = 40°$ and $M_3 = M_1 = 2.54$. One additional compression completes the rotation back to region 4 where $\nu_4 = 26.5°$, and $M_4 = M_e = 2$ and $p_4 = p_e = 0.1278\, p_0$.

The entire cycle repeats until the waves broaden, weaken, diffuse, and dissipate. The Mach angles can all be easily found from the Mach angle Prandtl-Meyer table, and by using the bisection rule the actual orientation of each wave may be found and from that the entire pattern constructed. For example, in the exit region $M_e = 2$ and $\mu = 29.9°$. In region 1, $M_1 = 2.54$ and $\mu_1 = 23.2°$. The wave between region e and 1 (line A) is oriented as shown in Fig. 8-29, bisecting the Mach lines. (Remember the Mach angle is the angle between the Mach line and the velocity vector.) The envelope of the region 1 (which is a slip-stream contact surface, line B) is simply parallel to the velocity in region 1.

8.7. Consider the two-dimensional nozzle of Problem 8.6. The design Mach number is 2 and the design back pressure is $p_e = 0.1278\ p_0$. Now let the actual back pressure be a bit higher than the design value (but not so high as to cause shocks to enter the nozzle). Now a pattern similar to that in Problem 8.6 occurs except the first wave from the corner is compressive. If the back pressure is set at $0.20\ p_0$ find the wave pattern external to the nozzle and the critical angles and properties in the flow. Use weak wave theory.

Referring to Fig. 8-30, the initial wave (with p_e and $v_e = 26.5°$) is compressive to region 1, where $p_1 = p_B = 0.20\ p_0$, $M_1 = 1.74$, and $v_1 = 19°$. The angle of flow turn is $(26.5° - 19°) = 7.5°$. In region 2, additional compression through a turn of $7.5°$ occurs to give $M_2 = 1.49$, $v_2 = 11.5°$, $p_2 = 0.276\ p_0$. Then an expansion occurs to region 3 of $7.5°$ to give $v_3 = 19°$, $M_3 = 1.74$, and $p_3 = 0.20\ p_0$. Finally an additional expansion through $7.5°$ occurs to region 4 which again is the same as the exit conditions, $M_e = M_4 = 2.0$, $v_e = v_4 = 26.5°$ and $p_e = 0.1278\ p_0$. Again the pattern repeats itself similar to that of Problem 8.6.

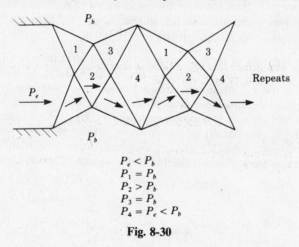

$$P_e < P_b$$
$$P_1 = P_b$$
$$P_2 > P_b$$
$$P_3 = P_b$$
$$P_4 = P_e < P_b$$

Fig. 8-30

Supplementary Problems

8.8. Air flows over a corner as shown in Fig. 8-31. Find the air properties and Mach number downstream. Draw in the expansion fan. What happens when the angle θ becomes larger and larger?

8.9. Air flows through a corner shown in Fig. 8-32. Find the properties and Mach number downstream. What happens as θ becomes larger and larger?

Fig. 8-31   **Fig. 8-32**

8.10. Find the shock pattern of the wedge of Fig. 8-33 with standard air flowing in at Mach 5.

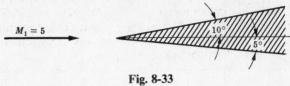

Fig. 8-33

8.11. Find the properties downstream and plot the Mach angle μ as a function of θ for the flow of Fig. 8-34.

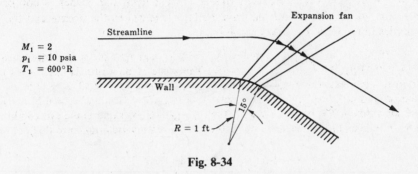

Fig. 8-34

8.12. Two-dimensional flow occurs through a bend as shown in Fig. 8-35. Locate the downstream bend so that no shock reflection obtains. What is the width D_2?

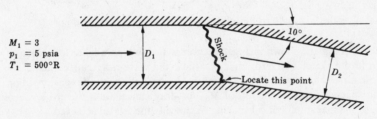

Fig. 8-35

8.13. Similar to Problem 8.12 is the expanding turn as shown in Fig. 8-36. Specify the shape of the top wall so that the expansion wave is not reflected. (*Hint.* The shape follows a streamline through the turn.) What is D_2 in terms of relevant parameters?

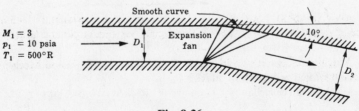

Fig. 8-36

8.14. What happens in Problem 8.5 where the shock is reflected when the angle of the first shock, β_1, becomes larger and larger? Can a simple reflection turn the flow parallel to the wall, or does something strange happen?

8.15. Flow occurs over a triangular shaped surface as shown in Fig. 8-37. For $M_1 > 1$ find the flow pattern over the wall by the linearized theory of small perturbations (assuming $\epsilon \ll \lambda$).

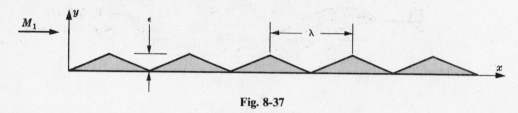

Fig. 8-37

8.16. A regular diamond shaped airfoil (Fig. 8-38) is inclined at an angle of $10°$ to a supersonic flow of Mach 2. The wedge angle of the airfoil is $20°$ so that the top and bottom surfaces are horizontal.

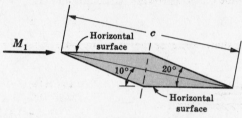

Fig. 8-38

Compute the lift on the airfoil by shock expansion theory and sketch in the shocks and waves. What is the wave drag?

Compute the lift by thin airfoil linearized theory and compare the results to those obtained by the more exact theory.

What does the flow pattern look like as indicated by linearized theory?

8.17. The Busemann biplane shown in Fig. 8-39 consists of two parallel airfoils, each of which is a half diamond as shown in the sketch. At zero angle of attack the wave drag can be made effectively zero by correctly spacing the airfoils for a given Mach number M_1. Assuming two-dimensional linear theory ($t \ll c$), find the maximum spacing G (as a function of M_1) for drag cancellation at zero angle of attack. (At certain sub-multiples of the maximum critical spacing G, cancellation is also achieved.)

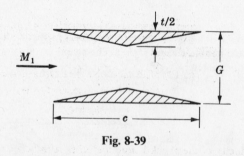

Fig. 8-39

Is there any lift at zero angle of attack?

Is there wave drag cancellation if the angle of attack is changed, keeping M_1 and G constant?

8.18 Find the lift and drag coefficient for the half diamond shaped (cross section) airfoil of Problem 8.4. Discuss the character of the contact surface in thin airfoil theory.

8.19. Using the theory of thin airfoils, find the flow pattern and the lift and drag coefficients over the two-dimensional airfoil shown in cross section in Fig. 8-40.

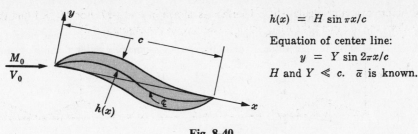

$h(x) = H \sin \pi x/c$

Equation of center line:
$$y = Y \sin 2\pi x/c$$
H and $Y \ll c$. $\bar{\alpha}$ is known.

Fig. 8-40

8.20. Using thin airfoil theory make a plot of C_L and C_D versus M for a two-dimensional thin flat plate airfoil inclined at an angle of attack of $5°$.

8.21. Make a similar plot to the one in Problem 8.20 but assume $M = 2$ and plot C_L and C_D versus the angle of attack.

8.22. Flow expands around a sharp corner as shown in Fig. 8-41. Locate the slipstream.

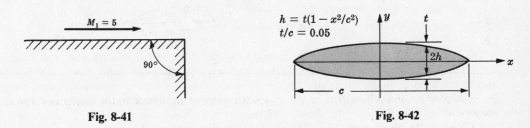

Fig. 8-41   **Fig. 8-42**

8.23. Make a plot of C_L versus C_D for a thin airfoil of the shape shown in Fig. 8-42. (Treat $\bar{\alpha}$ as an independent variable and M as a constant.) t is the maximum thickness and the airfoil is symmetrical, top and bottom.

8.24. As shown in Fig. 8-43, a piston is contained in a cylinder which has a closed end. The piston is suddenly set in motion at a constant velocity towards the closed end of the tube. A shock wave will then be initiated, travel to the end of the tube, and be reflected back toward the piston.

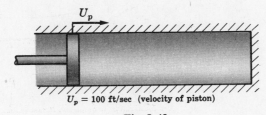

Fig. 8-43

Determine the velocity of the shock before and after reflection and the pressures in all regions of flow at all times. Qualitatively explain briefly what happens after the shock reflects off the piston face.

The air in the cylinder is initially at atmospheric pressure and at $520°R$. The piston is moved at a constant velocity of 100 ft/sec. *Hint.* Refer back to Chapter 7.

8.25. In Problems 8.1 and 8.2, find the velocity of the air in the various regions of flow.

8.26. A well-designed two-dimensional nozzle has an exit Mach number of 3. What is the necessary back pressure (in terms of stagnation pressure)? If the back pressure is reduced 10% what happens? Show a complete pattern of the waves that occur and indicate all relevant properties in each region of flow.

8.27. A straight walled two-dimensional nozzle to be designed for an exit Mach number of 2.5 is shown in Fig. 8-44. Using weak wave theory determine the divergence angle of the nozzle and nozzle length. Compare the value of A^*/A_e to the tabular value. Explain the error. Assume one wall reflection as shown in the figure.

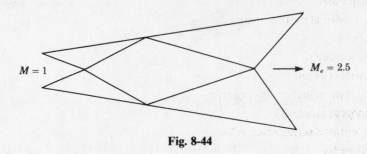

Fig. 8-44

8.28. Carry out the detail design for a nozzle as shown in Fig. 8-23 for an exit Mach number of $M_e = 1.5$. Use two straight initial expansion wall segments A and B as shown which give three initial expansion waves (and three further expansions about the three waves of the other family). The lengths A and B which may be taken the same are arbitrary. Try different lengths for A and B and notice the effect on the overall nozzle length. Compare the design A^*/A_e to the tabular value for $M_e = 1.5$.

8.29. A two-dimensional ramped inlet for a supersonic jet engine is shown in Fig. 8-45. The flight Mach number is 1.5. Find the pressure recovery, i.e., stagnation pressure, after the two oblique shocks. Further, assume a normal shock stands in the nozzle after the two oblique shocks. Find the stagnation pressure now as the air enters the compressor. Assume the normal shock occurs at an upstream Mach number equal to the Mach number of the flow just after the second oblique shock. Compare the results to the stagnation pressure which would exist after a single normal shock at the flight Mach number.

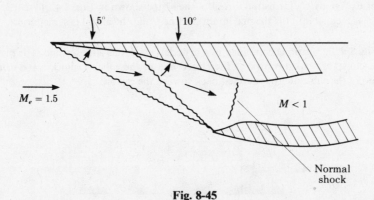

Fig. 8-45

NOMENCLATURE FOR CHAPTER 8

a = sonic speed

c = chord length (width of airfoil)

c_p = specific heat at constant pressure

c_v = specific heat at constant volume

C_D = drag coefficient

C_L = lift coefficient

C_P = pressure coefficient

D = drag

h = specific enthalpy

h_0 = total (or stagnation) specific enthalpy

k = ratio of specific heats, c_p/c_v

L = lift

M = Mach number

M_0 = Mach number of the free stream

p = pressure

s = specific entropy

u = x component of velocity

u' = x component of perturbation velocity

v = y component of velocity

v' = y component of perturbation velocity

$\mathbf{V}$ = velocity vector

V_0 = free stream velocity

w = z component of velocity

w' = z component of perturbation velocity

α = angle of attack

β = shock angle

θ = deflection angle

μ = Mach angle

ν = Prandtl-Meyer function

ρ = density

ϕ = velocity potential

Chapter 9

Incompressible Turbulent Flow

9.1 INTRODUCTION

Although we have previously considered flows which are turbulent (much of Chapter 5 was devoted to turbulent flows and many of the results of Chapters 7 and 8 are valid for turbulent flow), we will now consider the important problem of turbulent fluid motion in more depth in the present chapter.

The problems of fluid motion might be arranged in a sequence going from the least to the most difficult in the following way: potential flow → viscous laminar flow → turbulent flow. Historically, the early work in the field was primarily concerned with potential flow and laminar flow, with very little research effort devoted to the difficult problems of turbulent flow. This was unfortunate because most of the flows of engineering importance are turbulent. In recent years, however, the research effort in the area of turbulent flow has increased substantially. These problems, nevertheless, are far from being solved and will be with us for a long time.

Before we go further, let us answer two questions: What is turbulent flow? Where does it occur? Turbulent fluid motion is defined by Hinze (see reference 7) as "an irregular condition of flow in which various quantities (for example, velocity and pressure) show a random variation with time and space, so that statistically distinct average values can be discerned". One finds that this kind of motion predominates in most flows which occur in nature. When objects such as ships, automobiles, aircraft and re-entry vehicles move through fluids, the flow is nearly always turbulent. In these flows one finds that there is a fluctuating motion superimposed on the main or average flow.

Turbulence also occurs when a fluid moves through enclosures such as fans, pumps, ducts and pipes. The important criterion as to when the flow will be turbulent is the magnitude of the Reynolds number, as was previously indicated in Chapter 5.

Why does turbulent flow occur? What causes a laminar flow to become turbulent? The complete answer is not simple, but a basic physical understanding is not difficult. Essentially, small, even infinitesimal, disturbances are always present in a fluid. They may be due to small variations in properties, wall roguhness, variations in free surface effects, and any other small perturbation. Under certain conditions (usually at a low Reynolds number) these disturbances damp out and the flow remains laminar. As the Reynolds number becomes larger, infinitesimal disturbances tend to grow and the flow is said to be unstable. Because of non-linearities the final state of the fluid as the disturbances grow is often difficult to determine, and depends critically on the configuration of the flow. For some configurations, the flow will remain laminar but stabilize at a more complex flow where secondary flows and/or recirculations are present. For some configurations (and inevitably for all configurations as the Reynolds number is increased) the flow becomes turbulent, the detail structure of the turbulence depending on the geometrical configuration and the Reynolds number.

The study of the response of a fluid to disturbances or perturbations is known as stability theory and is a very important part of fluid mechanics but is rather beyond the scope of this book.

We are seeking (1) a physical understanding and (2) a quantitative description of turbulent fluid motion. The former will rely heavily (although, of course, not entirely) on experiment, while the latter will rely heavily on mathematical models. Both are important to the engineer.

There are two basically different approaches used in describing and seeking an understanding of turbulent motion. They are (1) phenomenological and (2) statistical. In the former, one establishes an expression for shear stress in terms of some empirical exchange coefficient and in the latter one studies the equations of motion in terms of time average quantities.

9.2 EQUATIONS OF MEAN VELOCITY

Equations of mean velocity distribution were obtained for turbulent flows in Section 5.2. They are summarized here for convenience. The reader may refer back to Chapter 5 for definitions of the symbols used here.

Power Law:

$$u/U = (y/\delta)^{m/(2-m)} = (y/\delta)^{1/n} \tag{5.25}$$

Here y is the distance from the wall, δ is the boundary layer thickness, U is the free stream velocity, and u is the velocity in the boundary layer. The power law applies for flow with a negligible pressure gradient over a flat plate. It also applies for fully developed flow in a pipe. The exponent varies weakly with Reynolds number.

Logarithmic Form of Law of the Wall:

$$u/u_\tau = 2.44 \ln (yu_\tau/v) + 4.9 \tag{5.28}$$

where u_τ is the friction velocity, $\sqrt{\tau_0/\rho}$, and τ_0 is the wall shear stress. The equation is valid over most of the boundary layer. It does not apply to a small region right at the wall $(0 < yu_\tau/v < 50)$ and to the outermost portion of the boundary layer. It does apply for a nonzero pressure gradient, but the larger the pressure gradient the more the region over which it applies is reduced.

Logarithmic Form of the Velocity Defect Law:

$$(U - u)/u_\tau = -2.44 \ln (y/\delta) + 2.5 \tag{5.29}$$

This equation applies to the intermediate portion of the boundary layer for zero or moderate pressure gradients. In the above equations (5.25), (5.28) and (5.29), u, as in Chapter 5, denotes the time average velocity in a steady turbulent flow, although in the remainder of this chapter we will use a bar to indicate mean values (i.e. $\bar{u}$).

9.3 STATISTICAL APPROACH

Turbulent Velocities and Averaging

Consider turbulent flow in a pipe. If we choose a particular point in the flow and observe the velocity at that point as a function of time, we obtain a random curve as shown in Fig. 9-1. We define the time average velocity $\bar{u}$ as

$$\bar{u} \equiv \frac{1}{T_1} \int_0^{T_1} u \, dt \tag{9.1}$$

where T_1 is a time large enough such that $\bar{u}$ is the same for any larger time for a steady mean flow. The bar over the quantity indicates the time average of the type indicated by equation (9.1). The instantaneous velocity u may be written in terms of a time average velocity $\bar{u}$ which we call the mean velocity and a fluctuating velocity u'. $(u = \bar{u} + u'.)$

We will have need to combine fluctuating quantities in various ways in the next section. The following set of rules will aid in these operations. If a and b are fluctuating quantities and c is a constant, we have the following set of rules (Reynolds' rules of averages):

$$\overline{a + b} = \bar{a} + \bar{b}$$

$$\overline{ca} = c\bar{a}$$

$$\overline{ab} = \bar{a}\bar{b} + \overline{a'b'}$$

$$\overline{\frac{\partial a}{\partial x}} = \frac{\partial \bar{a}}{\partial x}$$

where $a = \bar{a} + a'$, $b = \bar{b} + b'$. (The mean value of a fluctiation is zero, $\bar{a}' = \bar{b}' = 0$.)

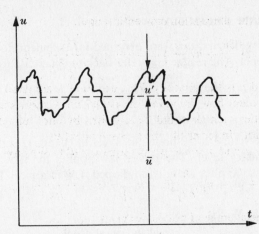

Fig. 9-1. Velocity as a function of time at a fixed point in a turbulent flow. u' is a random quantity superimposed on the mean velocity $\bar{u}$.

Equations of Motion for Turbulent Flow

We will now develop the equations of motion for turbulent flow. The method is basically the same in every case. First, the equations are written for the instantaneous quantities. Then we take the time average of both sides, noting that if the equality is valid instantaneously then it is also valid on the average for some period of time. Finally we simplify the equations such that only time average quantities appear.

Continuity Equation for Turbulent Flow

The differential form of the continuity equation as derived in Chapter 3 is (in cartesian tensor form)

$$\frac{\partial \rho}{\partial t} + \frac{\partial}{\partial x_i}(\rho u_i) = 0 \tag{3.30}$$

where ρ is the density and u_i is the ith component of velocity. This equation is valid for turbulent flow, and the dependent variables (ρ and u_i) represent instantaneous fluctuating quantities. If we take the time average of this equation, we obtain

$$\overline{\frac{\partial \rho}{\partial t} + \frac{\partial}{\partial x_i}(\rho u_i)} = 0$$

By replacing the instantaneous quantities with the time average plus the fluctuating part, and by applying Reynolds' rules of averages as given in the previous section, we obtain

$$\frac{\partial \bar{\rho}}{\partial t} + \frac{\partial}{\partial x_i}(\bar{\rho}\bar{u}_i) + \frac{\partial}{\partial x_i}\overline{(\rho' u_i')} = 0 \tag{9.2}$$

For incompressible flow, equation (9.2) becomes

$$\frac{\partial \bar{u}_i}{\partial x_i} = 0 \tag{9.3}$$

Momentum Equation for Turbulent Flow

The momentum equation in differential form was determined in Chapter 3 (equation (3.54)). We will assume that the flow is incompressible and the viscosity is constant. Then equation (3.54) becomes, in cartesian tensor notation with the summation convention used,

$$\rho\left(\frac{\partial u_i}{\partial t} + u_j \frac{\partial u_i}{\partial x_j}\right) = -\frac{\partial p}{\partial x_i} + \mu \frac{\partial^2 u_i}{\partial x_j \, \partial x_i} + B_i \qquad (9.4)$$

where B_i is the body force and μ is the viscosity. This equation is assumed to be valid for turbulent as well as laminar flow. For turbulent flow, however, the dependent variables all vary as functions of time. They are not averaged quantities such as would be indicated by pitot tube measurements. We will now proceed to establish the equations in terms of time average quantities.

Let us substitute $u_i = \bar{u}_i + u_i'$ and $p = \bar{p} + p'$ into equation (9.4). We have

$$\rho\left[\frac{\partial(\bar{u}_i + u_i')}{\partial t} + (\bar{u}_j + u_j') \frac{\partial}{\partial x_j}(\bar{u}_i + u_i')\right] = B_i - \frac{\partial(\bar{p} + p')}{\partial x_i} + \mu \frac{\partial^2(\bar{u}_i + u_i')}{\partial x_j \, \partial x_j}$$

Simplifying and taking the time average of both sides gives

$$\rho\left(\frac{\partial \bar{u}_i}{\partial t} + \bar{u}_j \frac{\partial \bar{u}_i}{\partial x_j} + \overline{u_j' \frac{\partial u_i'}{\partial x_j}}\right) = -\frac{\partial \bar{p}}{\partial x_i} + \mu \frac{\partial^2 \bar{u}_i}{\partial x_j \, \partial x_j} + \bar{B}_i \qquad (9.5)$$

The third term on the left side of equation (9.5) is usually written in a different form. From the continuity equation for incompressible flow, we have $\partial u_j'/\partial x_j = 0$. Thus

$$u_i' \frac{\partial u_j'}{\partial x_j} = 0 \qquad \text{and} \qquad \overline{u_i' \frac{\partial u_j'}{\partial x_j}} = 0$$

Then by adding $\overline{u_i' \dfrac{\partial u_j'}{\partial x_j}}$ (i.e. zero) to both sides of equation (9.5) we have

$$\overline{u_j' \frac{\partial u_i'}{\partial x_j} + u_i' \frac{\partial u_j'}{\partial x_j}} = \frac{\partial}{\partial x_j} \overline{u_i' u_j'}$$

and equation (9.5) becomes (with the turbulence term on the right side)

$$\rho\left(\frac{\partial \bar{u}_i}{\partial t} + \bar{u}_j \frac{\partial \bar{u}_i}{\partial x_j}\right) = -\frac{\partial \bar{p}}{\partial x_i} + \frac{\partial}{\partial x_j}\left(\mu \frac{\partial \bar{u}_i}{\partial x_j} - \rho \overline{u_i' u_j'}\right) + \bar{B}_i \qquad (9.6)$$

Equation (9.6) is the momentum equation for turbulent flow written in terms of time average quantities. It differs from the equation for instantaneous quantities, equation (9.4), only by the addition of the last term. This term is often referred to as the Reynolds stresses or turbulent stresses. When the turbulent flow equation is written in this form, it gives the appearance of being the same as the laminar flow equation except for a term added to the laminar stress term. Strictly speaking, this term is not a stress but is an inertia (or momentum exchange) effect (recall that it came from the left side of the equation) and is only referred to as a stress because of the way in which the laminar flow equation is modified.

Energy Equation for Turbulent Flow

Starting with the momentum equation for a constant viscosity and constant density fluid, equation (9.4), we multiply by u_i and simplify, obtaining

$$\frac{\partial}{\partial t}\left(\frac{u_i u_i}{2}\right) = -\frac{\partial}{\partial x_i}\left[u_i\left(\frac{p}{\rho} + \frac{u_j u_j}{2}\right)\right] + \nu \frac{\partial}{\partial x_j}\left[u_i\left(\frac{\partial u_i}{\partial x_j} + \frac{\partial u_j}{\partial x_i}\right)\right] - \nu\left(\frac{\partial u_i}{\partial x_j} + \frac{\partial u_j}{\partial x_i}\right)\frac{\partial u_i}{\partial x_j} \qquad (9.7)$$

Equation (9.7) has units of energy and is referred to as an energy equation. However, it is not to be confused with the first law of thermodynamics. The first law of thermodynamics is a statement of the conservation of energy of all forms. The turbulence energy equation comes from the conservation of momentum and does not involve any forms of thermal energy.

Now we will write (9.7) in terms of time average plus fluctuating quantities

$$u_i = \bar{u}_i + u_i'$$

$$p = \bar{p} + p'$$

$$u_i u_i = \bar{u}_i \bar{u}_i + 2\bar{u}_i u_i' + u_i' u_i'$$

and take time average of both sides. Then we combine (subtract) with equation (9.6) after it has been multiplied by $\bar{u}_i$ and obtain

$$\overbrace{\frac{\partial}{\partial t}\left(\frac{\overline{u_i' u_i'}}{2}\right)}^{1} + \overbrace{\frac{\partial}{\partial x_i}\left(\bar{u}_i \frac{\overline{u_j' u_j'}}{2}\right)}^{2} = -\overbrace{\frac{\partial}{\partial x_i}\overline{u_i'\left(\frac{p'}{\rho} + \frac{u_j' u_j'}{2}\right)}}^{3}$$

$$\underbrace{-\overline{u_i' u_j'}\frac{\partial \bar{u}_j}{\partial x_i}}_{4} + \underbrace{\nu \frac{\partial}{\partial x_i}\overline{u_j'\left(\frac{\partial u_i'}{\partial x_j} + \frac{\partial u_j'}{\partial x_i}\right)}}_{5} - \underbrace{\nu \overline{\left(\frac{\partial u_i'}{\partial x_j} + \frac{\partial u_j'}{\partial x_i}\right)\frac{\partial u_j'}{\partial x_i}}}_{6} \qquad (9.8)$$

Equation (9.8) is called the turbulence energy equation and each term is identifiable as a certain type of energy:

1. time rate of increase of turbulent kinetic energy.
2. convective diffusion of turbulent kinetic energy by the mean flow.
3. convective diffusion of total turbulent energy by turbulence.
4. production of turbulence (energy taken from the mean flow).
5. work done by the viscous shear stresses of the turbulent motion.
6. dissipation of turbulence by turbulent motion.

We will come back to this equation later in this chapter. In particular, we will look at the various terms for simple flows in order to gain some insight into the mechanism of turbulence.

9.4 PHENOMENOLOGICAL THEORIES

In the previous section we determined the equations for turbulent flow. The usefulness of these equations is somewhat restricted from an engineering point of view. There is no way to solve these equations even for very simple flows. Thus we must go to a more direct approach which involves models of the flow which are physically somewhat inexact but which allow approximate solutions for some flows of engineering interest.

The shear stress tensor τ_{ij} (laminar τ_{ij_l} plus turbulent τ_{ij_T}) is

$$\tau_{ij} = \mu\left(\frac{\partial \bar{u}_i}{\partial x_j} + \frac{\partial \bar{u}_j}{\partial x_i}\right) - \rho\overline{u_i' u_j'} = \tau_{ij_l} + \tau_{ij_T} \qquad (9.9)$$

$$\tau_{ij_l} = \mu\left(\frac{\partial \bar{u}_i}{\partial x_j} + \frac{\partial \bar{u}_j}{\partial x_i}\right), \qquad \tau_{ij_T} = -\rho\overline{u_i' u_j'}, \qquad \text{for } i \neq j$$

There is little hope in obtaining a solution for the Reynolds stresses. Thus the problem is to relate in some way the turbulent stresses to the mean velocity. This approach has been successful in describing free turbulent flows such as jets and wakes. All of these models are restricted to two dimensional flow.

Eddy Viscosity or Turbulent Viscosity

By writing the shear stress in terms of the laminar viscosity plus an additional term to account for the turbulent or macroscopic motion, we have (for two dimensional flow)

$$\tau_{ij} = (\mu + \rho\epsilon)\,\frac{\partial \bar{u}_i}{\partial x_j} \qquad\qquad (9.10)$$

where ϵ is referred to as the eddy viscosity and is related to the Reynolds stresses by

$$\epsilon = \frac{-1}{\partial \bar{u}_i/\partial x_j}\,\overline{u'_i u'_j}$$

The advantage of defining an eddy viscosity as in equation (9.10) is that if ϵ is a quantity which can be numerically determined (or expressed in terms of the mean velocity), then this form of the shear stress may be substituted into the momentum equation, thereby reducing the number of dependent variables. This procedure would lead to a substantial simplification of the problem. The difficulty is, however, that for most cases ϵ is different for each different flow condition and is not spatially constant for a given flow. We will consider the eddy viscosity for some actual flows later in this chapter.

Prandtl Momentum Mixing Length

Prandtl determined a length for transfer of momentum between layers of different mean velocity. This transfer process is illustrated in Fig. 9-2 and given by equation (9.11).

$$\tau = \rho l^2 \left|\frac{\partial \bar{u}}{\partial y}\right|\frac{\partial \bar{u}}{\partial y} \qquad\qquad (9.11)$$

The length l represents the distance that a fluid particle must travel in order to produce the apparent turbulent shear stress for the given velocity gradient. The momentum mixing length model of turbulence is analogous to kinetic theory where the microscopic or molecular viscosity is equal to a velocity (mean molecular velocity) times a length (mean free path). There are inconsistencies in applying the kinetic theory analogy to turbulence in that for turbulent flow, dissipation occurs, and bits of fluid do not maintain their identity. The mixing length may be applied to two dimensional flows such as pipe flow, flow over a flat plate, free jet and wake flow.

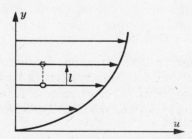

Fig. 9-2.   Illustration of Prandtl momentum mixing length.

The mixing length approach suffers from the same difficulties as the eddy viscosity approach for most flows, i.e. it is different for different flow conditions and is spatially variable for a given flow. In addition, in the derivation of the equations for mixing length the assumption is made that it is small. However, measurements have indicated that for many flows the mixing length is quite large.

Other Phenomenological Theories

There are other models of turbulent flow similar to the two previously mentioned. One of these is Taylor's vorticity transfer theory which assumes that the vorticity of each particle is constant. In this case one obtains a mixing length for vorticity in a manner similar to the momentum mixing length in Prandtl's model.

9.5　TURBULENCE CORRELATIONS

Bypassing the problem of solving the equations for the moment, let us ask the question: What would one need to know to have a complete description of a turbulent flow field? The instantaneous velocity components at every point in the flow (say $u_i(x, y, z, t)$) would yield much information. But, as we have previously noted, u_i is a random function of time and itself has limited practical utility. Also,

$\bar{u}_i$　describes mean velocity field.

$\overline{u_i'}$　does not help since it is zero.

$\overline{u_i'^2}$　gives information about the intensity of the fluctuations.

$\overline{u_i' u_j'}$　gives more information about the flow at a point (9 quantities which includes the 3 quantities $\overline{u_i'^2}$). These quantities appear in the equations of motion and are called one point double velocity correlations.

If all of these quantities were known, then the flow would still not be completely described. This is the same as saying that if we completely solved the momentum equation, we still would not have a complete description of the flow. The reason for this is that one point velocity correlations do not give information regarding the size of the turbulent eddies and the way in which energy is transferred from an eddy of one size to an eddy of a different size.

Two point velocity correlations do give information regarding the size of turbulent eddies. Here we want to establish the dependence of the velocity at one point, A, $(u_i')_A$ on the velocity at another point $(u_j')_B$, i.e.,

$$\overline{(u_i')_A (u_j')_B}$$

If there is no correlation, that is, no dependence of one velocity on the other, then this quantity is zero.

9.6　ISOTROPIC TURBULENCE

Isotropic turbulence requires that the various turbulence quantities be unaltered by rotation of the coordinate system, i.e. given quantities are the same (at a point) no matter which direction one takes in measuring this quantity. The condition of isotropy requires that the flow be homogeneous (i.e. the same at every point).

Isotropic turbulence is an idealization which can only be approximated by real flows. One example of nearly isotropic turbulence is flow with a uniform velocity downstream of a screen. Any flow having a mean velocity variation is a nonisotropic flow. Most real flows are far from isotropic and the only reason for studying isotropic flow is that to some extent it is a mathematically tractable problem which aids in understanding the more complicated nonisotropic flow. Much of the recent research has been in the area of isotropic turbulence. References 3 and 7 consider this subject in considerable depth.

9.7　WALL TURBULENCE

Wall turbulence may be described as turbulent flow which is influenced by a solid boundary. The flow is retarded by the solid boundary such that there is a mean velocity variation between the surface and the free stream. As a result of the mean velocity variation the flow must be anisotropic.

Wall turbulence is divided into two types: (1) external (boundary layer) and (2) internal (flow in pipes and ducts). In this chapter we will consider one simple example of each of these. First we will consider flow over a flat plate with zero pressure gradient and then we will consider fully developed pipe flow. While most flows are more complicated than these, they contain basically the same features and aid substantially in leading to a physical understanding of the more complicated boundary layer and channel flows.

Boundary Layer Flow Along a Flat Plate

Regions of the Turbulent Boundary Layer

We have considered methods of describing mean velocity distribution in Chapter 5. The results of the law of the wall apply especially well for the flat plate problem. For example, these results which are shown in Fig. 5-10 indicate that there are three distinct regions across the boundary layer. They are:

1. Wall Layer on Viscous Sublayer.

    This region is very small ($0 < yu_\tau/v < 5$). It is a region close to the wall where the viscous stresses are much larger than the turbulent stresses. The velocity distribution in this region is a linear function of distance from the wall, $u/u_\tau = yu_\tau/v$.

2. Wall Turbulent or Transition Region.

    This region is an intermediate region ($5 < yu_\tau/v < 30$) where viscous and turbulent stresses are of about the same magnitude.

3. Free Turbulent Region.

    This region is by far the largest of the three ($yu_\tau/v > 30$). The turbulent stresses are much larger than the viscous stresses and the mean velocity distribution is described by the logarithmic form of the law of the wall (equation (5.28)) except for the outermost part of the boundary layer.

Measurements of Turbulence Quantities in the Boundary Layer

The turbulence velocities have been measured with hot wire anemometers. As shown in Fig. 9-3, they increase rapidly from zero at the wall to a maximum value relatively close to the plate surface. Fig. 9-3 also shows that the degree of anisotropy increases in going from the boundary layer edge toward the wall. The component normal to the wall, v', has the smallest value. This is a result of the wall having a more restrictive influence in this direction.

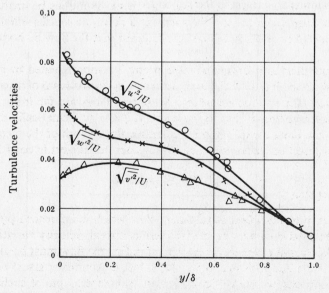

Fig. 9-3. Turbulence velocities for flow along a wall with zero pressure gradient (from reference 8).

Fig. 9-4 shows the distribution of Reynolds stress in the boundary layer. Here we note that there is a region close to the wall where the Reynolds stress is essentially constant. Also, the viscous stresses are important for only about 2% of the boundary layer.

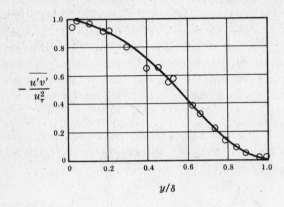
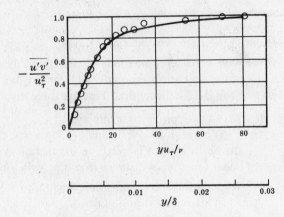

(a) Distribution of turbulent shear stress across the boundary layer.

(b) Distribution of turbulent shear stress close to the wall.

Fig. 9-4. Turbulent stress for flow over a flat plate with zero pressure gradient (from reference 8).

The energy equation for flow over a flat plate (except close to the wall where viscous stresses are important) is

$$\bar{u}\frac{\partial}{\partial x}\left(\frac{\overline{q^2}}{2}\right) + \bar{v}\frac{\partial}{\partial y}\left(\frac{\overline{q^2}}{2}\right) + \overline{u'v'}\frac{\partial \bar{u}}{\partial y} + \frac{\partial}{\partial y}\left[\overline{v'\left(\frac{p'}{\rho} + \frac{q^2}{2}\right)}\right]$$

$$-v\left[\frac{\partial^2}{\partial y^2}\left(\frac{\overline{q^2}}{2}\right) - \overline{\frac{\partial u'_j}{\partial x_i}\frac{\partial u'_j}{\partial x_i}}\right] = 0 \qquad (9.12)$$

where q^2 is the turbulent kinetic energy, $u'_i u'_i$. Fig. 9-5 shows the experimental results for the various terms of equation (9.12). Production and dissipation are seen to be the important terms over most of the boundary layer. These terms are largest in the region close to the wall.

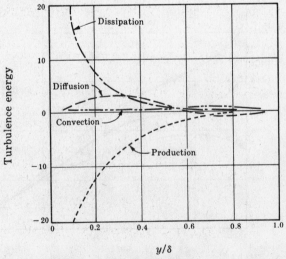

Fig. 9-5. Turbulence energy distribution in the boundary layer (from reference 7).

The preceding experimental results are given in terms of time average quantities. However, based on instantaneous hot wire anemometer measurements, the model for the boundary layer was postulated as shown in Fig. 9-6. There is a sharp, irregular boundary between the turbulent and nonturbulent flow regions. The figure shows an instantaneous picture, and the boundary moves irregularly with time between the limits $0.4 < y/\delta < 1.2$. Thus these are three distinct regions: a region close to the surface which is fully turbulent, $y/\delta < 0.4$; a region which is intermittently turbulent and nonturbulent, $0.4 < y/\delta < 1.2$; and a nonturbulent region, $y/\delta > 1.2$.

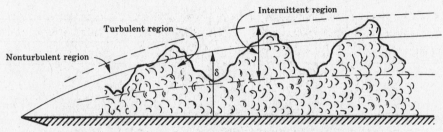

Fig. 9-6. Instantaneous picture of the turbulent boundary layer showing sharp division between turbulent and nonturbulent regions and intermittency.

Fully Developed Turbulent Flow in a Pipe

The meaning or definition of the length required, and other influencing factors regarding the attainment of fully developed flow were discussed in Chapter 5. The equations for this flow are greatly simplified since all terms (except pressure) of the form $\partial/\partial x$ are zero. The entrance length required to attain fully developed flow in terms of the mean velocity is between 20 and 100 diameters depending on entrance conditions, Reynolds number, wall roughness and inlet turbulence. The entrance length for satisfying the definition of fully developed flow in terms of the turbulence quantities is substantially longer.

Fully developed pipe flow is similar to the flat plate flow in some respects. For example, both are two dimensional flows and both have the three basic regions (viscous, wall turbulent and free turbulent) for the same range of y/δ distances from the wall.

The essential differences are that in pipe flow (1) there is no transverse (radial) mean velocity, $\bar{v}$, (2) the mean velocity in the main flow direction, $\bar{u}$, does not depend on the coordinate in that direction, (3) there is no free stream and thus no intermittency, and (4) the shear stress varies linearly with distance from the wall. Thus fully developed pipe flow is one of the simplest turbulent flows and for this reason it has received extensive experimental investigation.

The momentum equations for this case (in cylindrical coordinates) are, where u, v and w are respectively the velocities in the axial x direction, the radial direction v, and the azimuthal direction θ,

$$\frac{1}{\rho}\frac{\partial \bar{p}}{\partial x} + \frac{1}{r}\frac{d}{dr}\left(r\overline{u'v'}\right) - \nu\left(\frac{d^2\bar{u}}{dr^2} + \frac{1}{r}\frac{d\bar{u}}{dr}\right) = 0 \qquad (9.13)$$

$$\frac{1}{\rho}\frac{\partial \bar{p}}{\partial r} + \frac{1}{r}\frac{d}{dr}\left(r\overline{v'^2}\right) - \frac{\overline{w'^2}}{r} = 0 \qquad (9.14)$$

where r is the radial coordinate measured from the axis.

The energy equation is

$$\overline{u'v'}\frac{d\bar{u}}{dr} + \frac{1}{r}\frac{d}{dr}\left[r\overline{v'\left(\frac{p'}{\rho} + \frac{q^2}{2}\right)}\right] - \frac{\nu}{r}\frac{d}{dr}\left[r\frac{d}{dr}\left(\frac{\overline{q^2}}{2}\right)\right] + \nu\,\overline{\frac{\partial u_i'}{\partial x_j}\frac{\partial u_i'}{\partial x_j}} = 0 \qquad (9.15)$$

where q^2, the turbulence kinetic energy, is $(\overline{u'u'} + \overline{v'v'} + \overline{w'w'})$.

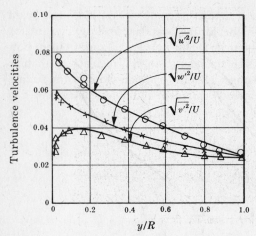

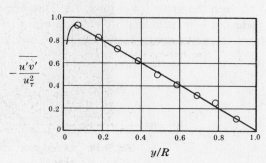

Fig. 9-7. Turbulence velocities for fully developed flow in a pipe (from reference 9). R is the pipe radius and y is the distance from the wall. U is the velocity (mean centerline velocity).

Fig. 9-8. Distribution of turbulence shear stress for fully developed flow in a pipe (from reference 9).

Fig. 9-7 shows the turbulence velocities and Fig. 9-8 the Reynolds stress distribution for fully developed flow in a pipe of radius R. Fig. 9-9 shows the distribution of the turbulence energy. y is the distance from the wall of the pipe and U is the mean centerline velocity.

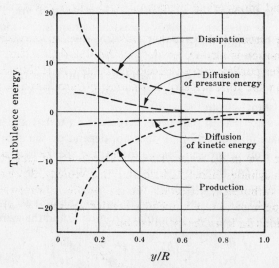

Fig. 9-9. Distribution of turbulence energy for fully developed pipe flow (from reference 9).

From these results we are able to construct a model of turbulent pipe flow. Many of the results of this model also apply to boundary layer flow. The degree of similarity between the two flows decreases with distance from the wall, mainly due to the intermittent character of the boundary layer.

The important characteristics of pipe flow are presented in Fig. 9-10. Production and dissipation are the largest terms over most of the boundary layer. The turbulent kinetic energy, production and dissipation all show a sharp maximum in the wall turbulent or transition region.

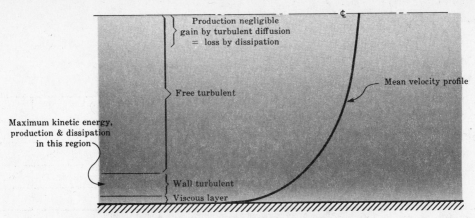

Fig. 9-10.   Description of fully developed turbulent flow in a pipe.

There is a transport of turbulent kinetic energy down the gradient toward the center of the pipe where the kinetic energy is a minimum. This energy is balanced by dissipation at the centerline.

9.8   FREE TURBULENCE

Turbulent fluid motion which is not being influenced directly by a solid boundary is called free turbulence. Examples of free turbulent flows are wakes behind objects in a moving fluid and jets emitting from a nozzle into a stagnant or slower moving fluid.

The problems of engineering interest are to determine the rate of spreading with distance in the flow direction, mean velocity and transport of momentum and energy during mixing with the surrounding fluid.

A characteristic of free turbulent flows is that viscosity effects have no controlling influence on the mean motion. The mean motion is dictated entirely by the turbulent eddies. In this sense, jet and wake flows are similar to the outer portion of the turbulent boundary layer. The only role that viscosity plays in free turbulent flows is in the final stage of dissipation of turbulent energy through the small scale eddies.

Wake Flows

Consider the turbulent flow in the wake of a circular cylinder of diameter d as shown in Fig. 9-11. The problem is: for a given cylinder diameter, fluid and fluid velocity, determine $u = u(x, y)$, $b = b(x)$ the wake half width, and the turbulence quantities. We are unable to determine a completely analytical solution, but with some experimental results useful generalizations may be established.

The momentum equations far from the cylinder become

$$U \frac{\partial \bar{u}}{\partial x} = - \frac{\partial}{\partial y} (\overline{u'v'}) \tag{9.16}$$

$$0 = \frac{\partial \bar{p}}{\partial y} + \rho \frac{\partial}{\partial y} (\overline{v'^2}) \tag{9.17}$$

where it has been assumed $\tau_T \gg \tau_l$, $\bar{u} \gg \bar{v}$, $\partial/\partial x \ll \partial/\partial y$, $\nabla \bar{p} = 0$. τ_T is the turbulent shear stress defined by equation (9.9) and τ_l is the laminar shear stress.

Even as simple as equations (9.16) and (9.17) are compared to the original equations, they are inadequate in yielding a solution. Thus the general approach in seeking solutions to jet and wake flow problems is to assume similar velocity profiles in conjunction with replacing the Reynolds stress with one of the phenomenological models such as eddy viscosity or Prandtl mixing length.

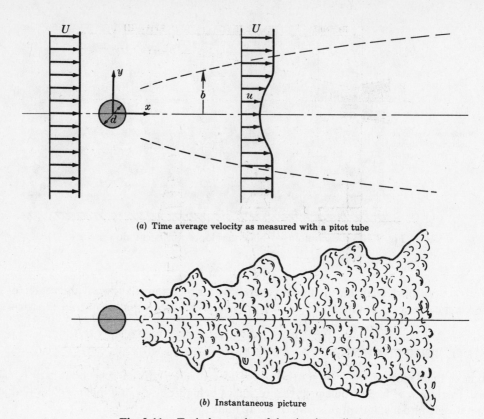

(a) Time average velocity as measured with a pitot tube

(b) Instantaneous picture

Fig. 9-11. Turbulent wake of the circular cylinder.

By a similar velocity profile we mean a velocity distribution having the form

$$\bar{u}/U = f(y/b) \qquad (9.18)$$

Schlichting (see reference 13) used the similarity argument in conjunction with the Prandtl momentum mixing length. By assuming that the mixing length is proportional to the wake width, he solved the differential equation and used the momentum integral equation to obtain

$$b = \sqrt{10}\,\beta(xC_D\,d)^{1/2} \qquad (9.19)$$

$$\frac{U - \bar{u}}{U} = \frac{\sqrt{10}}{18\beta}\,[x/C_D\,d]^{-1/2}[1 - (y/b)^{3/2}]^2 \qquad (9.20)$$

where $\beta = 0.18$, a constant determined from experimental measurements. C_D is the drag coefficient.

Equation (9.19) is in good agreement with experimental results for $x/d > 10$ and (9.20) for $x/d > 50$.

Jet Flows

Let us consider the turbulent circular jet as shown in Fig. 9-12. The flow may be separated into three regions, each having a distinctly different character. Region I consists of a potential core of uniform mean velocity U_P bounded by a shear layer which in turn is bounded by a fluid having a uniform mean velocity U_S. This region starts at the nozzle exit and has a length of 4 or 5 nozzle diameters. Region II is characterized by both the absence of a potential core and absence of similarity in the mean velocity distribution. This region is followed by region III which starts at about 8 nozzle diameters and is characterized by similar velocity distributions.

There have been a large number of investigations of turbulent jets mainly because of their engineering importance, simplicity and relevance to other turbulent flows. Two books have been published which are largely devoted to this subject (see references 1 and 11).

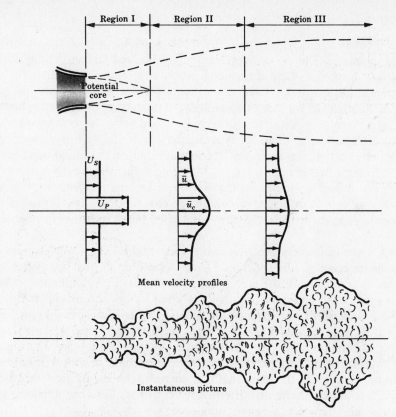

Fig. 9-12. Different regions and velocity profiles for a turbulent jet.

Hinze (reference 7) used the similarity method and an eddy viscosity to obtain a solution for the mean velocity distribution. Using the momentum equation (differential form) and the momentum integral equation, he found that for $(U_P - \bar{u})/U_S \gg 1.0$, b is proportional to $(x + a)$ and

$$\frac{U_P - \bar{u}}{U_P - \bar{u}_c} = \left[1 + \frac{(U_P - \bar{u}_c)r^2}{8\epsilon(x + a)} \right]^{-2}$$

For $(U_P - \bar{u})/U_S \ll 1.0$, b is proportional to $(x + a)^{1/2}$. Here ϵ is assumed constant and a is an experimental constant. U_P and U_S are shown on Fig. 9-12 and $\bar{u}_c$ is the centerline velocity.

The results for the rate of growth of the wake and jet and the centerline velocities are summarized in Table 9-1.

Table 9.1.   Wake Width and Centerline Velocity

	Width, Proportional to	Centerline velocity defect, $(\bar{u}_c - U_S)$, Proportional to
Plane two-dimensional jet	x	$x^{-1/2}$
Circular jet	x	x^{-1}
Plane two-dimensional wake	$x^{1/2}$	$x^{-1/2}$
Circular wake	$x^{1/3}$	$x^{-1/3}$

Of course, these results are not valid in the near wake and close to the nozzle.

9.9 RECENT DEVELOPMENTS

The complete theoretical description of turbulent flow is still one of the unsolved problems of modern physics. Many methods have been applied and have helped further the physical understanding of turbulence, but a full mathematical description seems still remote. Of particular interest are the method of renormalization borrowed from high energy physics and the concept of the theory of "chaos," which is based on the interaction of a multitude of interacting oscillating systems.

The difficulty of a theoretical description of turbulence can better be understood by viewing turbulence from the point of view of a "chaotic" system. Consider a simple linear two degree of freedom oscillator. The motion of one of the degrees of freedom depends on the initial conditions and becomes more complex than a single degree of freedom system, being in fact the superposition of two modes. As the number of degrees of freedom increases, the motion of any one becomes more complex and may repeat only after a long time. Further, the position (or coordinate of any degree of freedom at any particular time) becomes more sensitive to the initial conditions, particularly after a long time has elapsed.

If we imagine a system of a very large (virtually uncountable) number of degrees of freedom, the value of any coordinate corresponding to any degree of freedom at any time (after a long time has elapsed) is dependent so critically on the precise value of the initial conditions that it becomes impossible in practice to predict the performance of such a complex system no matter how accurately the initial conditions are prescribed. For example, in some systems of a large but still easily calculable number of degrees of freedom, particularly if they are nonlinear, the difference of a single digit in say the 16th place of an initial condition can completely change the state of the system at a given future time. Hence even with modern digital computers it becomes impossible to follow such a system, and its behavior is said to be "chaotic" in the sense that it is unpredictable in practice even though the system obeys deterministic Newtonian mechanics. The theory of "chaos" has been applied to turbulence and has led to new insights into the mechanisms of turbulence.

Also, new phenomenological theories have been developed which are particularly suited for computer analysis. Of particular importance is the so-called K-ε model which has been remarkably successful in describing turbulent flows, particularly in regions away from high shear flows which occur near walls.

The reader is referred to the references and current journals for further details and current thought on turbulence.

9.10 SUMMARY

We have considered the subject of incompressible turbulent flow in terms of "isotropic turbulence" and "shear turbulence". Isotropic turbulence is a highly idealized flow which requires that there be no mean velocity variation. While isotropic turbulence is rarely approximated by real flows of engineering importance, it does lend itself to some mathematical tractability and as a result it is the subject of extensive study in an attempt to gain some understanding of the more complicated nonisotropic flows.

Shear turbulence is further subdivided into "wall turbulence" such as boundary layer flows and "free turbulence" such as wakes and jets. The problems are to determine mean velocity, shear stress, and the turbulence quantities. There are two approaches: (1) the phenomenological approach of defining exchange coefficients such as eddy viscosity and mixing length to relate the turbulent stress to the mean velocity field and (2) the statistical approach of writing the basic differential equations in terms of time average quantities. Both approaches are useful from an engineering standpoint.

References

1. Abramovich, G. N., *The Theory of Turbulent Jets* (USSR), MIT Press, 1963.

2. Anderson, D. A., Tannehill, J. C., and Pletcher, R. H., *Computational Fluid Mechanics and Heat Transfer*, Hemisphere Publishing Corp., 1984.

3. Batchelor, G. K., *The Theory of Homogeneous Turbulence*, Cambridge University Press, 1960.

4. Chandrasekhar, S., *Hydrodynamic and Hydromagnetic Stability*, Oxford, 1961.

5. Drazin, P. G., and Reid, W. H., *Hydrodynamic Stability*, Cambridge University Press, 1981.

6. Essers, J. A., *Computational Methods for Turbulent, Transonic, and Viscous Flows*, Hemisphere Publishing Corp., 1983.

7. Hinze, J. O., *Turbulence*, 2nd ed., McGraw-Hill, 1975.

8. Klebanoff, P. S., "Characteristics of Turbulence in a Boundary Layer with Zero Pressure Gradient," NACA TN 3178, 1954.

9. Laufer, J., "The Structure of Turbulence in Fully Developed Pipe Flow," NACA R. 1174, 1954.

10. Moon, F. C., *Chaotic Vibrations*, John Wiley, 1987.

11. Pai, S. I., *Fluid Dynamics of Jets*, Van Nostrand, 1954.

12. Patankar, S. V., *Numerical Heat Transfer and Fluid Flow*, McGraw-Hill, 1980.

13. Schlichting, H., *Boundary Layer Theory*, 7th ed., McGraw-Hill, 1986.

14. Schubauer, G. B., and Tchen, C. M., *Turbulent Flow*, Princeton University Press, 1961.

15. Tennekes, H., and Lumley, J. L., *A First Course in Turbulence*, MIT Press, 1972.

16. Townsend, A. A., *The Structure of Turbulent Shear Flow*, Cambridge University Press, 1956.

For recent and current developments the reader is referred to journals; in particular the following are recommended: *Journal of Fluid Mechanics* and *Physics of Fluids*.

Supplementary Problems

9.1. Derive the equations of motion for constant density and viscosity for fully developed turbulent flow in concentric annuli. The flow is in the axial direction.

9.2. Show that the incompressible continuity equation is satisfied by the instantaneous deviations alone.

9.3. Consider two-dimensional fully developed flow between parallel walls, one stationary and one moving at a constant velocity U. Assume $\partial \bar{p}/\partial x = 0$ where x is measured in the direction of flow.

    (a) Determine the general shape of the mean velocity curve.

    (b) Determine the shear stress distribution.

    (c) Determine the approximate distribution of the eddy viscosity ϵ and mixing length l.

    (d) Determine approximately the pressure distribution.

9.4. Consider the flow of air ($p = 14.7$ psi, $T = 59°F$, $\rho = 0.00238$ slug/ft^3, $\mu = 0.0373(10)^{-5}$ slug/ft-sec, $v = 1.57 \times 10^{-4}$ ft^2/sec) with a free stream velocity of 50 ft/sec over a flat plate. At one section $\delta = 2.8$ in. and the wall shear stress $\tau_0 = 0.00813$ lb/ft^2. Assume that the velocity distribution for $20 < yu_\tau/v < 1000$ is given by

$$\bar{u}/u_\tau = 2.44 \ln yu_\tau/v + 4.9$$

Assuming further that the shear stress is constant, calculate (a) the Prandtl mixing length and (b) the ratio ϵ/v at $y = 0.05$ in. and at $y = 0.10$ in.

NOMENCLATURE FOR CHAPTER 9

    a = constant

    b = wake or jet half width

    B_i = body force per unit volume

C_D = drag coefficient

d = diameter of cylinder

f = functional notation

l = Prandtl momentum mixing length

m = exponent for power law velocity distribution

n = exponent for power law velocity distribution

p = pressure

q^2 = turbulence kinetic energy, $u_i' u_i'$

r = radial coordinate

R = radius of pipe

t = time

T_1 = length of time for averaging

U = free stream velocity, mean centerline velocity in a pipe

U_P = velocity of jet

U_S = velocity of secondary stream

$\bar{u}_c$ = mean centerline velocity

u_i = $u_1, u_2, u_3 = u, v, w$ velocity components

u_τ = friction velocity, $\sqrt{\tau_0/\rho}$

x_i = $x_1, x_2, x_3 = x, y, z$ cartesian coordinates

y = distance from wall

$\overline{u_i' u_j'}$ = one point double velocity correlation

$\overline{(u_i')_A (u_j')_B}$ = two point double velocity correlation

δ = boundary layer thickness

ϵ = eddy viscosity

μ = viscosity

v = kinematic viscosity, μ/ρ

ρ = density

τ_{ij} = shear stress tensor

τ_{ijT} = turbulent shear stress tensor = $-\rho \overline{u_i' u_j'}$

τ_{ij_l} = laminar shear stress tensor = $\mu \dfrac{\partial \bar{u}_i}{\partial x_j}$

τ = one component of shear stress relevant to a particular problem

τ_0 = shear stress at the wall

$(^-)$ = denotes mean value

$(\ )'$ = denotes fluctuating component

Chapter 10

Hypersonic Boundary Layer Flow

10.1 INTRODUCTION

Hypersonic flow has become very important because of interest in missiles and spacecraft. When an object enters or flies within the earth's atmosphere at very high Mach numbers, dissociation of molecules may occur as a result of the high temperature of the gas in the region surrounding the body. Hypersonic flow is defined as a high speed flow where the Mach number is so high that dissociation occurs ($M > 6$ for the earth's atmosphere).

Let us begin the study of hypersonic flow by considering an example, the different flow regimes around a vehicle entering the earth's atmosphere at very high speed as shown in Fig. 10-1. A *bow shock* wave forms ahead of and wraps around the body. The flow is undisturbed ahead of the shock. This shock is strongest on the axis of symmetry since it is normal to the flow there. Behind the shock there is a region where dissociation will occur. Immediately behind the shock is a region where viscous effects are not important, but close to the body surface is the boundary layer where viscous stresses are important and where the flow may be laminar or turbulent. As the fluid passes over the rear portion of the body, it has a tendency to turn in toward the axis as in a Prandtl-Meyer expansion. This is followed by a turning of the flow in the opposite direction (a straightening) as a result of the fluid from opposite sides coming together. The straightening of the flow results in an oblique shock called a *wake shock*. The flow following the wake shock consists of a wake containing a turbulent core surrounded by a non-turbulent flow region. In the wake the velocity is reduced (it is a minimum on the axis) but increases with distance behind the body until eventually the velocity is nearly uniform and equal to the free stream velocity. However, the turbulent core will persist for long distances behind the body (thousands of body diameters).

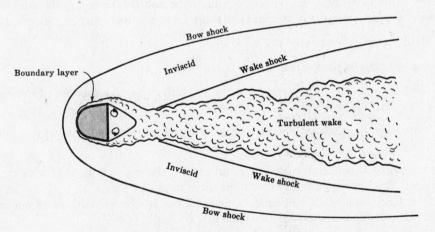

Fig. 10-1. Hypersonic flow regimes.

In this chapter we will consider the boundary layer and wake flows but will not discuss the determination of the shock shape and the inviscid flow regions, since they involve (ultimately) a numerical approach for solving the equations. We have mentioned the basic ideas of supersonic flow in Chapter 8 which will give some idea of the methods that must be used for these calculations.

The turbulent hypersonic boundary layer problem is one of the most complicated in fluid mechanics. It has all of the difficulties (and as we saw in Chapter 9 there were many) of incompressible turbulent flow, but in addition there are the added complications of large density and temperature variations, dissociation and recombination of molecules and diffusion.

We will approach the problem by writing the important equations and considering the way in which the results of simpler flows are used, and briefly discuss the methods of solving the equations.

10.2 BOUNDARY LAYER EQUATIONS

The boundary layer equations are the same as written in Chapter 5 except that now the effects of chemical reactions and diffusion must be included. The simple equations of Chapter 5 are not adequate for fluids which are not homogeneous in the concentrations of particles of various kinds. For hypersonic flow there are large temperature gradients and large diffusion gradients, and hence transport of mass, momentum and energy by diffusion will take place.

As we have mentioned before, there are two approaches to developing the equations of fluid mechanics. The first, and the one which has found more use, is the continuum approach. The equations of momentum and energy, for example, are derived assuming that the fluid is a continuum. Then, phenomenological equations relate shear and rate of strain, heat flux and temperature gradient, and mass diffusion and concentration gradient. These relationships are (to first order) for two-dimensional plane (xy) flow with velocity u in the x direction.

$$\tau_{yx} = \mu \frac{\partial u}{\partial y} \qquad \text{Stokes' law} \qquad (10.1)$$

$$q_y = -\kappa \frac{\partial T}{\partial y} \qquad \text{Fourier's law} \qquad (10.2)$$

$$V_{iy} = -D_{ij} \frac{1}{C_i} \frac{\partial C_i}{\partial y} \qquad \text{Fick's law} \qquad (10.3)$$

The coefficients, μ the viscosity, κ the thermal conductivity, and D_{ij} the binary diffusion coefficient (for diffusion of species i into a mixture of species i and j) are known as transport coefficients, and

τ_{yx} = shear stress

q_y = heat flux rate in the y direction

V_{iy} = diffusion velocity in y direction for species i diffusing into a mixture of species i and j [see equation (10.4)].

C_i = mass fraction of species i = ratio of mass of species i per unit volume to total mass density.

The second approach is from the particle point of view. Here one must deal with the dynamics of colliding particles. This approach is described in detail in books by Chapman and Cowling (reference 2) and Hirschfelder, Curtiss and Bird (reference 6), and will not be considered here except in that it does yield the conservation of species equation (see reference 1):

$$\rho \frac{DC_i}{Dt} = \dot{w}_i - \nabla \cdot (\rho C_i \mathbf{V}_i) \qquad (10.4)$$

where $\mathbf{V}_i = \mathbf{v}_i - \mathbf{V}$ = diffusion velocity of species i.

$\mathbf{v}_i$ = average velocity of species i.

$$\mathbf{V} = \text{mass average velocity of all species} = \frac{\sum_i \rho C_i \mathbf{v}_i}{\sum_i \rho C_i} = \frac{\sum_i \rho_i \mathbf{v}_i}{\sum_i \rho_i}$$

where ρ_i is the mass density of species i. $\mathbf{V}$ has components u and v.

$\dot{w}_i$ = mass rate of production of species i per unit volume by chemical reactions.

ρ = total mass density = $\sum_i \rho_i$.

Assuming plane two-dimensional steady boundary layer flow and neglecting diffusion terms in the x direction and using Fick's law, the *conservation* of *species equation* becomes

$$\rho \left[u \frac{\partial C_i}{\partial x} + v \frac{\partial C_i}{\partial y} \right] = \dot{w}_i + \frac{\partial}{\partial y} \left(\rho D_{12} \frac{\partial C_i}{\partial y} \right) \tag{10.5}$$

where u and v are the x and y components of the mass average velocity $\mathbf{V}$, and we have replaced D_{ij} by D_{12}, the binary diffusion coefficient. It can be shown that $D_{12} = D_{21}$ and for most systems of gases of interest the binary coefficient is adequate. We usually have two types of species, heavy ones and light ones. For example, in air we have the heavy particles O_2 and N_2 and the light ones O and N. D_{12} adequately describes diffusion of O or N through O_2 and N_2. The total continuity equation is

$$\frac{\partial(\rho u)}{\partial x} + \frac{\partial(\rho v)}{\partial y} = 0 \tag{10.6}$$

and the momentum equation is for the total fluid

$$\rho \left[u \frac{\partial u}{\partial x} + v \frac{\partial u}{\partial y} \right] = -\frac{\partial p}{\partial x} + \frac{\partial}{\partial y} \left(\mu \frac{\partial u}{\partial y} \right) \tag{10.7}$$

The total energy equation is for the total fluid (where the kinetic energy contribution $v^2/2$ is neglected compared to $u^2/2$),

$$\rho \left[u \frac{\partial}{\partial x} (h + u^2/2) + v \frac{\partial}{\partial y} (h + u^2/2) \right]$$

$$= \frac{\partial}{\partial y} \left[\frac{\mu}{P_r} \frac{\partial}{\partial y} (h + u^2/2) + \mu \left(1 - \frac{1}{P_r} \right) \frac{1}{2} \frac{\partial u^2}{\partial y} \right] - \frac{\partial}{\partial y} \left[\left(\frac{1}{L} - 1 \right) \rho D_{12} \sum_i h_i \frac{\partial C_i}{\partial y} \right] \tag{10.8}$$

where $\quad h = \sum_i C_i h_i$

$$h_i = \int_0^T c_{pi} \, dT + h_i^0$$

h_i^0 = heat of formation of species i

$P_r = \dfrac{\mu c_{pf}}{\kappa}, \quad$ the Prandtl number

$L = \dfrac{\rho D_{12} c_{pf}}{\kappa}, \quad$ the Lewis number

$c_{pf} = \sum_i C_i c_{pi}, \quad$ the frozen specific heat at constant pressure

For a derivation and discussion of the energy equation the reader may refer to reference 4. Since the energy equation is of importance in hypersonic flow, we will "outline" the derivation here.

Consider an infinitesimal control volume at a point in the flow as shown in Fig. 10-2. We will write the energy equation for this control volume. Consider the following fluxes.

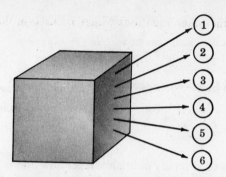

Fig. 10-2. Infinitesimal control volume for energy equation.

1. Net internal and kinetic energy flux out of control volume:

$$\nabla \cdot \left[(\rho \mathbf{V}) \left(\sum_i C_i e_i + u^2/2 \right) \right]$$

where $e_i = \int_0^T c_{vi}\, dT$, the specific internal energy for the ith species assuming it to be a perfect gas.

2. Net molecular transfer of energy out due to temperature gradient:

$$-\nabla \cdot (\kappa \nabla T)$$

3. Rate of work done by fluid per unit volume by pressure forces:

$$\nabla \cdot (p\mathbf{V})$$

4. Rate of work done by fluid per unit volume by viscous forces:

$$\nabla \cdot [\tau \cdot \mathbf{V}] = \frac{\partial}{\partial x_i} (\tau_{ij} u_j)$$

5. Energy carried out by mass diffusion of species i:

$$\nabla \cdot \left(\sum_i \rho \mathbf{V}_i C_i h_i \right)$$

where

$$\mathbf{V}_i = -\frac{D_{12}}{C_i} \nabla C_i$$

6. Energy added due to formation of a species:

$$-\sum_i \dot{w}_i h_i^0$$

where $h_i^0 = 0$, for molecules; $h_i^0 = $ a positive value for atoms.

Then the energy equation becomes, adding the six above effects,

$$\rho \mathbf{V} \cdot \nabla \left(\sum_i C_i e_i + u^2/2 \right) + \left(\sum_i C_i e_i + u^2/2 \right) \nabla \cdot (\rho \mathbf{V}) - \nabla \cdot (\kappa \nabla T)$$

$$+ \nabla \cdot (p\mathbf{V}) + \nabla \cdot (\tau \cdot \mathbf{V}) + \nabla \cdot \left(\sum_i \rho \mathbf{V}_i C_i h_i \right) + \sum_i \dot{w}_i h_i^0 = 0 \qquad (10.9)$$

In addition to equation (*10.9*), the other equations which are used in the derivation are the equation of state

$$p_i = \rho_i R_i T \tag{10.10}$$

where all species are assumed to be at the same temperature and $p = \sum_i p_i = \rho \bar{R} T$, $\bar{R} = \sum_i C_i R_i$,

$$h_i = e_i + R_i T \tag{10.11}$$

and continuity

$$\nabla \cdot \rho \mathbf{V} = 0 \tag{10.12}$$

Using equations (*10.4*), (*10.9*), (*10.10*), (*10.11*) and (*10.12*) and assuming $h_i = f(T)$, the energy equation is obtained in the form of equation (*10.8*).

The total energy equation for boundary layer flow in terms of the temperature becomes

$$c_{pf}\left[\rho u \frac{\partial T}{\partial x} + \rho v \frac{\partial T}{\partial y}\right] = \frac{\partial}{\partial y}\left(\kappa \frac{\partial T}{\partial y}\right) + u \frac{\partial p}{\partial x} + \mu\left(\frac{\partial u}{\partial y}\right)^2 - \sum_i \dot{w}_i h_i^0 + \sum_i c_{pi}\left(D_{ij}\rho \frac{\partial C_i}{\partial y}\right) \tag{10.13}$$

Equations (*10.4*) through (*10.8*) then, are the basic equations of hypersonic (laminar) boundary layer flow.

10.3 HYPERSONIC LAMINAR BOUNDARY LAYER

Although it is likely that transition will occur at some point on the body resulting in the flow going from laminar to turbulent, we will consider laminar flow only in this section and turbulent flow will be treated in the next section.

Consider an ideal dissociating gas. The molecule A_2 dissociates into the atom A

$$A_2 \rightleftarrows 2A$$

where $\rho_A/\rho = \alpha$, the mass fraction of atoms

$\rho_M/\rho = (1 - \alpha)$, the mass fraction of molecules.

The subscripts A and M refer to the atom and molecule, respectively. An alternative form of the energy equation (*10.8*) is

$$\rho u \frac{\partial h}{\partial x} + \rho v \frac{\partial h}{\partial y} = u \frac{\partial p}{\partial x} + \frac{\partial}{\partial y}\left(\kappa \frac{\partial T}{\partial y}\right) + \frac{\partial}{\partial y}\left(\rho D_{12} \sum_i h_i \frac{\partial C_i}{\partial y}\right) + \mu\left(\frac{\partial u}{\partial y}\right)^2 \tag{10.14}$$

Then, for the ideal dissociating gas, we have

$$\rho u \frac{\partial h}{\partial x} + \rho v \frac{\partial h}{\partial y} = u \frac{\partial p}{\partial x} + \frac{\partial}{\partial y}\left(\kappa \frac{\partial T}{\partial y}\right) + \frac{\partial}{\partial y}\left[\rho D_{12} h_A \frac{\partial \alpha}{\partial y} - \rho D_{12} h_M \frac{\partial \alpha}{\partial y}\right] + \mu\left(\frac{\partial u}{\partial y}\right)^2 \tag{10.15}$$

The conservation of species equation becomes

$$\rho u \frac{\partial \alpha}{\partial x} + \rho v \frac{\partial \alpha}{\partial y} = \frac{\partial}{\partial y}\left(\rho D_{12} \frac{\partial \alpha}{\partial y}\right) + \dot{w}_A \tag{10.16}$$

The conservation of species, continuity, momentum and energy equations, i.e. (*10.16*), (*10.6*), (*10.7*) and (*10.15*), along with property relationships define the problem if the particular boundary conditions are given. Also, one would need to determine the appropriate values for the transport coefficients.

The solution to these equations present enormous difficulties. Therefore let us consider a related problem of somewhat lesser difficulty. Consider the flow of a nonreacting gas on a flat plate with zero pressure gradient and constant properties (see Schlichting, reference 8). The equations become

$$\rho u \frac{\partial h}{\partial x} + \rho v \frac{\partial h}{\partial y} = \frac{\partial}{\partial y}\left(\kappa \frac{\partial T}{\partial y}\right) + \mu\left(\frac{\partial u}{\partial y}\right)^2$$

$$\rho u \frac{\partial u}{\partial x} + \rho v \frac{\partial u}{\partial y} = \mu \frac{\partial^2 u}{\partial y^2}$$

$$\frac{\partial u}{\partial x} + \frac{\partial v}{\partial y} = 0$$

The boundary conditions are

$$y = 0; \qquad u = v = 0, \qquad T = T_w$$

$$y = \infty; \qquad u = U_\infty, \qquad T = T_\infty$$

The result for the heat transfer at the wall is

$$q = -\kappa \frac{\partial T}{\partial y} = \frac{C_f}{2}(P_r)^{-2/3}\rho_\infty U_\infty\left[h_\infty + (P_r)^{1/2}\frac{U_\infty^2}{2} - h_w\right]$$

where $C_f = 1.328\sqrt{v/U_\infty x}$, $P_r = \mu c_p/\kappa$.

The additional complications of variable properties, dissociation and diffusion make the problem considerably more difficult. And, although we cannot obtain a solution, we can determine some of the physical characteristics of hypersonic flow.

The heat flux without dissociation is $q = -\kappa \partial T/\partial y$. The total heat flux with dissociation is [see equation (10.15)]

$$q = -\kappa \frac{\partial T}{\partial y} - \rho D_{12}(h_A - h_M)\frac{\partial \alpha}{\partial y} \qquad (10.17)$$

But

$$h_A - h_M = h_A^0 + \int_0^T (c_{pA} - c_{pM})\, dT$$

which is approximately h_A^0, and

$$q = -\kappa \frac{\partial T}{\partial y} - \rho D_{12} h_A^0 \frac{\partial \alpha}{\partial y}$$

Now consider an adiabatic wall, $q = 0$, and

$$\frac{\partial T}{\partial y} = -\frac{\rho D_{12} h_A^0}{\kappa}\frac{\partial \alpha}{\partial y}$$

Thus there may be a nonzero temperature gradient at the wall when the heat flux is zero as shown in Fig. 10-3.

Dorrance (reference 3) shows that heat transfer of the wall for a dissociating gas on a flat plate is

$$q = -\frac{C_f}{2}\rho_\infty U_\infty P_r^{-2/3}\left(h_\infty - h_w + P_r^{1/2}\frac{U_\infty^2}{2}\right)\left[1 + \frac{(L-1)(\alpha_\infty - \alpha_w)h_A^0}{h_\infty - h_w + P_r^{1/2}U_\infty^2/2}\right]^{2/3} \qquad (10.18)$$

We note in equation (10.18) that if $L = 1.0$ or $\alpha_\infty = \alpha_w$, the equation is the same as that for no dissociation. The subscripts ∞ and w refer to free stream and wall conditions, respectively. L is the Lewis number.

The previous discussion was for flow over a flat plate. There has been much interest in the heat transfer for flows over blunt nosed bodies, particularly in the region of the stagnation point where the heat flux is large. These problems have been considered in detail by Dorrance, by Lees, by Fay and Riddell and by others (see references 1, 7 and 4). We will not pursue this problem in depth here.

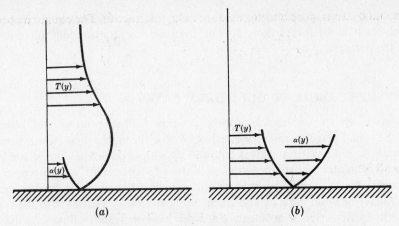

Fig. 10-3. Velocity and concentration distributions for an adiabatic wall. (*a*) Cold wall, (*b*) hot wall.

The method that is used is to set up coordinate transformations which result in changing the problem from one involving partial differential equations to one involving ordinary differential equations. The equations are then solved by numerical techniques for (1) frozen flow (i.e. for a sufficiently small recombination rate where the concentration of atoms is determined wholly by diffusive flow) and (2) equilibrium flow (i.e., for a sufficiently large recombination rate where the temperature would determine the concentration throughout). Results for particular values of Lewis and Prandtl numbers and wall temperature are shown in Fig. 10-4.

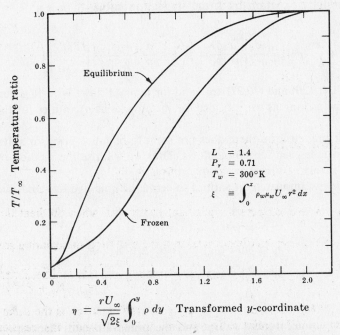

$$\eta = \frac{rU_\infty}{\sqrt{2\xi}} \int_0^y \rho \, dy \quad \text{Transformed } y\text{-coordinate}$$

Fig. 10-4. Temperature distribution in the boundary layer for hypersonic flow (from reference 4). The flow is assumed to occur over a cylindrical blunt-nosed object and r is the radius of the cylinder. y is the distance from the surface or wall.

For the same wall temperature and free stream temperature, the intermediate temperature of equilibrium flow is higher than for frozen flow. This results from the atoms recombining as they move into an area of lower temperature.

10.4 HYPERSONIC TURBULENT BOUNDARY LAYER

When an object enters the earth's atmosphere, the flow field surrounding the object goes through several phases. The initial stage of re-entry consists of flow through a low density medium where the continuum assumptions are not valid. This is followed by a more dense medium where the flow behaves as a continuum and is laminar. As the penetration of the atmosphere further progresses, random velocity fluctuations appear in the form of turbulence in the downstream wake. A further increase in density results in the boundary between laminar and turbulent flow moving upstream to the point where eventually most of the boundary layer is turbulent. When this has occurred the flow is as shown in Fig. 10-1.

We will not consider the problem of transition. Rather, we will assume that a turbulent boundary layer exists, then proceed to describe the flow. The method here will be quite similar to that of Chapter 9 on turbulence.

Consider the continuity, species, momentum and energy equations of the previous sections. If we multiply the continuity equation (10.6) by C_i and add the species equation (10.5), we obtain

$$\frac{\partial}{\partial x}(\rho u C_i) + \frac{\partial}{\partial y}(\rho v C_i) = -\frac{\partial}{\partial y}(\rho C_i V_{iy}) + \dot{w}_i \qquad (10.19)$$

Combining the continuity and momentum equation (10.7) gives

$$\frac{\partial}{\partial x}(\rho u^2) + \frac{\partial}{\partial y}(\rho v u) = \frac{\partial}{\partial y}\left(\mu \frac{\partial u}{\partial y}\right) \qquad (10.20)$$

Multiplying the continuity equation by I, the stagnation enthalpy $(h + u^2/2)$, and adding the energy equation gives

$$\frac{\partial}{\partial x}(\rho u I) + \frac{\partial}{\partial y}(\rho v I) = \frac{\partial}{\partial y}\left[\frac{\mu}{P_r}\frac{\partial I}{\partial y} + \mu\left(1 - \frac{1}{P_r}\right)\frac{1}{2}\frac{\partial(u^2)}{\partial y}\right] + \frac{\partial}{\partial y}\left[\left(\frac{1}{L} - 1\right)\sum_i V_{iy} h_i\right] \qquad (10.21)$$

Equations (10.19), (10.20) and (10.21) are valid for laminar flow. We will assume that they are valid also for turbulent flow as long as we consider the instantaneous quantities for the dependent variables in the equations.

However, in their present form there does not seem to be much hope for solving the equations as a function of space and time even for simple flows. Thus we will use the method of Chapter 9 to obtain associated equations involving time average quantities.

We will write the instantaneous quantities in terms of time average and fluctuating quantities as follows:

$$u = \bar{u} + u' \qquad\qquad C_i = \bar{C}_i + C_i'$$
$$\rho u = \overline{\rho u} + (\rho u)' \qquad\qquad h_i = \bar{h}_i + h_i'$$
$$\rho v = \overline{\rho v} + (\rho v)' \qquad\qquad \rho V_{iy} = \overline{\rho V_{iy}} + (\rho V_{iy})'$$
$$h = \bar{h} + h'$$

where the bar indicates time average values and the prime indicates fluctuating quantities. Now we substitute the above expressions into the continuity and equations (10.19), (10.20) and (10.21) and take the time average of each. We obtain:

Total Continuity:
$$\frac{\partial}{\partial x}(\overline{\rho u}) + \frac{\partial}{\partial y}(\overline{\rho v}) = 0 \qquad (10.22)$$

Species Continuity:
$$\overline{\rho u}\,\frac{\partial \bar{C}_i}{\partial x} + \overline{\rho v}\,\frac{\partial \bar{C}_i}{\partial y} = \frac{\partial}{\partial y}\left[\bar{\rho}D_{12}\,\frac{\partial \bar{C}_i}{\partial y} - \overline{(\rho v)'C_i'}\right] + \dot{w}_i \qquad (10.23)$$

where it was assumed $\dfrac{\partial}{\partial x}\,\overline{[(\rho u)'C_i']} \ll \dfrac{\partial}{\partial y}\,\overline{[(\rho v)'C_i']}$ and $\overline{[(\rho V_{iy})'C_i']} \ll \overline{[\rho V_{iy}\,C_i]}$.

Momentum:
$$\overline{\rho u}\,\frac{\partial \bar{u}}{\partial x} + \overline{\rho v}\,\frac{\partial \bar{u}}{\partial y} = \frac{\partial}{\partial y}\left[\mu\,\frac{\partial \bar{u}}{\partial y} - \overline{(\rho v)'u'}\right] \qquad (10.24)$$

where it is assumed $\dfrac{\partial}{\partial x}\,\overline{[(\rho u)'u']} \ll \dfrac{\partial}{\partial y}\,\overline{[(\rho v)'u']}$.

Energy:
$$\overline{\rho u}\,\frac{\partial \bar{h}}{\partial x} + \overline{\rho v}\,\frac{\partial \bar{h}}{\partial y} = \mu\left(\frac{\partial \bar{u}}{\partial y}\right)^2 - \overline{(\rho v)'u'}\,\frac{\partial \bar{u}}{\partial y}$$

$$+ \frac{\partial}{\partial y}\left[\kappa\,\frac{\partial \bar{T}}{\partial y} - \sum_i \overline{h_i\,\rho_i\,V_{iy}} - \sum_i \overline{C_i\,h_i'(\rho v)'} - \sum_i \overline{(\rho v)'C_i'\,h_i}\right] \quad (10.25)$$

It is interesting to note that there are additional terms which appear in the equations as a result of the turbulent fluctuations. These terms have a definite physical interpretation. For example,

$$-\overline{(\rho v)'C_i'} \qquad \text{is turbulent mass transfer}$$

$$-\overline{(\rho v)'u'} \qquad \text{is turbulent momentum transfer}$$

$$-\overline{(\rho v)'h_i'} \qquad \text{is turbulent energy transfer}$$

With some algebraic manipulations and after defining new terms such as

$$\epsilon = -\frac{\overline{(\rho v)'u'}}{\partial \bar{u}/\partial y}, \qquad \epsilon_T = -\frac{\overline{(\rho v)'h_i'}}{\partial \bar{T}/\partial y}, \qquad \bar{\rho}D_T = -\frac{\overline{(\rho v)'C_i'}}{\partial \bar{C}_i/\partial y},$$

$$P_{rT} = \frac{c_{pf}\,\epsilon}{\epsilon_T}, \qquad L_T = \frac{c_{pf}\,\bar{\rho}D_T}{\epsilon_T}$$

we have

Species Continuity:
$$\overline{\rho u}\,\frac{\partial \bar{C}_i}{\partial x} + \overline{\rho v}\,\frac{\partial \bar{C}_i}{\partial y} = \frac{\partial}{\partial y}\left[(\rho D_{12} + \rho D_T)\,\frac{\partial \bar{C}_i}{\partial y}\right] + \dot{w}_i \qquad (10.26)$$

Momentum:
$$\overline{\rho u}\,\frac{\partial \bar{u}}{\partial x} + \overline{\rho v}\,\frac{\partial \bar{u}}{\partial y} = \frac{\partial}{\partial y}\left[(\mu + \epsilon)\,\frac{\partial \bar{u}}{\partial y}\right] \qquad (10.27)$$

Energy:
$$\overline{\rho u}\,\frac{\partial \bar{h}}{\partial x} + \overline{\rho v}\,\frac{\partial \bar{h}}{\partial y} = (\mu + \epsilon)\left(\frac{\partial \bar{u}}{\partial y}\right)^2 + \frac{\partial}{\partial y}\left[\left(\frac{\mu}{P_r} + \frac{\epsilon}{P_{rT}}\right)\frac{\partial \bar{h}}{\partial y}\right]$$

$$+ \frac{\partial}{\partial y}\left\{\left[\frac{\kappa}{c_{pf}}\,(L - 1) + \frac{\epsilon_T}{c_{pf}}\,(L_T - 1)\right]\sum_i h_i\,\frac{\partial \bar{C}_i}{\partial y}\right\} \qquad (10.28)$$

The equations are now in terms of the microscopic and macroscopic transport coefficients. The solutions to the equations are considered in reference 3. The general procedure is to consider the simple case where

$$P_r = P_{rT} = L = L_T = 1.0$$

Then the results are extended to the more general case.

10.5 AERODYNAMIC HEATING

We close this chapter with a brief discussion of a problem that assumed great importance during the flourishing of the piloted space program. Vehicles traveling at very high speed through the atmosphere become heated because of viscous friction and the surface temperature, or, in effect, the adiabatic wall temperature may rise to a high level—beyond human endurance.

At supersonic speeds, say Mach number 1 to 2, the adiabatic wall temperature T_a gives us an idea of the temperature of the airplane surface in steady flight. At $M = 2$, for example, $T_a/T_0 \approx 1.8$ (for an air temperature of $0°F$, T_a is about $370°F$). These high temperatures do constitute a problem, and in commercial supersonic jet planes the problem of maintaining the inside temperature at a reasonable level is aggravated by the high outside surface temperature, even though the outside air may be quite cold. Proper insulation and air conditioning are vital.

A reentry vehicle may enter the upper atmosphere at hypersonic speed with a Mach number of about 10 to 30. At these speeds the temperature in the boundary layer is so high that dissociation of the air molecules and even ionization occur. The complete description of the boundary layer must take these chemical and atomic reactions into account.

However, we can apply some of the results of this chapter to the hypersonic heating problem and at least obtain some interesting qualitative information about how a reentry vehicle should be designed so that it does not burn up while passing down through the atmosphere. The conclusions are quite different from those for an airplane. For an airplane, the drag must be minimized and reasonably steady temperatures maintained. For a reentry vehicle, the problem is basically a transient one, and drag needs to be adjusted to minimize the heating—not to minimize power since a reentry vehicle is basically in free fall.

For an order-of-magnitude study we can express Newton's law of motion for the vehicle as

$$M \frac{dV}{dt} = -D = -C_D(\tfrac{1}{2}\rho V^2 A)$$

or

$$M \frac{d(V^2/2)}{dt} = -DV$$

where M is the mass of the vehicle, V is its velocity, D is the drag opposing the motion, and A is a characteristic surface area. (We assume gravity to be small compared to the drag.)

Now, the total rate of heat transfer dQ/dt to the vehicle at some bulk temperature T_w may be expressed approximately in terms of the adiabatic wall temperature and T_w as

$$\frac{dQ}{dt} \approx \bar{h}A(T_a - T_w) \approx \frac{\bar{h}AV_\infty^2}{2c_p}$$

For air, $T_a \approx T_\infty + (rV_\infty^2/2c_p)$, where T_∞ and V_∞ are the free-stream values, r is the "recovery factor," and $\bar{h}$ is the average film coefficient. We assume that $r \approx 1$ and conservatively we assume that $T_w \approx T_\infty$. Further, it follows that

$$\bar{h} \propto \tfrac{1}{2}\rho c_p C_{Df} V_\infty$$

and the drag due to friction D_f is

$$D_f = C_{Df} \frac{\rho V_\infty^2}{2} A$$

which defines C_{Df} as the friction coefficient. The total drag D may be written

$$D = C_{Df} \frac{\rho V_\infty^2}{2}$$

$$C_D = C_{Df} + C_{Dp}$$

where C_{Dp} is the pressure or form drag coefficient. Combining these relationships, we have

$$\frac{dQ}{dt} \approx -\frac{1}{2} \frac{C_{Df}}{C_D} M \frac{d(V_\infty^2/2)}{dt}$$

and hence Q, the total heat transferred to the object during deceleration, is the integral of the above equation and may be expressed in terms of the initial velocity $V_{\infty i}$ as

$$Q \approx \frac{1}{2} \frac{C_{Df}}{C_D} \left(\frac{MV_{\infty i}^2}{2} \right) \qquad (10.29)$$

where the final velocity is neglected. We see that in order to minimize the heat input to the object or spacecraft, the ratio C_{Df}/C_D should be small. In other words, the object should be blunt, with the form drag making up the main contribution to the total drag. Even then, ablative materials are used on the surface of reentry vehicles to help absorb the heat and prevent the interior temperature from rising too high. In the Soviet Union, typical reentry vehicles have been spherical, whereas American craft have generally been blunt but conical in shape. The sphere has a lower value of C_{Df}/C_D, but the cone shape has better directional stability during the reentry flight.

10.6 SUMMARY AND DISCUSSION

The turbulent reacting boundary layer is one of the most complicated problems in fluid mechanics, one which will probably receive considerable attention as we progress into the space age.

There are two kinds of transport coefficients in the equations. The transport coefficients μ, κ and D_{12} are due to microscopic effects and are properties of the fluid. They do not depend on the flow conditions. However, the transport coefficients ϵ, ϵ_T and D_T are the result of macroscopic effects and they are not properties of the fluid. They depend on the flow conditions and should be considered as unknowns in the equations just as are velocity, entropy, etc. Determination of the transport coefficients (microscopic and macroscopic) remains one of the major problems in turbulent hypersonic boundary layer flow.

Hypersonic aerodynamics and the analysis of reentry problems have become reasonably well understood branches of fluid mechanics. The aerodynamics design of the space shuttle was based on hypersonic flow theory and reacting boundary layer flow. The fact that the shuttles have performed so well, aerodynamically, attests to the maturation of this area of fluid mechanics. Presently, a great deal of effort is being devoted to the hypersonic transport program, which is for the development of a high speed, high altitude, jet engine powered transport plane.

References

1. Anderson, J. D., *Hypersonic and High Temperature Gas Dynamics*, McGraw-Hill, 1989. (A useful bibliography of recent journal articles is provided.)

2. Chapman, S., and Cowling, T. G., *The Mathematical Theory of Non-Uniform Gases*, Cambridge University Press, 1958.

3. Dorrance, W. H., *Viscous Hypersonic Flow*, McGraw-Hill, 1962.

4. Fay, J. A., and Riddell, F. R., "Theory of Stagnation Point Heat Transfer in Dissociated Air," *J. Aero. Sci.*, 25, No. 2, pp. 73–85, 1958.

5. Hall, J. G., Editor, *Fundamental Phenomena in Hypersonic Flow*, Cornell University Press, 1966.

6. Hirschfelder, J. O., Curtiss, C. F., and Bird, R. B., *Molecular Theory of Gases and Liquids*, John Wiley, 1954.

7. Lees, L., "Laminar Heat Transfer over Blunt Nosed Bodies at Hypersonic Flight Speeds," *Jet Propulsion*, April 1956.

8. Schlichting, H., *Boundary Layer Theory*, McGraw-Hill, 1960.

9. Vincenti, W. G., and Kruger, C. H., Jr., *Introduction to Physical Gas Dynamics*, John Wiley, 1965. (Reprinted by Krieger Publishing Co.)

The reader who desires further study in hypersonic flow will find that references 1 and 3 will provide an excellent introduction and discussion of the problems. Both of these books list many of the important publications in hypersonic flow.

The following journals are particularly useful for the reader interested in following current research in the area of hypersonic flow: *Physics of Fluids* and *Transactions of the A.I.A.A.*

NOMENCLATURE FOR CHAPTER 10

C_f = skin friction coefficient

C_i = species mass fraction

c_p = specific heat at constant pressure

c_{pi} = specific heat at constant pressure for species i

c_{pA} = specific heat at constant pressure for atoms

c_{pf} = the frozen specific heat, $\sum_i C_i c_{pi}$, at constant pressure

c_{pM} = specific heat at constant pressure for molecules

c_{vi} = specific heat at constant volume for species i

D_{12} = binary diffusion coefficient

D_T = turbulent diffusion coefficient

e_i = specific internal energy of species i

h = specific enthalpy

h_i = specific enthalpy of species i

h_i^0 = specific enthalpy of formation of species i

h_w = specific enthalpy at the wall

h_∞ = specific enthalpy of free stream

I = $u^2/2 + h$, total or stagnation specific enthalpy

L = Lewis number

L_T = turbulent Lewis number

M = Mach number

p = pressure

p_i = partial pressure of species i

P_r = Prandtl number = $\mu c_{pf}/\kappa$

P_T = turbulent Prandtl number

q = heat flux in the y direction

$\bar{R}$ = gas constant for total gas = $\sum_i C_i R_i$

R_i = gas constant for species i

t = time

T = temperature

T_∞ = temperature of the free stream

u = mass average velocity in x direction

U_∞ = velocity of the free stream

v = mass average velocity in y direction

$\mathbf{V}$ = mass average velocity = $\dfrac{\sum_i C_i \rho_i \mathbf{v}_i}{\sum_i C_i \rho_i}$

V_{iy} = diffusion velocity of species i in y direction

$\mathbf{V}_i$ = diffusion velocity of species i

$\mathbf{v}_i$ = average velocity of species i

w = mass average velocity in z direction

$\dot{w}_i$ = mass rate of production of species i per unit volume, by chemical reaction

x, y, z = coordinates

α = mass fraction of atoms

α_∞ = mass fraction of atoms for free stream

α_w = mass fraction of atoms at the wall

ϵ = turbulent viscosity

ϵ_T = turbulent energy transfer coefficient

κ = thermal conductivity

μ = viscosity

ρ = density

ρ_A = density of atoms

ρ_M = density of molecules

τ_{yx} = shear stress

$(\ )_A$ refers to the property of an atom

$(\ )_M$ refers to the property of a molecule

$(\ )_w$ refers to the property at the wall

$(\ )_\infty$ refers to the property in the free stream

Chapter 11

Magnetohydrodynamics

11.1 INTRODUCTION

Magnetohydrodynamics (MHD) is a relatively new but important branch of fluid dynamics. It is concerned with the interaction of electrically conducting fluids and electromagnetic fields. When a conducting fluid moves through a magnetic field, an electric field and consequently a current may be induced and, in turn, the current interacts with the magnetic field to produce a body force on the fluid.

Such interactions occur both in nature and in new man-made devices. MHD flow occurs in the sun, the earth's interior, the ionosphere, and the stars and their atmosphere, to mention a few. In the laboratory many new devices have been made which utilize the MHD interaction directly, such as propulsion units and power generators; or which involve fluid-electromagnetic field interactions, such as electron beam dynamics, traveling wave tubes, electrical discharges, and many others.

There are two basic approaches to this problem, the macroscopic, fluid continuum model known as MHD, and the microscopic statistical model known as plasma dynamics. We will be concerned here only with MHD, that is, electrically conducting liquids and fairly dense ionized gases.

11.2 ELECTRODYNAMICS OF MOVING MEDIA

Before we begin the study of MHD itself, a thorough understanding of classical electromagnetic field theory is necessary. A brief review is presented here, although it is assumed that the reader has some familiarity with Maxwell's equations.

The basic laws of electromagnetic theory may be presented in a mathematical form known as the Maxwell equations. These equations relate the basic field quantities and show how they are produced. In addition to the Maxwell equations, for a complete description of an electromagnetic system, we must relate the field quantities seen by observers in relative motion. These laws are all contained in the Special Theory of Relativity, but we will confine ourselves here to nonrelativistic problems, that is, we will always assume that all velocities are small compared to the speed of light, c (mathematically, $V^2/c^2 \ll 1$). We will use RMKS units throughout.

The Maxwell Equations

The four basic Maxwell equations are the following. The displacement field $\mathbf{D}$ is given by the source equation

$$\int_{\mathcal{V}} \rho_e \, d\mathcal{V} = \int_A \mathbf{D} \cdot d\mathbf{A} \qquad (11.1a)$$

which states that the true space charge density ρ_e generates the total $\mathbf{D}$ field. $\mathbf{D}$ lines must always terminate on charges and cannot just end in space. In differential form,

$$\nabla \cdot \mathbf{D} = \rho_e \qquad (11.1b)$$

The source equation for the magnetic field $\mathbf{H}$, which is known as Ampere's law, is

$$\oint \mathbf{H} \cdot d\mathbf{l} = \int_A \mathbf{J} \cdot d\mathbf{A} + \int_A \frac{\partial \mathbf{D}}{\partial t} \cdot d\mathbf{A} \qquad (11.2a)$$

where $\mathbf{J}$ is the true current density and $\partial\mathbf{D}/\partial t$ is the displacement current. In differential form,

$$\nabla \times \mathbf{H} = \mathbf{J} + \frac{\partial\mathbf{D}}{\partial t} \qquad (11.2b)$$

Faraday's law states

$$\oint \mathbf{E} \cdot d\mathbf{l} = -\int_A \frac{\partial\mathbf{B}}{\partial t} \cdot d\mathbf{A} \qquad (11.3a)$$

or

$$\nabla \times \mathbf{E} = -\frac{\partial\mathbf{B}}{\partial t} \qquad (11.3b)$$

Finally, for the magnetic induction field $\mathbf{B}$,

$$\int_A \mathbf{B} \cdot d\mathbf{A} = 0 \qquad (11.4a)$$

which states that all $\mathbf{B}$ lines must form closed loops. In differential form,

$$\nabla \cdot \mathbf{B} = 0 \qquad (11.4b)$$

These Maxwell equations hold for any observer, regardless of his motion, so long as he measures all field quantities in his frame of reference. Such equations which have the same form for all observers, regardless of their relative motion, are said to be covariant. The field quantities as measured by the various observers are related by the Lorentz transformations. Further, the quantities of mass, length, and time and their derivatives must also be related by the Lorentz transformations among various observers in relative motion. For low velocities (compared to that of light) the Galilean transformations relate distance (time and mass are then the same), but even then the electromagnetic field quantities must be related by the low velocity limit of the Lorentz transformations. These transformations may be derived from the principles of special relativity. Beginning with the postulate of the invariance of the speed of light and the covariance of the laws of nature all the appropriate transformations follow, but they will not be derived here.

Consider two frames of reference S and S' shown in Fig. 11-1. Frame S' moves with velocity $\mathbf{V}$ with respect to frame S. We will use S' throughout to denote the rest frame of material media so that the observer in S' moves locally with the medium (fluid in MHD). In the rest frame (and only in the rest frame, in general) can the necessary constitutive equations be written. These equations relate $\mathbf{D}'$ to $\mathbf{E}'$, $\mathbf{B}'$ to $\mathbf{H}'$, and usually are considered to include Ohm's law which relates $\mathbf{J}'$ to $\mathbf{E}'$. We will use primes

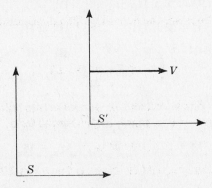

Fig. 11-1. Observers in relative motion. Frame S' moves with velocity $\mathbf{V}$ relative to frame S.

throughout to designate quantities measured in the rest frame. Hence we can write

$$\mathbf{D'} = \epsilon \mathbf{E'} = \epsilon_0 \kappa \mathbf{E'} = \mathbf{P'} + \epsilon_0 \mathbf{E'}$$

$$\mathbf{B'} = \mu \mathbf{H'} = \mu_0 \kappa_m \mathbf{H'} = \mu_0(\mathbf{H'} + \mathbf{M'}) \qquad (11.5)$$

$$\mathbf{J'} = \sigma \mathbf{E'}$$

where ϵ is the permittivity, ϵ_0 the permittivity of free space, and κ the relative permittivity. Similarly, μ is the permeability,[1] μ_0 the permeability of free space, and κ_m the relative permeability. σ is the scalar electrical conductivity. $\mathbf{P}$ is the polarization and $\mathbf{M}$ the magnetization. The first two equations of (11.5) are generally true, ϵ and μ being only functions of temperature and pressure for linear media and also functions of the fields if the material is nonlinear. Ohm's law is a phenomenological equation and is not generally true. The form given here for a scalar σ is good for liquids and dense gases, but in rarefied gases the conductivity becomes a tensor.

In any frame of reference we can write

$$\mathbf{D} = \mathbf{P} + \epsilon_0 \mathbf{E}$$

$$\mathbf{B} = \mu_0(\mathbf{M} + \mathbf{H}) \qquad (11.6)$$

but (11.5) is only true in the rest frame. In free space or in media where $\kappa = \kappa_m = 1$ we see that $\mathbf{P} = \mathbf{M} = 0$ and thus

$$\mathbf{D} = \epsilon_0 \mathbf{E}$$

$$\mathbf{B} = \mu_0 \mathbf{H} \qquad (11.7)$$

in any frame of reference.

In any MHD problem we usually need all four Maxwell equations and the constitutive equations. However, it is often convenient not to use the equation $\nabla \cdot \mathbf{D} = \rho_e$, but instead the equation

$$\frac{\partial \rho_e}{\partial t} + \nabla \cdot \mathbf{J} = 0 \qquad (11.8)$$

(or in steady state or in cases where ρ_e is negligible in this equation, $\nabla \cdot \mathbf{J} = 0$ as is the case in MHD). Equation (11.8) follows directly from Maxwell's equations and is not independent.

The Lorentz Transformations

The Lorentz transformations are now listed below. $\mathbf{V}$ is the velocity of the rest frame S' with respect to S and $\perp$ and $\parallel$ indicate the components of the quantities perpendicular to and parallel to the vector $\mathbf{V}$ respectively.

$$\mathbf{E'}_\perp = \beta(\mathbf{E} + \mathbf{V} \times \mathbf{B})_\perp \qquad\qquad \mathbf{E'}_\parallel = \mathbf{E}_\parallel$$

$$\mathbf{D'}_\perp = \beta(\mathbf{D} + \mathbf{V} \times \mathbf{H}/c^2)_\perp \qquad\qquad \mathbf{D'}_\parallel = \mathbf{D}_\parallel$$

$$\mathbf{H'}_\perp = \beta(\mathbf{H} - \mathbf{V} \times \mathbf{D})_\perp \qquad\qquad \mathbf{H'}_\parallel = \mathbf{H}_\parallel$$

$$\mathbf{B'}_\perp = \beta(\mathbf{B} - \mathbf{V} \times \mathbf{E}/c^2)_\perp \qquad\qquad \mathbf{B'}_\parallel = \mathbf{B}_\parallel$$

$$\mathbf{J'}_\perp = \mathbf{J}_\perp \qquad\qquad \mathbf{J'}_\parallel = \beta(\mathbf{J} - \rho_e \mathbf{V})_\parallel \qquad (11.9)$$

$$\rho'_e = \beta(\rho_e - \mathbf{V} \cdot \mathbf{J}/c^2)$$

$$\mathbf{P'}_\perp = \beta(\mathbf{P} - \mathbf{V} \times \mathbf{M}/c^2)_\perp \qquad\qquad \mathbf{P'}_\parallel = \mathbf{P}_\parallel$$

$$\mathbf{M'}_\perp = \beta(\mathbf{M} + \mathbf{V} \times \mathbf{P})_\perp \qquad\qquad \mathbf{M'}_\parallel = \mathbf{M}_\parallel$$

where $\beta = 1/\sqrt{1 - V^2/c^2}$.

---

[1] In this chapter the viscosity is denoted as μ_f to avoid confusion with permeability.

In the nonrelativistic limit, which is the case of interest in MHD, the transformations reduce to:

$$\mathbf{E}' = \mathbf{E} + \mathbf{V} \times \mathbf{B} \qquad\qquad \mathbf{J}' = \mathbf{J} - \rho_e \mathbf{V}$$

$$\mathbf{D}' = \mathbf{D} + \mathbf{V} \times \mathbf{H}/c^2 \qquad\qquad \rho_e' = \rho_e - \mathbf{V} \cdot \mathbf{J}/c^2$$

$$\mathbf{H}' = \mathbf{H} - \mathbf{V} \times \mathbf{D} \qquad\qquad \mathbf{P}' = \mathbf{P} - \mathbf{V} \times \mathbf{M}/c^2 \qquad\qquad (11.10)$$

$$\mathbf{B}' = \mathbf{B} - \mathbf{V} \times \mathbf{E}/c^2 \qquad\qquad \mathbf{M}' = \mathbf{M} + \mathbf{V} \times \mathbf{P}$$

The Boundary Conditions

The boundary conditions on the field quantities at an interface are the same as in stationary media. Across an interface the following conditions hold in any frame of reference. We need only write the conditions with *all* the field quantities measured in any arbitrary frame of reference. (See Fig. 11-2.)

$$\mathbf{n} \times (\mathbf{E}_2 - \mathbf{E}_1) = 0 \qquad\qquad \mathbf{n} \times (\mathbf{H}_2 - \mathbf{H}_1) = \mathbf{J}_s \qquad\qquad (11.11)$$

$$\mathbf{n} \cdot (\mathbf{D}_2 - \mathbf{D}_1) = \rho_s \qquad\qquad \mathbf{n} \cdot (\mathbf{B}_2 - \mathbf{B}_1) = 0$$

where ρ_s is the surface charge density on the interface and $\mathbf{J}_s$ is the surface current on the interface. Physically then, the tangential component of $\mathbf{E}$ and the normal component of $\mathbf{B}$ are continuous across the interface. The tangential component of $\mathbf{H}$ jumps if a surface current is present, and the normal component of $\mathbf{D}$ jumps if a surface charge exists.

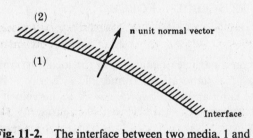

Fig. 11-2.   The interface between two media, 1 and 2.

11.3 THE INDUCED EMF AND TERMINAL VOLTAGE

In MHD problems we are often interested in the terminal potential or voltage between two electrodes. The question is: given two electrodes, what is the terminal voltage between them if an electrically conducting fluid flows between them?

In general we define the emf about a closed loop (in a laboratory frame of reference in which we locate the terminals) as

$$\text{emf} = \oint \mathbf{E}' \cdot d\mathbf{l} \qquad\qquad (11.12)$$

where $\mathbf{E}'$ is the local electric field seen by an observer moving locally with the fluid, but $d\mathbf{l}$ is fixed in the laboratory frame. Using Maxwell's equations and the nonrelativistic Lorentz transformations, we obtain

$$\text{emf} = \oint (\mathbf{V} \times \mathbf{B}) \cdot d\mathbf{l} - \int_A \frac{\partial \mathbf{B}}{\partial t} \cdot d\mathbf{A} \qquad\qquad (11.13)$$

where $\partial \mathbf{B}/\partial t$ must *not* be taken outside the integral. $\mathbf{V}$ is the velocity of the conductor instantaneously coincident with $d\mathbf{l}$.

The terminal potential V_{AB} is then the emf minus the IR drop between the electrodes A and B:

$$V_{AB} = \text{emf} - \int_A^B \mathbf{E}' \cdot d\mathbf{l} \qquad (11.14)$$

where V_{AB} is the voltage at B with respect to A and the line integral is taken between the electrodes, through the fluid. It should be noted that the emf may be dependent on the integration path used to determine it (if eddy currents are present in AC systems), but the terminal potential is always independent of the path of integration.

In steady MHD problems such as generators, pumps, etc., we obtain the simplified form of (11.14) which follows directly by using the fact that $\partial/\partial t = 0$ and the Lorentz transformations. We get

$$V_{AB} = - \int_A^B \mathbf{E} \cdot d\mathbf{l} \qquad (11.15)$$

Note that in (11.15) $\mathbf{E}$ is measured in the laboratory frame of reference, and (11.15) also follows from the fact that (since $\nabla \times \mathbf{E} = 0$ in steady state) $\mathbf{E} = -\nabla\phi$, and hence

$$\phi_{AB} = - \int_A^B \mathbf{E} \cdot d\mathbf{l}$$

and ϕ is the potential, identical with V_{AB}.

11.4 THE ELECTROMAGNETIC BODY FORCE

The electromagnetic body force is derivable from the Coulomb force law, since basically this is the only interaction that occurs.

The electromagnetic force on a particle of charge q in the local rest frame of the particle is $\mathbf{F} = q\mathbf{E}'$. Now, if we transform to the laboratory frame, $\mathbf{E}' = \mathbf{E} + \mathbf{V} \times \mathbf{B}$ (nonrelativistic) so that $\mathbf{F} = q(\mathbf{E} + \mathbf{V} \times \mathbf{B})$ which is the familiar Lorentz force law. $\mathbf{V}$ is the velocity of the particle with respect to the laboratory frame.

The Body Force in a Fluid

The complete electromagnetic body force in a conducting fluid may be derived, microscopically, from the Coulomb law, or by a virtual work and energy method. We will only present the result here. The body force density $\mathbf{f}'_e$ in the rest frame (because the constitutive equations have been used) is

$$\mathbf{f}'_e = \rho'_e \mathbf{E}' + \mathbf{J}' \times \mathbf{B}' - \tfrac{1}{2}\epsilon_0 E'^2 \nabla' \kappa - \tfrac{1}{2}\mu_0 H'^2 \nabla' \kappa_m$$

$$+ \tfrac{1}{2}\epsilon_0 \nabla'\left(E'^2 \frac{\partial \kappa}{\partial \rho} \rho \right) + \tfrac{1}{2}\mu_0 \nabla'\left(H'^2 \frac{\partial \kappa_m}{\partial \rho} \rho \right) \qquad (11.16)$$

However, in MHD, only the first two terms are important. These two terms are covariant and the expression holds in any frame of reference. In MHD then,

$$\mathbf{f}_e = \rho_e \mathbf{E} + \mathbf{J} \times \mathbf{B} \qquad (11.17)$$

and nonrelativistically, $\mathbf{f}_e = \mathbf{f}'_e$ so that we can compute $\rho'_e \mathbf{E}' + \mathbf{J}' \times \mathbf{B}'$ or $\rho_e \mathbf{E} + \mathbf{J} \times \mathbf{B}$ and obtain the same numerical result. As we will see, $\rho_e \mathbf{E}$ is usually negligible compared to $\mathbf{J} \times \mathbf{B}$.

The Stress Tensor

An alternative but equivalent description of the body force is the electromagnetic stress tensor T_{ij}. The body force is then $f_j = \dfrac{\partial T_{ij}}{\partial x_i} - \dfrac{\partial g_j}{\partial t}$ where g_i is the electromagnetic momentum flux vector. A covariant expression for T_{ij} (neglecting striction effects) is

$$T_{ij} = -\tfrac{1}{2}(\mathbf{D} \cdot \mathbf{E} + \mathbf{B} \cdot \mathbf{H})\delta_{ij} + D_i E_j + B_i H_j \qquad (11.18)$$

For all practical purposes the term $\partial g_i/\partial t$ is negligible and is of no concern in MHD. The most important use of the stress tensor concept is the physical interpretation of the body force which it permits. If the stress tensor is diagonalized (the principal stresses determined) we get an interesting result. If we make the assumptions, which are valid in MHD, that the electric terms in the stress tensor are negligible, and that B and H are collinear (they will be collinear in the rest frame or in any frame in MHD since there $\mathbf{B} = \mathbf{B}'$ and $\mathbf{H} = \mathbf{H}'$ for nonrelativistic speeds), we obtain the result that the three principal stresses, λ_i, are

$$\lambda_1 = \tfrac{1}{2}\mathbf{H} \cdot \mathbf{B}, \qquad \lambda_2 = \lambda_3 = -\tfrac{1}{2}\mathbf{H} \cdot \mathbf{B} \qquad (11.19)$$

and the principal axes are oriented so that λ_1 is a tension along the magnetic lines of force, and λ_2 and λ_3 represent a compression normal to the field lines. Alternatively, we can say that there is a hydrostatic compression of $\mathbf{H} \cdot \mathbf{B}/2$ with a tension of $\mathbf{H} \cdot \mathbf{B}$ (along the field lines) superposed on the hydrostatic compression.

The concept of magnetic pressure is based on these results. It should be remembered, however, that this body force is not a physical tension or pressure in the fluid, but enters into the momentum equation as a body force and consequently may generate mechanical stresses. In static equilibrium, the pressure gradient must be balanced out by the tensor divergence of the electromagnetic stress tensor (which, physically, is a body force). The situation is similar to pressure in a fluid being generated in a static fluid because of a gravitational field.

Another way to arrive at the idea of magnetic pressure is to write the body force $\mathbf{J} \times \mathbf{B}$ as $(\nabla \times \mathbf{H}) \times \mathbf{B}$ (where we have used the Maxwell equation $\nabla \times \mathbf{H} = \mathbf{J} + \dot{\mathbf{D}}$ and $\dot{\mathbf{D}}$ is negligible as we shall show). $(\nabla \times \mathbf{H}) \times \mathbf{B}$ may be written (using $\mathbf{B} = \mu\mathbf{H}$)

$$(\nabla \times \mathbf{H}) \times \mathbf{B} = -\nabla\left(\frac{B^2}{2\mu}\right) + \frac{1}{\mu}(\mathbf{B} \cdot \nabla)\mathbf{B} \qquad (11.20)$$

Hence the first term on the right is the irrotational part of the body force and adds directly to pressure in the momentum equation, and the second term is the rotational part (corresponding to the tension along the magnetic field lines) and is generally not zero in MHD.

11.5 BASIC CONCEPTS OF MHD FLOW

We will now review the basic assumptions of MHD and the relevant principles and equations.

Assumptions

The MHD assumptions which are usually made are the following:

1. All velocities are small compared to that of light, so that $V^2/c^2 \ll 1$.
2. The electric field is of order $\mathbf{V} \times \mathbf{B}$. Any $\mathbf{E}$ fields involved are induced or of the order of the induced field. We must always write $\mathbf{E}' = \mathbf{E} + \mathbf{V} \times \mathbf{B}$ and distinguish between $\mathbf{E}'$ and $\mathbf{E}$.
3. $\mathbf{H} \cong \mathbf{H}'$ and $\mathbf{B} \cong \mathbf{B}'$, since

$$\mathbf{B}' = \mathbf{B} - \frac{\mathbf{V} \times \mathbf{E}}{c^2} \cong \mathbf{B} - 0\left[\frac{V^2}{c^2}\mathbf{B}\right]$$

Liquid metals and ionized gases have permeability μ_0, so that we write $\mathbf{B} = \mu_0\mathbf{H}$ in any frame of reference.

4. The displacement current $\dot{\mathbf{D}}$ is negligible with respect to $\mathbf{J}$.

5. Ohm's law is $\mathbf{J} = \sigma\mathbf{E}' = \sigma(\mathbf{E} + \mathbf{V} \times \mathbf{B})$, and $\mathbf{J}' \cong \mathbf{J}$ since $\rho_e \mathbf{V}$ is neglected compared to $\sigma(\mathbf{E} + \mathbf{V} \times \mathbf{B})$.

6. The space charge ρ_e' is zero, but $\rho_e \neq 0$. We must write $\nabla \cdot \mathbf{D} = \rho_e$, and *not* set $\nabla \cdot \mathbf{D}$ to zero. Since ρ_e is not usually known, $\nabla \cdot \mathbf{D} = \rho_e$ is not a useful equation. It is better and more convenient to use $\nabla \cdot \mathbf{J} = 0$ (even in unsteady problems, since space charge transport is negligible compared to $\mathbf{J}$).

7. The force $\rho_e \mathbf{E}$ is negligible compared to $\mathbf{J} \times \mathbf{B}$. The electric stress and energy, proportional to $\mathbf{E} \cdot \mathbf{D}$, is negligible compared to $\mathbf{H} \cdot \mathbf{B}$.

8. For high conductivity, $\sigma \to \infty$, Ohm's law indicates that for finite $\mathbf{J}$, $\mathbf{E}' = 0$, and $\mathbf{E} = -\mathbf{V} \times \mathbf{B}$. The current is then determined by $\nabla \times \mathbf{H} = \mathbf{J}$ and not by Ohm's law.

Equations

The basic equations of MHD can be written then as the Maxwell equations, Ohm's law, the equation of continuity, the equation of motion with the $\mathbf{J} \times \mathbf{B}$ body force, and the energy equation with Joule heating. In addition, we must use the nonrelativistic Lorentz transformations. The equation of state for a perfect gas can still be used under the MHD approximations. We can list the equations as:

Maxwell's:
$$\left.\begin{aligned} \nabla \times \mathbf{H} &= \mathbf{J} \\ \nabla \cdot \mathbf{B} &= 0 \\ \nabla \times \mathbf{E} &= -\partial\mathbf{B}/\partial t \\ \nabla \cdot \mathbf{J} &= 0 \end{aligned}\right\} \tag{11.21}$$

Ohm's Law:
$$\mathbf{J} = \sigma(\mathbf{E} + \mathbf{V} \times \mathbf{B}) \tag{11.22}$$

Motion:
$$\rho\left[\frac{\partial\mathbf{V}}{\partial t} + (\mathbf{V} \cdot \nabla)\mathbf{V}\right] = -\nabla p + \nabla \cdot \tau + \mathbf{J} \times \mathbf{B} - \rho\nabla\psi$$
$$= -\nabla(p + B^2/2\mu) + (\mathbf{B} \cdot \nabla)\mathbf{B}/\mu + \nabla \cdot \tau - \rho\nabla\psi \tag{11.23}$$

where τ is the mechanical stress tensor and ψ is the gravitational potential.

Energy:
$$\rho\frac{Du}{Dt} = -p\nabla \cdot \mathbf{V} + \kappa_T\nabla^2 T + \Phi + \mathbf{E}' \cdot \mathbf{J}' \tag{11.24}$$

We use κ_T for thermal conductivity here to avoid confusion with relative permittivity. u is the specific internal energy. The Joule heat $(\mathbf{E}' \cdot \mathbf{J}' \cong \mathbf{E}' \cdot \mathbf{J} = J^2/\sigma)$ is the only interaction if the material is linear, that is, ϵ and μ are not functions of temperature. The reader is referred to reference 5 for a detailed discussion of this equation.

State:
$$p = \rho RT \tag{11.25}$$

Parameters

These equations may be put into dimensionless form, as discussed in Chapter 4, and the relevant dimensionless parameters obtained. We will not go through the calculations here, but will list the parameters. In addition to all the parameters of ordinary flow (of Chapter 4) we have two additional independent parameters. We list several (not independent) below.

R_m = Magnetic Reynolds number = $V_0 L\sigma\mu$, which is a measure of the ratio of magnetic convection to magnetic diffusion. If $R_m \ll 1$, it can be shown that the induced magnetic field is small compared to the applied magnetic field.

M_m = Magnetic Mach number = V_0/A, where A is the Alfvén speed, defined as $\sqrt{\mathbf{B} \cdot \mathbf{H}/\rho}$ which will be discussed presently.

M = Hartmann number = $\sqrt{\sigma B_0^2 L^2/\mu_f} = \sqrt{N_R R_m/M_m}$, which is a measure of the ratio of the magnetic body force to the viscous force.

P_m = Magnetic Prandtl number = $\nu/\eta = R_m/N_R$, which is a measure of the ratio of vorticity diffusion to magnetic diffusion.

N = The interaction parameter, which is a measure of the ratio of the magnetic body force to the inertia force. For finite conductivity N is taken as $\sigma B_0^2 L/\rho V_0 = R_m/M_m^2$, and for very high or infinite conductivity $N = M_m^{-2}$.

Magnetic Transport

A very important equation may be derived by combining Maxwell's equations and Ohm's law. $\nabla \times \mathbf{H} = \mathbf{J} = \sigma(\mathbf{E} + \mathbf{V} \times \mathbf{B})$, and $\mathbf{E}$ may be eliminated by taking the curl of the above equation and using $\nabla \times \mathbf{E} = -\dot{\mathbf{B}}$. Then making use of a few vector identities and $\nabla \cdot \mathbf{B} = 0$, we arrive at

$$\frac{\partial \mathbf{B}}{\partial t} = \eta \nabla^2 \mathbf{B} + \nabla \times (\mathbf{V} \times \mathbf{B}) \qquad (11.26)$$

which is the equation of transport, diffusion and convection, of the magnetic field. η is the magnetic diffusivity, $1/\sigma\mu$. This equation may be used simultaneously with the equation of motion [with the body force written as $(\nabla \times \mathbf{H}) \times \mathbf{B}$] in certain problems where it is not convenient to introduce the electric field.

By introducing the dimensionless variables

$$\mathbf{B}^* = \mathbf{B}/\mathbf{B}_0, \qquad t^* = tV_0/L, \qquad \mathbf{r}^* = \mathbf{r}/L$$

the transport equation becomes

$$\frac{\partial \mathbf{B}^*}{\partial t^*} = -\frac{1}{R_m} \nabla^* \times (\nabla^* \times \mathbf{B}^*) + \nabla^* \times (\mathbf{V}^* \times \mathbf{B}^*) \qquad (11.27)$$

which indicates the significance of R_m. The first term on the right side is a diffusion term and the second a convection. If no motion is present, $\mathbf{V} = 0$, and (11.26) becomes

$$\frac{\partial \mathbf{B}}{\partial t} = \eta \nabla^2 \mathbf{B} \qquad (11.28)$$

a diffusion equation where η is the diffusivity. On a laboratory scale the diffusion usually dominates, ($R_m \ll 1$), but on an astronomical scale (for example in the sun) convection dominates ($\sigma \to \infty$, and $R_m \gg 1$). In fact, as $\sigma \to \infty$, the magnetic field is transported entirely by convection and there is no diffusion of the field lines through the conducting media. Hence it is said that then the magnetic field lines are "frozen" in the fluid and move along with it.

There is an exact analogy between the transport of the magnetic field here and the transport of vorticity in ordinary fluid mechanics. The classical Kelvin theorems, etc., may be applied here to the magnetic field.

We may think of the field lines as being elastic, and the flowing fluids drag them until they are in static equilibrium. This is the case when a magnetic field is induced (which then adds to the original applied magnetic field). The induced field is caused by the distortion of the applied field lines because of the fluid convecting them. As $\sigma \to \infty$ the lines lose their ability to resist being dragged along by the fluid.

Alfvén Waves

As we have shown earlier, stress tensor considerations show that the body force is equivalent to a hydrostatic pressure plus a tension of $\mathbf{H} \cdot \mathbf{B}$ along the field lines. As we will show later, these lines behave like taut strings and transverse waves will propagate along them at what is called Alfvén speed, $\sqrt{\mathbf{H} \cdot \mathbf{B}/\rho}$. Coupled with this transverse field disturbance must be a transverse velocity wave which is 90° out of phase with the magnetic wave. These waves, which are called Alfvén or magnetohydrodynamic waves, travel without dispersion or attenuation if $\sigma \to \infty$ and viscosity is neglected.

The Bernoulli and Kelvin Theorems

The equation of motion (*11.23*) may be written for an inviscid fluid and may be integrated along a streamline to yield (for a compressible fluid in steady flow)

$$\frac{V_2^2 - V_1^2}{2} + \psi_2 - \psi_1 + \int_1^2 \frac{d(p + B^2/2\mu)}{\rho} = \frac{1}{\mu} \int_1^2 \frac{(\mathbf{B} \cdot \nabla)\mathbf{B} \cdot ds}{\rho} \qquad (11.29)$$

If $(\mathbf{B} \cdot \nabla)\mathbf{B} = 0$, the classical Bernoulli theorem is valid with the pressure replaced by $(p + B^2/2\mu)$ which is then the total (mechanical plus magnetic) pressure.

By forming the circulation, the equation of motion yields the MHD form of Kelvin's theorem as

$$\frac{D\Gamma}{Dt} = \frac{D}{Dt} \int_A (\nabla \times \mathbf{V}) \cdot d\mathbf{A} = \int_A \nabla \times \left[\frac{1}{\mu} (\mathbf{B} \cdot \nabla)\mathbf{B} \right] \cdot d\mathbf{A} \qquad (11.30)$$

The force $(\mathbf{B} \cdot \nabla)\mathbf{B}$ is rotational and may generate vorticity even though the fluid is inviscid.

11.6 INCOMPRESSIBLE VISCOUS MHD FLOW

As one of the simplest examples of MHD flow we will now discuss the steady flow of an electrically conducting, viscous, incompressible fluid between parallel plates with an applied transverse magnetic field. The configuration is shown in Fig. 11-3. The details of this problem provide an excellent illustration of method and in particular of the construction of an equivalent electric circuit. We assume the flow to be fully developed so that only pressure varies in the x direction. The channel extent in the z direction is much greater than that in the y direction so that no variations occur with z. The electrodes are assumed to be perfect conductors, and the fluid has conductivity σ. The applied magnetic field $\mathbf{B}_0$ is steady and uniform. This problem is known as the Hartmann problem, after the Danish physicist.

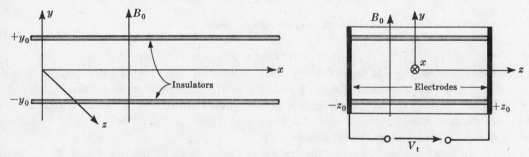

Fig. 11-3. Flow between parallel plates. The Hartmann problem.

On a laboratory scale the magnetic Reynolds number for such a problem is very small and the induced magnetic field would be small and could be thought of as produced by a slight dragging of the $\mathbf{B}_0$ lines by the flowing fluid. However, we will see that for this problem the induced field (which will be entirely in the x direction) does not couple into the equations of motion at all and may be found as the final step in the problem. In many problems (which are not one-dimensional) the induced field may couple into the basic equations and make a solution very difficult. Even then it is often possible, if $R_m \ll 1$, to assume that the induced field is negligible compared to $\mathbf{B}_0$.

We begin the Hartmann problem by writing the Maxwell equations, Ohm's law, and the equations of motion. From continuity, $v = w = 0$. From $\nabla \cdot \mathbf{J} = 0$ we conclude that J_z must only depend on y and be equal to (by Ohm's law)

$$J_z = \sigma(E_z + uB_0)$$

From $\nabla \times \mathbf{E} = 0$ we conclude that E_x is a constant which we arbitrarily take as zero since no electric field is applied in the x direction. Also, since the electrodes have high conductivity, $E_y = 0$ in the electrodes and hence everywhere. It follows that E_z is a function only of y.

The pressure gradient $\partial p / \partial x$ is a constant, but a gradient $\partial p / \partial y$ will also exist due to a "pinch" effect.

The equation of motion in the x direction becomes

$$0 = -\frac{\partial p}{\partial x} + \mu_f \frac{d^2 u}{dy^2} - \sigma(E_z + uB_0)B_0 \qquad (11.31)$$

which can be integrated immediately with the boundary conditions

$$u = 0, \qquad y = \pm y_0$$

to give the velocity profile as

$$u = \frac{y_0^2}{M^2}\left(\frac{1}{\mu_f}\frac{\partial p}{\partial x} + \frac{M}{y_0}\sqrt{\frac{\sigma}{\mu_f}}E_z\right)\left(\frac{\cosh My/y_0}{\cosh M} - 1\right) \qquad (11.32)$$

where M, the Hartmann number is $\sqrt{B_0^2 y_0^2 \sigma / \mu_f}$.

The y equation of motion is

$$0 = -\frac{\partial p}{\partial y} + J_z B_x$$

and allows determination of $\partial p / \partial y$ once B_x, the induced magnetic field, is found later. From (11.32) and Ohm's law we find J_z as

$$J_z = \sigma E_z \frac{\cosh My/y_0}{\cosh M} + \frac{y_0}{M}\sqrt{\frac{\sigma}{\mu_f}}\frac{\partial p}{\partial x}\left(\frac{\cosh My/y_0}{\cosh M} - 1\right) \qquad (11.33)$$

The total current (per unit channel length in the x direction) is given by

$$\mathfrak{I} = \int_{-y_0}^{+y_0} J_z \, dy = 2\sigma y_0 E_z \frac{\tanh M}{M} + \frac{2y_0^2}{M}\sqrt{\frac{\sigma}{\mu_f}}\frac{\partial p}{\partial x}\left(\frac{\tanh M}{M} - 1\right) \qquad (11.34)$$

The terminal voltage V_t (voltage of electrode at $+z_0$ with respect to the one at $-z_0$) is defined as

$$V_t = -\int_{-z_0}^{+z_0} E_z \, dz = -2z_0 E_z$$

and using (11.34) to solve for E_z in terms of $\mathfrak{I}$, we find

$$V_t = -\frac{z_0 M \mathfrak{I}}{\sigma y_0 \tanh M} + \frac{2y_0 z_0}{M\sqrt{\sigma \mu_f}}\frac{\partial p}{\partial x}\left(1 - \frac{M}{\tanh M}\right) \qquad (11.35)$$

Now we can construct an equivalent circuit of the device by adding the open circuit voltage in series with the effective internal resistance R_i which is defined as the ratio of the open circuit voltage $V_{t_{oc}}$ to the short circuit current I_{sc}. From (11.35) V_{toc} is found by setting $\mathfrak{J} = 0$ to get

$$V_{toc} = \frac{2y_0 z_0}{M\sqrt{\sigma\mu_f}} \frac{\partial p}{\partial x}\left(1 - \frac{M}{\tanh M}\right) \qquad (11.36)$$

(which is a negative number) and I_{sc} is found by setting $E_z = 0$ in (11.34) to get

$$\mathfrak{J}_{sc} = \frac{2y_0^2}{M}\sqrt{\frac{\sigma}{\mu_f}}\frac{\partial p}{\partial x}\left(\frac{\tanh M}{M} - 1\right) \qquad (11.37)$$

which is a negative number, which indicates current flows from the $+z_0$ to the $-z_0$ terminal through the channel. Hence for a device of length L, I_{sc} (the total short circuit current for the entire device) is $L\mathfrak{J}_{sc}$ so that R_i is

$$R_i = \frac{V_{toc}}{I_{sc}} = \frac{z_0}{L\sigma y_0}\frac{M}{\tanh M} \qquad (11.38)$$

The equivalent circuit then is, by Thévenin's theorem, as shown by Fig. 11-4. A load resistor R_L may be attached to the terminals as indicated, and by using Kirchhoff's law we get

$$V_{toc} + I(R_L + R_i) = 0 \qquad (11.39)$$

Note that R_L may also be replaced by an external generator. The sign on V_t and I depends on whether a load R_L or an external generator is attached.

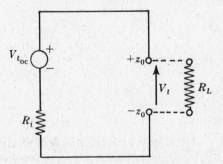

Fig. 11-4. The equivalent circuit of an MHD flow device with sign conventions. For the Hartmann problem, V_{toc} is negative and hence the polarity of V_{toc} is the reverse of that shown.

The above circuit method may be used to analyze generators, pumps, and meters in incompressible or compressible flow for any channel shape and lays the groundwork for the MHD generator analysis.

Now the induced magnetic field H_x may be found from $\nabla \times \mathbf{H} = \mathbf{J}$, since $\mathbf{J}$ is known. From (11.33),

$$\frac{dH_x}{dy} = -J_z = -\sigma E_z \frac{\cosh My/y_0}{\cosh M} - \frac{y_0}{M}\sqrt{\frac{\sigma}{\mu_f}}\frac{\partial p}{\partial x}\left(\frac{\cosh My/y_0}{\cosh M} - 1\right)$$

which may be integrated to find $H_x(y)$. The boundary condition depends on the geometric configuration of the return circuit, and can be determined by applying Ampere's circuital law. If the return circuit is symmetric in y (consisting of two plates as shown in Fig. 11-3), the value of $H_x(y = 0) = 0$ and H_x is antisymmetric in y. H_x has a constant value between the insulating walls and the return path plates, then it drops to zero through the plates. Integration of the above equation gives

$$H_x = \frac{y_0^2}{M}\sqrt{\frac{\sigma}{\mu_f}}\frac{\partial p}{\partial x}\cdot\frac{y}{y_0} - \left(\frac{y^2}{M}\sqrt{\frac{\sigma}{\mu_f}}\frac{\partial p}{\partial x} + \frac{I}{2L}\right)\frac{\sinh My/y_0}{\sinh M} \qquad (11.40)$$

In Hartmann flow, the flow rate is reduced as the external load R_L is increased (for a given Hartmann number). If an external voltage source is applied, the flow rate may be further reduced (and even made to flow against the pressure gradient) or it may be increased, depending on the polarity of the applied voltage. For a fixed value of R_L, the profile is flattened as the Hartmann number is increased, as shown in Fig. 11-5. However, if an external voltage is applied, the curve may not flatten but steepen. In general, an increasing Hartmann number means an increasing interaction.

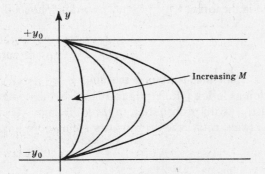

Fig. 11-5.   The velocity profile in Hartmann flow for a load
resistor R_L and a constant pressure gradient.

This analysis of the Hartmann problem may be applied to generators and pumps, the device acting as either depending on the terminal conditions and pressure gradient. These methods will be applied to the compressible flow in an MHD generator channel in Section 11.7. The extension to pipes and channels of finite cross section will not be made here, but the reader is referred to reference 5.

11.7 WAVES AND SHOCKS IN MHD

Earlier, we mentioned a new type of wave that occurs in MHD, the Alfvén wave. Now we will investigate general plane wave motion and see that the Alfvén wave is just one of several new phenomena that arise because of electromagnetic field-fluid interactions. We will confine our attention here to plane waves, since these are the only type that are amenable to simple description, but the conclusions that may be reached by this analysis suffice to give a good picture of the waves that propagate in MHD.

If we assume that a wave propagates in the x direction, a plane wave is characterized by the fact that no variations of any variable occur in the y or z directions. True plane waves do not exist in nature; but at least over small regions of space, waves may exhibit plane wave behavior. And, in any case, more complex wave geometries may always be thought of as a superposition of waves which behave like plane waves.

There are two types of plane waves, transverse and longitudinal. Transverse waves, such as shear waves and TEM (electromagnetic waves), involve only y and z components of vector variables in the wave (assuming propagation in the x direction). Pressure and density variations do not accompany transverse waves. Longitudinal waves are those that have only x components of variables, or pure scalars, entering into the wave motion. A common example is the acoustic or sound wave which carries with it a pressure and density disturbance. In MHD we will find that the transverse shear waves couple with the field quantities to give a new type of wave (the Alfvén wave), and the ordinary acoustic wave is split into fast, slow and intermediate magnetoacoustic waves.

In our study we will assume that the fluid is a continuum and that Ohm's law holds. In rarefied gases, a more complex fluid model must be used and the results, particularly at high frequency, are different. The study of radio waves in the ionosphere is based on these rarefied ionized gas (or plasma) oscillations, but we cannot discuss them here. Another assumption which we will make which is valid for liquids and dense gases is that the displacement current is negligible compared to the conduction current. This assumption amounts to saying that $\sigma/\omega\epsilon \gg 1$ where ω is the angular frequency and ϵ is the dielectric constant. For mercury (in RMKS units), $\sigma \cong 10^6$, $\epsilon \cong 10^{-12}$, so that for frequencies even into the microwave region the assumption is quite good. Even for poorly ionized gases at $\sigma \cong 1$, the frequency would have to be of order 10^{12} hertz before displacement current would become significant.

The approach here will be to assume that the waves consist of small disturbances or perturbations in the variables. We perturb the governing equations, linearize, and assume phasor solutions. Then a characteristic value problem may be set up and a dispersion equation derived. We assume that all disturbances may be expressed in the form $e^{j(\omega t - kx)}$ for propagation in the x direction. j is $\sqrt{-1}$.

Plane Waves in Gases

Neglecting displacement current and assuming the fluid to be initially at rest, we can derive the appropriate dispersion equation as follows. Referring to Fig. 11-6, the applied magnetic field is assumed to have components H_{0x} and H_{0y}, with propagation in the x direction. There is no loss of generality in leaving H_{0z} out, since we can always rotate the coordinate system about the x axis so that H_{0z} is zero.

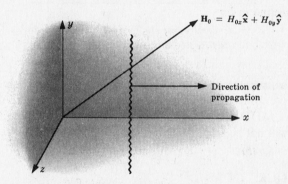

Fig. 11-6.   Plane wave propagation in the presence of an applied magnetic field.

The relevant equations are three motion, three magnetic transport, continuity, energy, and state, a total of nine equations. We assume perturbation quantities are represented by small letters and the perturbation pressure, density and temperature are represented by primes:

$$\mathbf{H} = \mathbf{H}_0 + \mathbf{h} \qquad \rho = \rho_0 + \rho'$$
$$\mathbf{V} = \mathbf{v}, (\mathbf{V}_0 = 0) \qquad T = T_0 + T'$$
$$p = p_0 + p'$$

The components of $\mathbf{v}$ are u, v, and w.

The continuity equation is
$$\frac{\partial p}{\partial t} + \nabla \cdot (\rho \mathbf{V}) = 0$$

which becomes
$$\frac{\partial}{\partial t}(\rho_0 + \rho') + \nabla \cdot [(\rho_0 + \rho')\mathbf{v}] = 0$$

The first order linearized equation is
$$\frac{\partial \rho'}{\partial t} + \rho_0 \frac{\partial u}{\partial x} = 0$$

Then the algebraic form (assuming a phasor solution $\rho' = \rho'^* e^{j(\omega t - kx)}$ and $u = u^* e^{j(\omega t - kx)}$, where ρ'^* and u^* are phasors) is

$$u^* \rho k - \rho'^* \omega - u^* \rho_0 k = 0 \qquad (11.41)$$

Similarly, we can write these algebraic forms of the remaining eight linearized equations. The first order equations become:

State:
$$\frac{p'}{p_0} = \frac{T'}{T_0} + \frac{\rho'}{\rho_0} \qquad (11.42)$$

Motion:

(a) $\qquad \rho_0 \frac{\partial u}{\partial t} = -\frac{\partial p'}{\partial x} + (\zeta + \tfrac{4}{3}\nu\rho_0)\frac{\partial^2 u}{\partial x^2} - \mu H_{0y}\frac{\partial h_y}{dx}$

(b) $\qquad \rho_0 \frac{\partial v}{\partial t} = \nu\rho_0 \frac{\partial^2 v}{\partial x^2} + \mu H_{0x}\frac{\partial h_y}{\partial x}$ $\qquad\qquad (11.43)$

(c) $\qquad \rho_0 \frac{\partial w}{\partial t} = \nu\rho_0 \frac{\partial^2 w}{\partial x^2} + \mu H_{0x}\frac{\partial h_z}{\partial x}$

Energy (the dissipation is second order):

$$\rho_0 c_p \frac{\partial T'}{\partial t} = \frac{\partial p'}{\partial t} + \kappa_T \nabla^2 T' \qquad (11.44)$$

Magnetic Transport:

(a) $\qquad \dfrac{\partial h_x}{\partial t} = \eta \dfrac{\partial^2 h_x}{\partial x^2}$

(b) $\qquad \dfrac{\partial h_y}{\partial t} = \eta \dfrac{\partial^2 h_y}{\partial x^2} + H_{0x}\dfrac{\partial v}{\partial x} - H_{0x}\dfrac{\partial u}{\partial x}$ $\qquad\qquad (11.45)$

(c) $\qquad \dfrac{\partial h_z}{\partial t} = \eta \dfrac{\partial^2 h_z}{\partial x^2} + H_{0x}\dfrac{\partial w}{\partial x}$

Now we can obtain the algebraic forms. There result nine algebraic equations for the nine phasor quantities. However, these equations are not linearly independent and the determinant of the coefficients of the phasors must be zero. This equation is then the dispersion equation. The determinental form is

$$
\begin{array}{c}
\quad\begin{array}{ccccccccc} h_x^* & h_z^* & w^* & u^* & v^* & h_y^* & p'^* & \rho'^* & T'^* \end{array}\\[4pt]
\left.
\begin{array}{l}
(12.45a)\\
(12.43c)\\
(12.45c)\\
(12.41)\\
(12.43a)\\
(12.43b)\\
(12.45b)\\
(12.42)\\
(12.44)
\end{array}
\right|
\begin{array}{c}
\\
\boxed{2 \times 2}\\
\\
\\
6 \times 6\\
\\
\\
\end{array}
\right| = 0
\end{array}
\qquad (11.46)
$$

There are three uncoupled terms. The first term is not important and merely indicates that $h_x = 0$. The 2×2 determinant represents two transverse waves and the 6×6 represents a set of coupled longitudinal and transverse waves. We will not go into the details of these general waves, but will look at some special cases.

Transverse Waves

The 2×2 determinant may be written out as

$$\nu \eta k^4 + [A_x^2 + j(\nu + \eta)\omega]k^2 - \omega^2 = 0 \qquad (11.47)$$

which represents two transverse modes since it is quadratic in k^2. These modes are due to a coupling between viscous and magnetic diffusion and the Alfvén wave. (A $+$ or $-$ sign indicates a forward or backward wave.) A_x is the x component of Alfvén velocity, $\sqrt{B_{0x} H_{0x}/\rho_0}$. If $A_x = 0$, then we get two uncoupled waves given by

$$(\eta k^2 + j\omega)(\nu k^2 + j\omega) = 0 \qquad (11.48)$$

which represents pure viscous diffusion and pure magnetic diffusion (since k^2 is imaginary).

If $\nu = \eta = .0$ (no dissipation), then we get $A_x^2 k^2 - \omega^2 = 0$ so that the phase velocity, $v_p = \omega/(\text{Re } k)$, is the Alfvén speed, that is, $v_p = \pm A_x$. Then the k is real and no attenuation occurs.

The transverse waves are identical for gases or liquids.

Coupled Longitudinal Waves

The 6×6 determinant of (11.46) written out explicitly is

$$\left[\kappa_T \left\{ \frac{1}{\rho_0} + j\, \frac{\omega}{\rho_0 p_0} \left(\tfrac{4}{3}\nu\rho_0 + \zeta \right) \right\} k^4 - \left\{ \frac{\omega^2 \kappa_T}{p_0} - j\omega c_p + \frac{\omega^2 c_p}{a_0^2 \rho_0} \left(\tfrac{4}{3}\nu\rho_0 + \zeta \right) \right\} k^2 - j\, \frac{\omega^3 c_p}{a_0^2} \right]$$

$$\cdot \left[\nu\eta k^4 + \left\{ A_x^2 + j(\nu + \eta)\omega \right\} k^2 - \omega^2 \right] - k^2 A_y^2 (j\omega + \nu k^2)\left(\frac{\omega^2 c_p}{a_0^2} - \frac{j\omega \kappa_T k^2}{p_0} \right) = 0 \qquad (11.49)$$

where $A_x = \sqrt{B_{0x} H_{0x}/\rho_0}$, $A_y = \sqrt{B_{0y} H_{0y}/\rho_0}$, and $a_0 = $ the sonic speed, $\sqrt{kRT_0}$.

We will not discuss the general case, but will look at some special cases of particular interest.

If we set the field $\mathbf{H}_0 = 0$ ($A_x = A_y = 0$) then we get ordinary acoustic waves in a viscous, heat conducting fluid. If we further set $\nu = \zeta = \kappa_T = 0$ we get ordinary simple acoustic waves and (11.49) reduces to

$$a_0^2 k^2 - \omega^2 = 0$$

so that $v_p = \pm a_0$, the ordinary sonic velocity.

If $\nu = \eta = \kappa_T = \zeta = 0$, then there is no dissipation and we get "ideal" magnetoacoustic waves. These waves may be examined in detail and are of extreme importance in MHD. We note that (11.49) reduces to two factors, one identical to the Alfvén wave of the previous section, and the other quadratic in k^2:

$$[(a_0^2 k^2 - \omega^2)(A_x^2 k^2 - \omega^2) - k^2 A_y^2 \omega^2](A_x^2 k^2 - \omega^2) = 0 \qquad (11.50)$$

The two roots of the quadratic expression represent two longitudinal waves (magnetoacoustic waves), a fast and slow wave. The phase velocities are:

Longitudinal:

$$v_{p_{\text{fast}}} = \pm\left[\tfrac{1}{2}(A_x^2 + A_y^2 + a_0^2) + \sqrt{\tfrac{1}{4}(A_x^2 + A_y^2 + a_0^2)^2 - a_0^2 A_x^2} \right]^{1/2}$$

$$v_{p_{\text{slow}}} = \pm\left[\tfrac{1}{2}(A_x^2 + A_y^2 + a_0^2) - \sqrt{\tfrac{1}{4}(A_x^2 + A_y^2 + a_0^2)^2 - a_0^2 A_x^2} \right]^{1/2} \qquad (11.51)$$

Transverse: $$v_p = \pm A_x$$

The character of these waves can best be seen physically by looking at a plot of the phase velocity as a function of direction of propagation relative to the magnetic field. In Fig. 11-7 we show a plot of the phase velocity in a three-dimensional polar diagram. We see that, in general, there are the three waves, and for propagation at right angles to $\mathbf{H}_0$ ($H_{0x} = 0$, then) there is only one wave which is a modified acoustic wave.

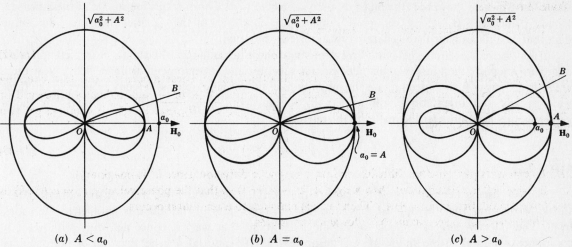

Fig. 11-7. Phase velocity of ideal magnetoacoustic waves. The radius vector OB represents the phase velocity of plane wave propagation in that direction relative to the applied magnetic field H_0. The diagram should be imagined rotated about H_0 to form three-dimensional surfaces. $A^2 = A_x^2 + A_y^2$.

From a practical standpoint, A is usually much smaller than a_0 in liquid metals and the effect on acoustic behavior is difficult to measure in the laboratory. On an astronomical and planetary scale, however, such is not the case and these waves become important.

Another type of diagram of interest is the ray-normal diagram for the ideal magnetoacoustic waves. These are called Friedrichs diagrams and are shown in Fig. 11-8. These diagrams are the envelope of plane waves traveling outward from the origin and hence represent the unit time development in space of a pulse at the origin. In ordinary acoustics such a diagram would be a sphere of radius a_0. (The Mach cone is the development of a sound pulse emanating from a supersonically moving source.) These diagrams may be constructed geometrically from Fig. 11-7. The perpendiculars are drawn to the rays

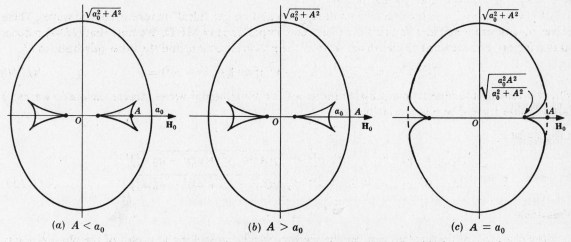

Fig. 11-8. The ray-normal, or Friedrichs diagrams for ideal magnetoacoustic waves. A sound pulse at the origin develops into these surfaces in unit time. They should be rotated about H_0 to form three-dimensional surfaces.

(such as OB) where they cross the phase velocity locus curve. We form a continuous set of these perpendicular planes as OB sweeps through all angles. The envelopes of these planes then form the sound disturbance surfaces which are generated by a pulse at the origin.

The disturbance surface generated by a moving source (as an aircraft) becomes rather complicated, and there are more than just two distinct cases (as subsonic and supersonic in ordinary flow). Now there are four regions of flow depending on the magnitude of the moving objects' velocity relative to the three wave speeds and orientation with respect to the magnetic field. We will not discuss magneto-aerodynamics here but refer the reader to the references.

MHD Shock Waves

We will discuss here only a special case of the MHD shock, which we will call the normal MHD shock. We assume that the applied magnetic field on both sides of the shock is parallel to the shock front. The velocity *may be oblique* to the shock front, so that we are concerned only with velocity components normal to the shock front. The normal MHD shock is a generalization of the ordinary oblique shock. The more general case of an arbitrary magnetic field will not be discussed here.

We will carry out the analysis in the shock frame of reference and assume that B is the same in all frames. Referring to Fig. 11-9 we can write the following conservation equations across the shock.

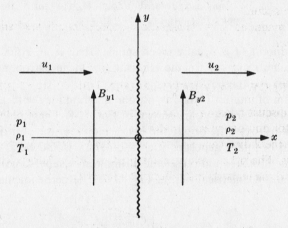

Fig. 11-9. The normal MHD shock. We orient the coordinate
system so that $B_{z1} = 0$, without loss of generality. B_{z2}
must also be zero to satisfy magnetic transport.

Continuity:
$$[\rho u]_1^2 = 0$$

Momentum:
$$\left[\rho u^2 + p + \frac{1}{2\mu} B_y^2\right]_1^2 = 0$$

Magnetic Transport:
$$[uB_y]_1^2 = 0, \qquad B_{z1} = B_{z2} = 0$$

Also, v and w, the y and z velocity components, must be continuous.

Energy:
$$\left[\frac{p}{\rho}\left(\frac{k}{k-1}\right) + \frac{u^2}{2} + \frac{B_y^2}{\rho\mu}\right]_1^2 = 0$$

The symbol $[\ \ ]_1^2$ is used to represent the jump condition. That is, $[A]_1^2 = A_2 - A_1$, where A is any quantity. These four equations above allow a complete description of the shock and are analogous to the Rankine-Hugoniot equations.

A current sheet exists in the shock to effect the change in B_y.

Solving, we find the shock speed a_s into still air (the same as u_1 in Fig. 11-9) as

$$a_s = \frac{2}{k+1} \cdot \frac{a_1^2 + A_1^2[1 + (1 - k/2)(\rho_2/\rho_1 - 1)]}{\rho_1/\rho_2 - (k-1)/(k+1)} \qquad (11.52)$$

where a_1 is the sonic speed upstream and A_1 the upstream Alfvén speed, $\sqrt{B_{y1}^2/\mu\rho_1}$. As A_1 goes to zero we get the shock speed of ordinary gas dynamics. The shock speed is increased by the presence of the magnetic field. As the strength becomes small, $\rho_2/\rho_1 \to 1$, $a_s^2 = a_1^2 + A_1^2$, which gives the magnetosonic velocity in a perpendicular field.

The pressure jump $[p]_1^2 = p_2 - p_1$ is given by

$$p_2 - p_1 = a_s^2 \rho_1(1 - \rho_1/\rho_2) + \frac{B_{y1}^2}{2\mu}(1 - \rho_2^2/\rho_1^2) \qquad (11.53)$$

and a magnetosonic Mach number, $M_t^2 = a_s^2/(a_1^2 + A_1^2)$, can be expressed as

$$M_t^2 = \frac{2}{k+1} \cdot \frac{1 + [A_1^2/(a_1^2 + A_1^2)](1 - k/2)(\rho_2/\rho_1 - 1)}{\rho_1/\rho_2 - (k-1)/(k+1)} \qquad (11.54)$$

As $\rho_2/\rho_1 \to 1$, $M_{t1} \to 1$. For $M_{t1} < 1$, no shock can occur; and for $M_{t1} > 1$ and $a_s^2 > a_1^2$ there can exist a shock. For $a_s^2 < a_1^2$ no shock can occur for any value of A_1. Hence as the field is increased beyond a critical value, no shock can occur.

For a general oblique magnetic field the situation is more complicated and a multiplicity of shocks and discontinuities can occur.

11.8 COMPRESSIBLE FLOW—MAGNETOGASDYNAMIC CHANNEL FLOW

In this section we will discuss quasi-one-dimensional compressible channel flow which is vital to the study of the MHD generator and propulsion systems. As shown in Fig. 11-10 we assume the channel to vary in the cross section in the x direction, and the electrodes to be parallel perfect conductors and form two opposite channel walls. The other walls are assumed to be insulators. The velocity components, v and w, are assumed $\ll u$, and the induced field effects are neglected.

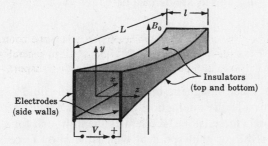

Fig. 11-10. The one-dimensional channel with compressible flow.

The terminal potential V_t is related to the electric field E_z (E_y and E_x are zero) by

$$E_z = -V_t/l$$

so that E_z is constant throughout the channel, and $J_z = \sigma(E_z + uB_0)$ is a function only of x (since u is a function of x). We could let l vary with x, then E_z would be a function of x but we confine our discussion here to parallel electrodes.

Under open circuit conditions the total external current I is zero, but $\mathfrak{J}(x)$ (the current density per unit length) is not necessarily zero and currents can circulate in the channel. We cannot determine $V_{t_{oc}}$

without knowing the function $u(x)$. However, on a local scale, (at a given value of x) the local open circuit conditions ($J_z = 0$) are given by $E_z = -uB_0$. Under local short circuit, $E_z = 0$. Hence any given local portion of the channel operates as a generator if $-uB_0 < E_z < 0$. In a full channel this condition must be satisfied at every value of x. By considering the sign of E_z, we can show that E_z must also be negative for the channel to act as a pump or flow meter. We will confine our study to negative E_z.

Basic One-Dimensional Flow

The basic channel flow equations can be obtained by averaging the differential equations across the channel. We assume steady flow, σ to be a scalar constant, a perfect gas, and no variations across the channel, and adiabatic flow.

Continuity:
$$\rho u A = m \text{ (a constant)} \tag{11.55}$$

Motion:
$$\rho u \frac{du}{dx} = -\frac{dp}{dx} - \sigma(E_z + uB_0)B_0 \tag{11.56}$$

Energy:
$$\rho u\left[c_p \frac{dT}{dx} + u \frac{du}{dx}\right] = \sigma(E_z + uB_0)E_z \tag{11.57}$$

State:
$$p = \rho RT \tag{11.58}$$

Here, A is the cross sectional area of the channel.

These equations cannot be solved explicitly for a general area variation $A(x)$, but a useful first integral may be obtained. By suitably combining the above equations we obtain

$$\frac{du}{dx} = \frac{(u/A)(dA/dx) - (\sigma B_0^2/p)(u - u_3)(u - u_1)}{M^2 - 1} \tag{11.59}$$

$$\frac{dM}{dx} = \frac{1}{(M^2 - 1)}\left\{\left(1 + \frac{k-1}{2} M^2\right) \frac{M}{A} \frac{dA}{dx} - \left(1 + \frac{k-1}{2} M^2\right) \frac{\sigma B_0^2}{ap} (u - u_3)(u - u_2)\right\} \tag{11.60}$$

where the critical velocities u_1, u_2, and u_3 are defined as

$$u_1 = -\frac{k-1}{k} \cdot \frac{E_z}{B_0} = \frac{k-1}{k} u_3, \qquad u_2 = \frac{1 + kM^2}{2 + (k-1)M^2} u_1, \qquad u_3 = -\frac{E_z}{B_0} = \frac{V_t}{B_0 l} \tag{11.61}$$

and M is the local Mach number.

Now the net effect of the electromagnetic interaction with the fluid is not merely the $\mathbf{J} \times \mathbf{B}$ force, but the Joule heat also enters into the interaction. The net effect then can be determined from (11.59) and (11.60). For $u < u_3$ the body force, alone, accelerates the flow, and for $u > u_3$ the body force, alone, decelerates the flow. The speed u_1 is that speed at which the body force and Joule heat cancel each other. The net effect then is: if $M > 1$ the flow is accelerated if u lies between u_3 and u_1, and decelerated if u lies outside that region. If $M < 1$, the flow follows the converse behavior. The net effect on M may be seen from (11.60). If $M > 1$, M increases with x if u lies between u_2 and u_3 and decreases with x if M lies outside this region. For $M < 1$, the reverse is true.

For operation at $E_z < 0$ and for various values of u_0 (inlet velocity at $x = 0$), we can plot qualitatively the behavior of u and M along the channel. This behavior is shown in Fig. 11-11 and 11-12. For generator operation, $u > u_3$, and from the figures we see that in only cases $1A$ and $2A$ is $u > u_3$. In $2A$, u is $>u_3$ throughout and in $1A$ there is the possibility for the channel to act as a generator only over a portion of its length (case $1Ab$). We must remember, however, that $u > u_3$ is the local requirement for generator operation, and the overall channel behavior ($V_{t_{oc}}$, I_{sc} and R_i) can only be calculated after $u(x)$ is known.

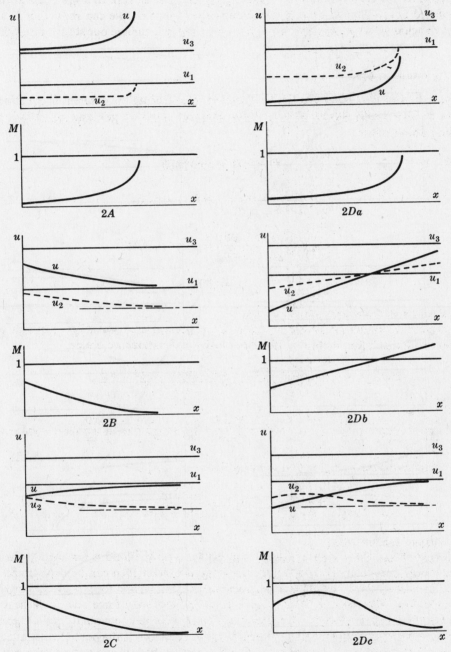

Fig. 11-11. The qualitative behavior of u and M along a constant area channel for various subsonic initial conditions. 2A: $u_2 < u_1 < u_3 < u_0$. 2B: $u_2 < u_1 < u_0 < u_3$. 2C: $u_2 < u_0 < u_1 < u_3$. 2Da, 2Db, 2Dc: $u_0 < u_2 < u_1 < u_3$.

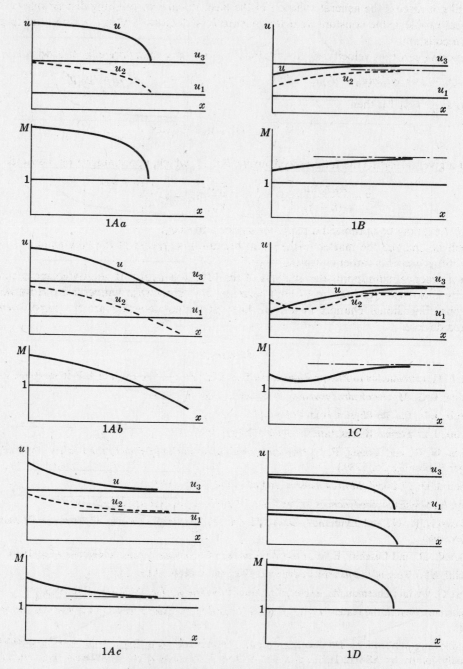

Fig. 11-12. The qualitative behavior of u and M along a constant area channel for various supersonic initial conditions. $1Aa$, $1Ab$, $1Ac$: the three possibilities when $u_1 < u_2 < u_3 < u_0$. $1B$: $u_1 < u_2 < u_0 < u_3$. $1C$: $u_1 < u_0 < u_2 < u_3$. $1D$: $u_0 < u_1 < u_2 < u_3$.

Generator Operation

We will not discuss the general solution of the basic equations, but only discuss one special simple but practical problem, the constant velocity generator. $A(x)$ must vary in a special way to give rise to $u(x)$ being a constant.

By imposing constant velocity, u_0, on the basic equations we can derive the following relationship:

$$p_0 - p = x B_0^2 \sigma (u_0 - u_3), \qquad T/T_0 = (p/p_0)^{(k-1)u_3/ku_0}, \qquad A/A_0 = (p_0/p)^{1-(k-1)u_3/ku_0} \qquad (11.62)$$

The total current I is then

$$I = \int_0^L \sigma(-V_t + u_0 B_0) \frac{A}{l} \, dx \qquad (11.63)$$

so that for a given V_t (or u_3) the total power output P is IV_t which, upon integration, becomes

$$P = \frac{A_0 p_0 k u_0}{(k-1)} \left[1 - \left\{ 1 - \frac{B_0 \sigma}{p_0} (u_0 - u_3) L \right\}^{(k-1)u_3/ku_0} \right] \qquad (11.64)$$

Equation (11.62) may be optimized directly for u_3 (for a fixed u_0).

We will not pursue the matter further here because integration of the flow equations under arbitrary conditions becomes rather complex.

From a practical standpoint the analysis of the MHD generator is somewhat more complex. The conductivity is a tensor (giving rise to Hall currents) and varies with temperature. However, the principles of one-dimensional channel flow have been presented here and are the same, even for more complicated devices.

References

1. Chen, F. F., *Introduction to Plasma Physics and Controlled Fusion*, 2nd ed., Vol. I, Plenum Press, 1984.

2. Cowling, T. G., *Magnetohydrodynamics*, Interscience, 1957.

3. Delcroix, J. L., *Plasma Physics*, John Wiley, 1965.

4. Delcroix, J. L., *Plasma Waves*, Interscience Publisher, 1963.

5. Hughes, W. F., and Young, F. J., *The Electromagnetodynamics of Fluids*, John Wiley, 1966. (Reprinted by Krieger Publishing Co., 1989.)

6. Jackson, J. D., *Classical Electrodynamics*, 2nd ed., John Wiley, 1975.

7. Jeffrey, A., *Magnetohydrodynamics*, Interscience (John Wiley), Oliver and Boyd, 1966.

8. Kulikovskiy, A. G., and Lyubimov, G. A., *Magnetohydrodynamics* (translated from the Russian), Addison-Wesley, 1965.

9. Landau, L. D., and Lifschitz, E. M., *Electrodynamics of Continuous Media*, Addison-Wesley, 1960.

10. Shercliff, A., *A Text of Magnetohydrodynamics*, Pergamon Press, 1965.

11. Sutton, G. W., and Sherman, A., *Engineering Magnetohydrodynamics*, McGraw-Hill, 1965.

For current research the reader is referred to the following journals: *Journal of Fluid Mechanics* and *Physics of Fluids*.

The following publications are also of interest: *The Proceedings of Engineering Aspects of MHD*, published annually, jointly by ASME, IEEE, and AIAA. *The Proceedings of the Beer-Sheva International Seminar on MHD Flows and Turbulence*. Published periodically by AIAA in the series *Progress in Astronautics and Aeronautics*.

Solved Problems

11.1. The Faraday disk generator is shown in Fig. 11-13. Consider a conducting thin disk rotating with uniform angular velocity ω in the presence of a uniform steady magnetic field $\mathbf{B}_0$ in the

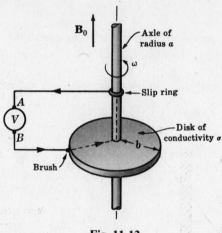

Fig. 11-13

axial direction. The perfectly conducting axle (of radius a) and periphery of the disk are connected by sliding contacts to a meter. What open circuit voltage does the meter read?

Since the motion is steady, there are no eddy currents and the emf is the same as the terminal voltage. Hence

$$V_{AB} = -\int_a^b E_r \cdot dr$$

and under open circuit, $J_r = 0 = \sigma(E_r + r\omega B_0)$. Hence

$$V_{AB} = \int_a^b r\omega B_0 \, dr = \tfrac{1}{2}(b^2 - a^2)\omega B_0$$

Alternatively we can say

$$\text{emf} = V_{BA} = \oint \mathbf{V} \times \mathbf{B} \cdot d\mathbf{l} - \int \frac{\partial \mathbf{B}}{\partial t} \cdot d\mathbf{A}$$

and taking a path through the fixed leads, disk, and axle, we get only a $\mathbf{V} \times \mathbf{B}$ contribution in the disk. It gives

$$V_{AB} = -V_{BA} = \int_a^b (r\omega)B_0 \, dr = \tfrac{1}{2}(b^2 - a^2)\omega B_0$$

which agrees with the result obtained above.

11.2. What are the boundary conditions of the induced magnetic field at the wall of a flow channel?

Consider fully developed flow in a rectangular channel of constant area. The flow region near a wall is shown in Fig. 11-14. At a wall that is a perfect conductor, $\sigma = \infty$, the tangential electric field, E_y, is zero. Also, the fluid velocity is zero at the wall. Hence $J_y = \sigma(E_y + \mathbf{V} \times \mathbf{B}) = 0$ at the wall. But from $\nabla \times \mathbf{H} = \mathbf{J}$, we have $\partial H_x/\partial z = 0$ where z is the coordinate normal to the wall, and the flow is in the x direction. For an insulating wall, $J_z = 0$ at the wall, so that we find $\partial H_x/\partial y = 0$. Hence we conclude that at a perfectly conducting wall

$$\partial H_x/\partial n = 0$$

and at an insulating wall

$$\partial H_x/\partial t = 0$$

where t and n are the tangential and normal directions to the wall, and both are normal to the flow direction x.

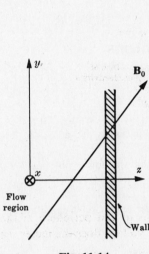

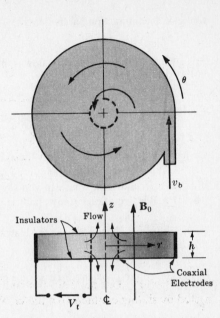

Fig. 11-14   Fig. 11-15

In a finite conductivity wall, more complex relationships may be derived. The reader is referred to reference 5 for details.

11.3. Calculate the terminal characteristics of the compressible vortex generator shown in Fig. 11-15.

m is the flow rate and v_b the inlet tangential velocity. B_0 is steady, uniform and applied along the z axis. Cylindrical screen electrodes are used. We assume: $v \gg u$ (where v is the θ component of velocity and u the radial component), no variations with z, and $R_m \ll 1$ (induced field negligible). The θ equation of motion is

$$\rho u(\partial v/\partial r + v/r) = -J_r B_0$$
$$= -\sigma(E_r + vB_0)B_0$$

Then, from $\nabla \cdot \mathbf{J} = 0$, $J_r = I/2\pi rh$ and $m = 2\pi u\rho rh$ so that the above equation reads

$$\frac{m}{2\pi rh}\left(\frac{dv}{dr} + \frac{v}{r}\right) = -\frac{IB_0}{2\pi rh}$$

which integrates to

$$\frac{v}{v_b} = \frac{b}{r} + \frac{IB_0 r}{2mv_b}\left(\frac{b^2}{r^2} - 1\right)$$

V_t is found from

$$V_t = -\int_a^b E_r\, dr$$

where E_r is given in terms of I from

$$I = J_r \cdot 2\pi rh = \sigma(E_r + vB_0)2\pi rh$$

We find after integrating and setting $I = 0$,

$$V_{t_{oc}} = v_b bB_0 \ln (b/a)$$

and setting $V_t = 0$,

$$I_{sc} = 2\pi h\sigma v_b bB_0\left[1 - \frac{N}{2}\left(1 - \frac{b^2/a^2 - 1}{2(b/a)^2 \ln (b/a)}\right)\right]^{-1}$$

where N, the interaction parameter is

$$N = 2\pi\sigma B_0^2 \, b^2 h/m$$

m is a negative number since the flow is inward (u is negative) and r is positive outward.

Supplementary Problems

11.4. Under what conditions is the body force $\mathbf{J} \times \mathbf{B}$ irrotational?

11.5. If $\mathbf{J} \times \mathbf{B}$ is irrotational, what can you say about the flow pattern? In two-dimensional incompressible flow when $\mathbf{J} \times \mathbf{B}$ is irrotational, show that the velocity distribution is exactly the same as for potential flow.

11.6. In the above problem, what are the restrictions on $\mathbf{B}_0$ such that the flow is potential? What is the orientation of $\mathbf{B}_0$?

11.7. In the Hartmann problem discuss the sign of E_z and I. Under open circuit conditions is $V_{t_{oc}}$ a positive or negative number?

11.8. In the Hartmann problem show that $\partial p/\partial x$ is a constant throughout the channel.

11.9. In the Hartmann problem solve for the pressure variation across the channel. ($\partial p/\partial y$ is not zero.)

11.10. Show that in two-dimensional flow where the velocity and $\mathbf{B}_0$ are coplanar, $\nabla^2 \phi$ is proportional to the only component of $\nabla \times \mathbf{V}$, the vorticity vector. ϕ is the electric potential.

11.11. Consider a missile traveling through the ionosphere with velocity $\mathbf{V}_0$ in the presence of a magnetic field. What is the boundary condition on $\mathbf{E}$, relative to the missile, at infinity?

11.12. Show that if $\epsilon = \epsilon_0$ and $\mu = \mu_0$ for a conducting liquid, then in any frame of reference, we can write the constitutive equations as $\mathbf{B} = \mu_0 \mathbf{H}$ and $\mathbf{D} = \epsilon_0 \mathbf{E}$.

11.13. Show that the above equations can be written exactly in any frame only if both ϵ and μ are ϵ_0 and μ_0 respectively. (However, under the MHD approximation, $\mathbf{B} = \mu\mathbf{H}$ is valid in any frame.)

11.14. Find the dispersion equation for transverse waves in a fluid moving in the direction of propagation with speed V_0. The fluid is conducting and a magnetic field is applied.

11.15. Why is the magnetic transport equation not needed in the Hartmann problem? Could you use it to work the problem in a different manner?

11.16. If the transport equation is used instead of the Maxwell's equations and Ohm's law, how does the external electric circuit enter into the formulation? *Hint.* Show how the boundary conditions of the induced field depend on the external electric circuit.

11.17. Solve the Couette MHD problem (which is the Hartmann problem with the top plate moving with velocity V_0 in the direction of flow).

11.18. Suppose the top and bottom insulating walls of the Hartmann problem have finite conductivity. Show that these walls just act as external resistances in parallel and do not change the solution.

11.19. How is the induced magnetic field affected in the above problem?

11.20. Is there such a thing as a moving field?

11.21. Calculate I_{sc} and R_i for the Faraday generator of Problem 11.1.

11.22. In Problem 11.2, under what conditions is $H_x = 0$ at the channel wall?

11.23. Solve the vortex generator for a radial field of the form $B_r = B_0/r$ (see Problem 11.3).

11.24. What is the appropriate energy equation for Problem 11.23? Can it be integrated to find $T(r)$? What about $p(r)$? *Hint.* Assume $v^2/2 \gg u^2/2$.

NOMENCLATURE FOR CHAPTER 11

a = sonic velocity

a_s = shock velocity

A = Alfvén velocity; area

$\mathbf{B}$ = magnetic induction field

$\mathbf{B}_0$ = applied magnetic field

c = velocity of light

c_p = specific heat at constant pressure

$\mathbf{D}$ = displacement field

$\mathbf{E}$ = electric field

$\mathbf{H}$ = magnetic field

$\mathtt{J}$ = current per unit length of channel

I = total current for a device

I_{sc} = short circuit current

$\mathbf{J}$ = current flux

k = ratio of specific heats, c_p/c_v; propagation constant

L = characteristic length

m = mass rate of flow

$\mathbf{M}$ = magnetization vector

M = Hartmann number

M_m = magnetic Mach number

N_R = Reynolds number

N = Interaction parameter

p = pressure

P = power

$\mathbf{P}$ = polarization vector

P_m = magnetic Prandtl number

R_m = magnetic Reynolds number

T_{ij} = electromagnetic stress tensor

u = x component of velocity, internal energy

v = y component of velocity

v_p = phase velocity

$\mho$ = volume

$\mathbf{V}$ = velocity vector

V_t = terminal potential

$V_{t_{oc}}$ = open circuit terminal potential

w = z component of velocity

β $\quad = 1/\sqrt{1 - V^2/c^2}$

ϵ $\quad =$ permittivity

ϵ_0 $\quad =$ permittivity of free space

ζ $\quad =$ second coefficient of viscosity

η $\quad =$ magnetic diffusivity

κ $\quad =$ relative permittivity

κ_m $\quad =$ relative magnetic permeability

κ_T $\quad =$ thermal conductivity

μ $\quad =$ magnetic permeability

μ_0 $\quad =$ magnetic permeability of free space

μ_f $\quad =$ fluid viscosity

ν $\quad =$ kinematic viscosity

ρ $\quad =$ fluid density

ρ_e $\quad =$ charge density

σ $\quad =$ electrical conductivity

τ $\quad =$ mechanical stress tensor

Φ $\quad =$ dissipation function

ϕ $\quad =$ electric potential

ψ $\quad =$ gravitational potential

ω $\quad =$ frequency

Chapter 12

Non-Newtonian Fluids

12.1 INTRODUCTION

There are fluids that do not obey the simple relationship between shear stress and shear strain rate given by equation (3.52) for a Newtonian fluid. These fluids have been given the general name non-Newtonian fluids. Many common fluids are non-Newtonian. Examples are: paints, solutions of various polymers, food products such as apple sauce and ketchup, emulsions of water in oil or oil in water, and suspensions of various solids and fibers in a liquid paper pulp or coal slurries and the drilling mud used in well drilling. It is not for lack of applications that many textbooks and courses have failed to give any consideration to non-Newtonian fluids. Although the properties of non-Newtonian fluids do not lend themselves to the elegant and precise analysis that has been developed for Newtonian fluids, the flow of non-Newtonian fluids does possess some interesting, useful and even exciting characteristics. For example: in the fracturing treatment of oil wells, materials have been developed which when added to water will make a fluid so thick that it will suspend sand, glass or metal pellets. Yet the same fluid can be pumped down a well through tubing at enormous rates with less than half the friction loss of water. In a typical fracturing job, over 100 barrels (4200 gallons) might be pumped per minute through several thousand feet of three-inch ID tubing. (Fracturing of oil and gas wells is a process used to increase the production of the well. A horizontal crack is initiated in the producing zone. Fluid pumped down the well under high pressure greatly extends this crack. The sand, or pellets, act as propping agents to hold the fracture open after the treatment. Fluid must be delivered at a rapid rate to overcome the loss by diffusion of fluid within the pores of the fractured rock.)

Coal slurries having consistency of over 80% by volume of powdered or crushed coal in water possess non-Newtonian properties and can be pumped long distances with much less power required for pumping than pure water.

12.2 CHARACTERISTICS AND CLASSIFICATION OF NON-NEWTONIAN FLUIDS

General

As was illustrated in Section 3.3, the shearing stress τ for a Newtonian fluid is linearly related to the shearing rate $\dot{\gamma}$ by the viscosity μ,

$$\tau = \mu\dot{\gamma} = \mu \frac{\partial u}{\partial y} \qquad (12.1)$$

(where $\dot{\gamma}$ is the rate of shear strain);[1] or conversely, the shearing rate is linearly related to shearing stress,

$$\dot{\gamma} = \frac{1}{\mu}\tau \qquad (12.2)$$

In what follows, non-Newtonian fluids which do not obey this linear relationship will be grouped and discussed in three general classifications:

1. The simplest of these is the time independent non-Newtonian fluid in which the shear rate is a unique but non-linear function of the shear stress.

---

[1] In Chapter 3 we used the notation γ_{ij} for the shear strain rate. Here we use γ for shear strain and $\dot{\gamma}$ for shear strain rate, since both terms may be considered in a non-Newtonian fluid.

2. Time dependent non-Newtonian fluids have more complex shearing stress strain rate relationships. In these fluids the shearing rate is not a single valued function of the shear stress. The shear rate depends on shearing time or on the previous shear stress rate history of that fluid.

3. In viscoelastic fluids, shear strain as well as strain rate are related in some way to shear stress. Unlike a truly viscous fluid in which all its energy of deformation is dissipated, some of the energy of deformation of a viscoelastic fluid may be recoverable as it is in the deformation of an elastic solid.

Time Independent Fluids

For the time independent non-Newtonian fluid,

$$\dot{\gamma} = f(\tau) \qquad\qquad (12.3)$$

A Newtonian fluid is simply a special case of the above where the function $f(\tau)$ is linear, and thus is also termed as being purely a viscous non-Newtonian fluid. The majority of the non-Newtonian fluids that we encounter probably fall into this category; and in some cases flow of fluids not in this category, such as time dependent fluids, may be approximated in this category for simple cases such as steady flow in a pipe or Couette flows.

The time independent non-Newtonian fluids have been commonly represented by three distinct types as shown by Fig. 12-1. These are:

1. Bingham plastics, curve A
2. pseudoplastic fluids, curve B
3. dilatant fluids, curve C.

The Newtonian fluids are indicated by straight lines as shown by line D.

Bingham Plastics

Bingham plastics exhibit a yield stress at zero shear rate, followed by a straight line relationship between shear stress and shear rate. The characteristics of these fluids are defined by two constants: the

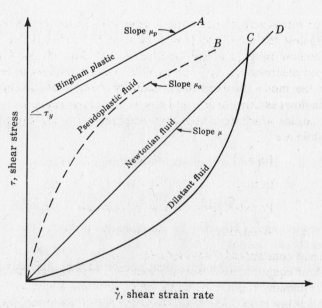

Fig. 12-1. Typical shear stress strain rate relationships for non-Newtonian fluids.

yield stress τ_y which is the stress that must be exceeded for flow to begin, and the plastic viscosity μ_p which is the slope of the straight line portion of curve A in Fig. 12-1. The equation for a Bingham plastic is then

$$\tau = \tau_y + \mu_p \dot{\gamma} \qquad (12.4)$$

The Bingham plastic concept has been found to closely approximate the behavior of many real fluids such as slurries, plastics, emulsions such as paint, and suspensions of finely divided solids in a liquid. An important example of the latter are drilling muds, which consist primarily of clays suspended in water.

Because of its simple straight line relationship between shear stress and shear rate, the Bingham plastic concept is convenient for analysis.

Pseudoplastic Fluids

Pseudoplastic fluids (curve B of Fig. 12-1), as well as dilatant fluids (curve C), do not have a yield stress. The pseudoplastic fluid is also characterized by a progressively decreasing slope of shear stress versus shear rate. This slope has been defined as apparent viscosity

$$\mu_a = \tau/\dot{\gamma} \qquad (12.5)$$

At very high rates of shear in real fluids the apparent viscosity becomes constant and equal to μ_∞, and the shear stress versus shear rate becomes linear.

There are a number of empirical relations that have been used to describe pseudoplastic fluids. The simplest of these is the power law due to Ostwald. It may be written as

$$\tau = k\dot{\gamma}^n \qquad \text{where} \qquad n < 1 \qquad (12.6)$$

k and n are constant for a particular fluid. k is a measure of the consistency of the fluid and n, the exponent, is a measure of how the fluid deviates from a Newtonian fluid. Note that for a Newtonian fluid $n = 1$ and $k = \mu$, the viscosity of the fluid.

Defining apparent viscosity as

$$\mu_a = \tau/\dot{\gamma} \qquad (12.7)$$

equation (*12.6*) gives
$$\mu_a = k\dot{\gamma}^{(n-1)} \qquad (12.8)$$

Note that when the shear rate is zero, the apparent viscosity is infinite. This is one of several objections that have been raised against the use of the power law model. Another is that for real fluids n is not constant over the entire flow range, and still another is that the constant k has dimensions which depend on n. As has been mentioned, n approaches 1, i.e. Newtonian flow, at very high rates of shear. Nevertheless the power law model has been found desirable because of its simplicity, as well as being adequate to analyze such flows as Couette flow and flow in pipes and channels.

Other empirical equations which have been proposed and which also overcome some of the objections to the power law fluid are

Prandtl	$\tau = A \sin^{-1}(\dot{\gamma}/C)$	(*12.9*)
Eyring	$\tau = \dot{\gamma}/B + C \sin(\tau/A)$	(*12.10*)
Powell-Eyring	$\tau = A\dot{\gamma} + B \sinh^{-1} C\dot{\gamma}$	(*12.11*)
Williamson	$\tau = A\dot{\gamma}/(B + \dot{\gamma}) + \mu_\infty \dot{\gamma}$	(*12.12*)

where A, B and C represent constants (different for each model).

Because the latter four equations lead to much greater complexity of analysis than the power law model which is adequate in many engineering applications, further discussion of the pseudoplastic fluids will be limited to the power law model. For cases where the power law model does not give an adequate representation of the fluid, it might be more practical to use a computer program that works from the actual measured properties of the fluid rather than resort to another empirical relationship.

Dilatant Fluids

Dilatant fluids are similar to pseudoplastic fluids in having no yield stress. They differ from pseudo-plastic fluids in that the apparent viscosity increases with increasing shear rate. These fluids are far less common than the pseudoplastic fluids. As with the pseudoplastic fluids, they may be represented by the power law model where the exponent n is now greater than unity.

The power law fluids can be represented graphically most simply by plotting the logarithms of shear stress versus the logarithms of shear rate, as illustrated in Fig. 12-2. Taking the logarithm of both sides of equation (12.6),

$$\log \tau = \log k + n \log \dot{\gamma} \qquad (12.13)$$

which is the equation of a straight line where the slope is n and the intercept of the curves with $\log \dot{\gamma} = 0$ or $\dot{\gamma} = 1$ gives the value of $\log k$, the consistency constant. A Newtonian fluid, $n = 1$, is just a special case of a power law fluid.

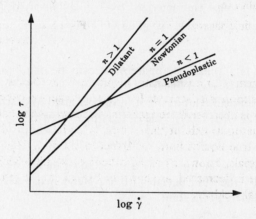

Fig. 12-2.   Log-log plot of power law fluids.

Time Dependent Fluids

Some fluids are more complex than those just described and the apparent viscosity depends not only on the shear rate but also on the time the shear has been applied. There are two general classes of such fluids:

1. thixotropic fluids
2. rheopectic fluids.

The shear stress decreases with time as the fluid is sheared for a thixotropic fluid and increases with time for a rheopectic fluid. A common example of a thixotropic fluid is printer's ink which is usually worked over several rolls before being applied to a plate.

Thixotropic Fluids

The consistency or apparent viscosity of thixotropic fluids depends on the length of time of shearing as well as on the shear rate. As the fluid is sheared from a state of rest, it breaks down (on a molecular scale), but then the structural reformation will increase with time. An equilibrium situation is eventually reached where the breakdown rate is equal to the buildup rate. If allowed to rest, the fluid builds up slowly and eventually regains its original consistency. Thixotropy is then a reversible process.

Fig. 12-3 shows a plot of stress versus strain rate for a thixotropic fluid immediately after shearing and after the fluid has rested for varying times. The initial curve is shown as Newtonian in Fig. 12-3 but could well be non-Newtonian.

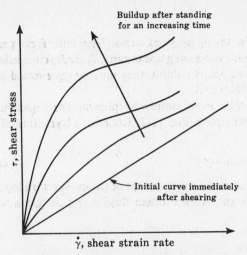

Fig. 12-3. Thixotropic fluid sheared at different times.

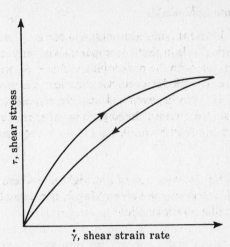

Fig. 12-4. Hysteresis loops for a thixotropic fluid.

If a thixotropic fluid is sheared at a constantly increasing rate, then at a constantly decreasing rate, a curve similar to a hysteresis loop is generated. In Fig. 12-4 is such a curve for a pseudoplastic type of thixotropic fluid. As the shear is decreased, the apparent viscosity is less than for the increasing shear.

Some Bingham plastic materials exhibit thixotropic behavior but, if the stress is high enough, will break down and behave like true liquids until the structure reforms. This behavior is indicated in Fig. 12-5(a). However, some materials, known as as false-bodies, exhibit a yield stress even after shearing, although the yield stress value is decreased, as shown in Fig. 12-5(b). It generally takes a long time for a false-body to regain its original yield strength.

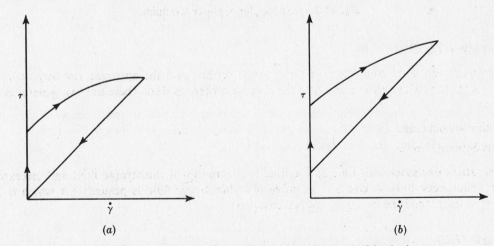

Fig. 12-5. (a) True thixotropic Bingham plastic, (b) false-body
behavior.

Rheopectic Fluids

In rheopectic fluids, molecular structure is formed by shear and the behavior is opposite to that of thixotropy. A simple example of the formation of structure by shear is the beating and thickening of egg whites, although egg white is probably not a true rheopectic fluid. Many substances lose their rheopectic property at extremely high shear rates and may then even behave as thixotropic fluids.

Viscoelastic Fluids

A viscoelastic material exhibits both elastic and viscous properties. The simplest type of such a material is one which is Newtonian in viscosity and obeys Hooke's law for the elastic part. We can write

$$\dot{\gamma} = \tau/\mu_0 + \dot{\tau}/\lambda \qquad (12.14)$$

where λ is a rigidity modulus. Under steady flow then, $\dot{\gamma} = \dot{\tau}/\mu_0$ and the fluid behaves like a simple Newtonian fluid. However, if the shear stress is changed, an elastic effect is noticed.

Maxwell first proposed (12.14) in the form

$$\tau + (\mu_0/\lambda)\dot{\tau} = \mu_0 \dot{\gamma} \qquad (12.15)$$

and liquids which obey this law are known as *Maxwell liquids*. The constant $(\mu_0/\lambda)^{-1}$ is known as the relaxation time and is, physically, the time constant for exponential decay of stress at a constant strain. If the motion is stopped the stress relaxes as $e^{-t\lambda/\mu_0}$.

Rather complex models of viscoelastic materials have been developed in which higher time derivatives of τ and γ appear. For time varying processes, the elastic constants may be complex functions of frequency. A very readable introduction to these models is given by Wilkinson (see reference 10).

12.3   LAMINAR FLOW IN PIPES

Power Law Fluids

To illustrate an application of the power law model, the equations for the velocity profile and pressure drop versus flow rate will be derived for fully developed laminar flow of a power law model fluid in a pipe.

The equation of shear stress for steady one dimensional flow in cylindrical coordinates (which is merely the steady state equation of motion for fully developed flow) is

$$\frac{d}{dr}(r\tau) = r\left(\frac{dp}{dz}\right) \qquad (12.16)$$

which for pipe flow integrates to

$$\tau = \frac{r}{2}\left(\frac{dp}{dz}\right) \qquad (12.17)$$

where τ is the shearing stress defined positive when acting in the downstream direction on the surface of a cylindrical element, and dp/dz is the downstream pressure gradient as shown in Fig. 12-6.

For a power law fluid, the rheological equation is

$$\tau = -k\left(-\frac{du}{dr}\right)^n \qquad (12.18a)$$

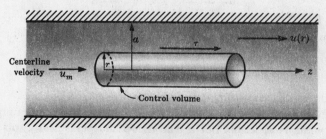

Fig. 12-6.   One-dimensional flow in a pipe.

Equation ($12.18a$) is somewhat awkward and in general for a power-law fluid (with a one-dimensional velocity profile) τ may be expressed as

$$\tau = k\varepsilon \left| \frac{du}{dy} \right|^n$$

where ε is a sign factor given by

$$\varepsilon = \frac{du/dy}{|du/dy|}$$

so that

$$\tau = k \left| \frac{du}{dy} \right|^{(n-1)} \frac{du}{dy} \qquad (12.18b)$$

Hence the term $k|du/dy|^{(n-1)}$ plays the role of an apparent viscosity. Now equation (12.17) may be written explicitly (if we assume du/dy is negative, corresponding to flow in the positive z direction).

$$k\left(-\frac{du}{dr} \right)^n = -\frac{r}{2}\left(\frac{dp}{dz} \right) \qquad (12.19)$$

or using the notation $-dp/dz = G$ and transposing (12.19),[1]

$$-\frac{du}{dr} = \left(\frac{G}{2k} \right)^{1/n} r^{1/n} \qquad (12.20)$$

Integrating (12.20),

$$\int_u^0 -du = \int_r^a \left(\frac{G}{2k} \right)^{1/n} r^{1/n}\, dr \qquad (12.21)$$

which gives

$$u = \frac{n}{n+1}\left(\frac{G}{2k} \right)^{1/n} (a^{(n+1)/n} - r^{(n+1)/n}) \qquad (12.22)$$

The total flow is

$$Q = \int_0^a 2\pi u r\, dr \qquad (12.23)$$

Substituting (12.22) into (12.23) and integrating,

$$Q = \frac{n\pi}{(3n+1)}\left(\frac{G}{2k} \right)^{1/n} a^{(3n+1)/n} \qquad (12.24)$$

The mean velocity is $V = Q/\pi a^2$, giving

$$V = \frac{n}{(3n+1)}\left(\frac{G}{2k} \right)^{1/n} a^{(n+1)/n} \qquad (12.25)$$

Setting $G = (\Delta p)/L$ for pipe flow (L = length of pipe, Δp = pressure drop and $D = 2a$, the pipe diameter), equation (12.25) can be written

$$\frac{2(3n+1)}{n}\frac{V}{D} = k^{-1}\left(\frac{D\,\Delta p}{4L} \right)^{1/n} \qquad (12.26)$$

For a Newtonian fluid ($k = \mu$, $n = 1$), equation (12.26) reduces to the familiar equation for Poiseuille flow,

$$\frac{8V}{D} = \frac{D\,\Delta p}{4\mu L} \qquad (12.27)$$

---

[1] G is taken as a positive number in what follows. Flow is taken positive in the positive z direction and dp/dz would be a negative value (and G positive) for positive flow.

Substituting $(G/2k)^{1/n}$ from (12.25) into (12.22) gives for the velocity profile in terms of the mean velocity,

$$\frac{u}{V} = \left(\frac{3n+1}{n+1}\right)\left[1 - \left(\frac{r}{a}\right)^{(n+1)/n}\right] \tag{12.28}$$

Fig. 12-7 shows velocity profiles given by (12.28) plotted for various values of n.

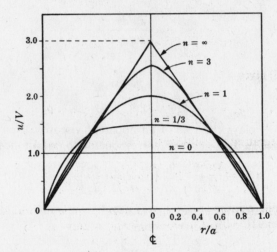

Fig. 12-7. Velocity profiles in a circular pipe for a power law fluid for various values of n.

The lower the value of n the flatter the velocity profile is, until eventually for very low values of n the flow assumes a plug-like form with only the fluid adjacent to the pipe walls being sheared, so that the velocity profile is almost indistinguishable from the profile for a Bingham plastic (see Fig. 12-8).

In some cases a fluid may be successfully treated as either a Bingham plastic or a power law fluid. This has been the case with certain drilling muds.

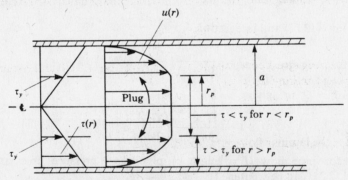

Fig. 12-8. Flow of a Bingham fluid in a circular pipe.

Bingham Plastics

For the laminar flow of a Bingham plastic, the equation for shear stress to be substituted into equation (12.17) is

$$\tau = \tau_y + \mu_p \dot{\gamma} \tag{12.29}$$

where $\tau > \tau_y$; if τ is less than τ_y, then $\dot{\gamma} = 0$ and the fluid moves as a plug of radius r_p, as shown in Fig. 12-8. We have for flow in a pipe of radius a,

$$\tau = \tfrac{1}{2}rG = \tau_y - \mu_p \frac{du}{dr} \tag{12.30}$$

for $r_p < r < a$; and for $r < r_p$, $du/dr = 0$. At $r = r_p$, $du/dr = 0$, giving

$$r_p = 2\tau_y/G \tag{12.31}$$

Integrating (12.30) for velocity gives

$$\int_0^u du = \frac{1}{\mu_p} \int_a^r (\tau_y - \tfrac{1}{2}rG)\, dr \tag{12.32}$$

so that for $r_p < r < a$,

$$u = \frac{G}{4\mu_p}(a^2 - r^2) - \frac{\tau_y}{\mu_p}(a - r) \tag{12.33}$$

Setting $r = r_p = 2\tau_y/G$, (12.33) gives

$$u_p = \frac{\tau_y^2}{\mu_p G}(a/r_p - 1)^2, \qquad r_p > r > 0 \tag{12.34}$$

The flow rate found from integrating the velocity over the cross sectional area of the pipe is

$$Q = \frac{\pi a^4 G}{8\mu_p}\left[1 - \frac{4}{3}\left(\frac{2\tau_y}{aG}\right) + \frac{1}{3}\left(\frac{2\tau_y}{aG}\right)^4\right] \tag{12.35}$$

which reduces to the Poiseuille expression for $\tau_y = 0$. The above equation, called Buckingham's equation, gives the pressure gradient $G = (\Delta p/L)$ as an implicit function.

12.4 GENERALIZED METHOD FOR FLOW IN PIPES

Generalized Reynolds Number and Friction Factor for Pseudoplastic Fluids

A method has been given by Metzner and Reed (see reference 5) which defines a generalized Reynolds number which can be used to find friction factors for both laminar and turbulent flow of time independent fluids.

As in the case of a Newtonian fluid, the generalized Reynolds number can be defined as

$$Re' = \frac{8\rho V^2}{\tau_{0L}} \tag{12.36}$$

where τ_{0L} is the wall shearing stress corresponding to laminar flow, and V is the mean velocity.

Defining the Fanning friction factor[1] as

$$f = \frac{2\tau_0}{\rho V^2} \tag{12.37}$$

in laminar flow leads to the laminar flow relationship

$$f = \frac{16}{Re'} \tag{12.38}$$

between the friction factor and Reynolds number for Newtonian fluids.

The next step in the method is to use measurements made in capillary tubes or other devices to determine rheological constants that can be used to evaluate the generalized Reynolds number.

The pressure drop Δp and mean velocity V are measured. Then two constants for the fluid are defined by the relation

$$\frac{\Delta p}{4L} = \tau_0 = K'(8V/D)^{n'} \tag{12.39}$$

[1] The Fanning friction factor differs from the one used previously, the Darcy friction factor (in Chapters 4 and 5), by a factor of 4: $(f, \text{Darcy}) = 4 \times (f, \text{Fanning})$.

where K^r is the consistency index and n' is the flow behavior index. $8V/D$ would be the wall shear rate for Poiseuille flow of a Newtonian fluid. n' can be found from the tangent slope of $\log \tau_0$ plotted versus $\log 8V/D$:

$$n' = \frac{d(\log \tau_0)}{d(\log 8V/D)} \tag{12.40}$$

Equation (12.39) is based on an expression by Mooney for the wall shear rate at the pipe wall:

$$-\left(\frac{du}{dr}\right)_w = \left(\frac{3n' + 1}{4n'}\right)\left(\frac{8V}{D}\right) \tag{12.41}$$

Equations (12.39), (12.40) and (12.41) are applicable to other fluids besides power law fluids, i.e. n' and K' are not necessarily constants. For the special case of power law fluids, $n = n'$ and we have

$$K' = k\left(\frac{3n + 1}{n}\right)^n \tag{12.42}$$

Substituting τ_0 from (12.39) into (12.36), the expression for the generalized Reynolds number gives

$$Re' = \frac{D^{n'}V^{(2-n')}\rho}{K'(8)^{(n'-1)}} \tag{12.43}$$

For a Newtonian fluid $n' = 1$ and $Re' = \rho VD/K'$, so that $K' = \mu$, the viscosity.

Turbulent Flow

Dodge and Metzner (see reference 2) have produced a relationship based on the logarithmic resistance formula of von Karman for Newtonian fluids relating friction factor to the generalized Reynolds number Re'. Analogous to this equation for Newtonian fluids, the equation proposed by Dodge and Metzner for a power law type of fluid flowing in smooth pipes is

$$\frac{1}{f} = \frac{4.0}{(n')^{0.75}} \log_{10}\left[Re' f^{1-n'/2}\right] - \frac{0.4}{(n')^{1.2}} \tag{12.44}$$

Fig. 12-9 is a graph of this equation which does not give the friction factor f explicitly.

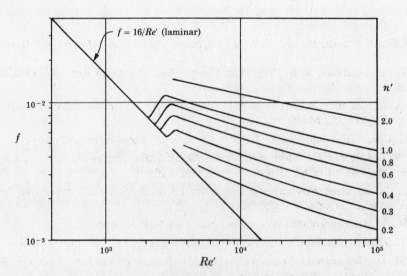

Fig. 12-9. Fanning friction factor f versus Reynolds number Re', equation (12.44).

The data of Dodge and Metzner show that the critical generalized Reynolds number corresponding to the onset of turbulence increases with decreasing values of the flow behavior index n', Re' critical increasing from 2100 at $n' = 1$ to 3100 at $n' = 0.38$. However, no exact criterion has been established for the critical Reynolds number.

Anomalous (Turbulent) Flow of Non-Newtonian Fluids

While a great number of non-Newtonian fluids have been found to satisfy the generalized turbulent flow criteria of Metzner, solutions of certain high molecular weight organic compounds have been found that have friction factors much lower than those predicted by the type of relationship given by (12.44) for turbulent flow. Instead of the friction factor leveling out after the critical generalized Reynolds number is reached, the friction factor continues to drop almost as fast as the curve for laminar flow, making it appear that either turbulence has been suppressed or the onset of turbulence delayed. For certain fluids this trend may continue until friction factors are reached that are an order of magnitude lower than that for pure water.

Guar gum and certain polymers supplied commercially to reduce friction in pumping fluids possess these characteristics. One application of these fluids is in oil well treatment where huge quantities of fluid may have to be pumped down a well in a very short time. One unfortunate disadvantage of many of these high molecular weight compounds is their susceptibility to physical degradation, i.e. tearing apart of the very large molecules due to high fluid shearing rates so that they lose their desirable friction reducing properties when fluid is reused.

Slurries of solids in water, where the solid phase may not be tightly dispersed by the water, may also produce a lower friction factor in turbulent flow than water in pipes. At the pipe walls the water or liquid phase separates from the solid phase to form a low viscosity layer. The more rigid core of the slurry tends to move as a plug. Because there is less turbulent momentum transfer from the boundary layer to the core, the laminar boundary layer is actually thicker and the friction factor less than it would be for the flow of pure water. Highly concentrated coal slurries show a much lower friction factor than water. The friction factor for turbulent flow of paper pulp in water also decreases below that of water as the concentration of pulp increases.

References

1. Coleman, B. D., Markovitz, H., and Noll, W., *Viscometric Flow of Non-Newtonian Fluids*, Springer-Verlag, 1966.

2. Dodge, D. W., and Metzner, A. B., "Turbulent Flow of Non-Newtonian Systems," *A.I.Ch.E. Journal*, 5, pp. 189–204, 1959.

3. Fredricson, A. G., and Bird, R. Byron, "Non-Newtonian Flow in Annuli," *Industrial and Engineering Chemistry*, 50, No. 3, pp. 347–352, March 1958.

4. Larson, R. G., *Constitutive Equations for Polymer Melts and Solutions*, Butterworth, 1987.

5. Metzner, A. B., and Reed, J. C., "Flow of Non-Newtonian Fluids—Correlation of the Laminar Transition and Turbulent-flow Regions," *A.I.Ch.E. Journal*, 1, pp. 434–440, 1955.

6. Savins, J. G., "Generalized Newtonian (Pseudoplastic) Flow in Stationary Pipes and Annuli," *Petroleum Transactions*, *A.I.M.E.*, 213, pp. 325–332, 1958.

7. Showalter, W. R., *Mechanics of Non-Newtonian Fluids*, Pergamon Press, 1978.

8. Streeter, V. L. (Editor), *Handbook of Fluid Dynamics*, Chapter 7, by A. B. Metzner, McGraw-Hill, 1961.

9. Vaughn, R. D., and Bergman, D., "Laminar Flow of Non-Newtonian Fluids in Concentric Annuli," *Industrial and Engineering Chemistry—Process Design and Development*, 5, No. 1, pp. 44–47, 1966.

10. Wilkinson, W. L., *Non-Newtonian Fluids*, Pergamon Press, 1960.

Solved Problems

12.1. A clay slurry, $K' = 1.693$ (units of K' are in lb, ft and sec, the power depending on n') and $n' = 581$, is to be pumped at 1150 gallons per minute in 3.826″ ID pipe. What will be the pressure loss in the pipe?

   1150 gal/min corresponds to 2.56 ft³/sec. The pipe cross sectional area is 0.0798 ft², giving a velocity of 32 ft/sec. From equation (*12.43*) the modified Reynolds number is $N'_R = 19,900$, well in the turbulent regime. From equation (*12.44*) the Fanning friction factor is found to be 0.00442. This value of f cannot be determined explicitly from (*12.44*) but may easily be found by assuming a trial value of f to substitute in the term $\log_{10}(N'_R f^{1-n'/2})$. Then (*12.44*) is solved for a new f to be used again in solving for f from (*12.44*), etc. f soon converges to the correct value to satisfy (*12.44*). Also, f can be found from Fig. 12-9. The pressure gradient $(\Delta p)/L$ is found in the usual way from the formula

$$\tau_0 = \frac{D\,\Delta p}{4L} = \frac{f\rho V^2}{2}$$

giving $(\Delta p)/L = 0.396$ psi/ft of pipe.

12.2. Paint approximates a Bingham plastic. A paint of yield stress τ_y and plastic viscosity μ_p is applied to a wall. What is the maximum thickness h of paint that will adhere to the wall and not run?

   Referring to Fig. 12-10 the paint will build up until the weight causes the shear stress at the wall to exceed τ_y, at which point the paint flows off to maintain that critical thickness. By considering the equilibrium of a unit square of paint of thickness $(h-x)$ the weight balances the shear face. Hence τ varies linearly from zero on the surface to ρgh at the wall. The minimum thickness is then given by $(\rho gh_{max}) = \tau_y$ and hence $h_{max} = \tau_y/\rho g$.

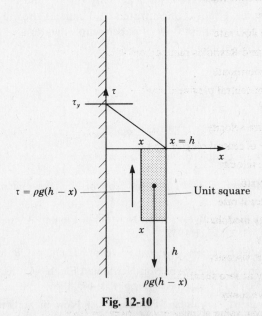

Fig. 12-10

Supplementary Problems

12.3. Capillary tube rheometer measurements of a fluid have shown $\log \tau_0$ versus $\log 8V/D$ to be a straight line between $8V/D = 88$ sec^{-1} and 5400 sec^{-1} corresponding to τ_0 varying from $\tau_0 = 1.20$ to 2.50. What is the pressure drop for the fluid flowing through a 3.820″ ID tube at 400 gallons/minute? Is the flow laminar or turbulent?

12.4. Show that the velocity profile for flow of a power law fluid between parallel plates can be expressed as

$$u = \frac{n}{1+n} \left(\frac{G}{k}\right)^{1/n} [h^{(n+1)/n} - y^{(n+1)/n}]$$

where h is the half width of the channel and y is the coordinate measured from the centerline.

12.5. Derive the velocity profile for Bingham plastic flow between stationary parallel plates.

12.6. A stepped slider bearing (see Problem 5.11) is used with a Bingham plastic. Describe the flow as the velocity of the bearing increases.

12.7. The hydrostatic thrust bearing of Problem 5.13 uses a Bingham plastic as a lubricant. Describe the flow through the bearing as the reservoir pressure is increased from zero to a high value.

NOMENCLATURE FOR CHAPTER 12

a　　　　= radius of pipe
A, B, C = constants used in various stress-strain rate relationships
D　　　　= diameter
f　　　　= friction factor (Fanning)
G　　　　= $-dp/dz$, pressure gradient in a pipe
k　　　　= proportionality constant for a power law fluid
K'　　　= consistency index
L　　　　= length
n　　　　= exponent for a power law fluid
n'　　　= flow behavior index
p　　　　= pressure
Q　　　　= volume flow rate
Re'　　= generalized Reynolds number, $8\rho V^2/\tau_{0L}$
r　　　　= radial coordinate
r_p　　　= radius of central plug in a pipe
u　　　　= velocity
u_m　　　= maximum velocity
u_p　　　= velocity of central plug in a pipe
V　　　　= average velocity
γ　　　= shear strain
$\dot{\gamma}$　　　= shear strain rate
λ　　　= a rigidity modulus
μ　　　= viscosity
μ_a　　　= apparent viscosity
μ_0　　　= viscosity at zero shear rate
μ_p　　　= plastic viscosity
μ_∞　　　= asymptotic value of apparent viscosity for high rates of shear
τ　　　= shear stress
$\dot{\tau}$　　　= shear stress rate
τ_0　　　= wall shear stress
τ_{0L}　　= wall shear stress corresponding to laminar flow
τ_y　　　= yield stress

Appendix A

Some Properties of Fluids

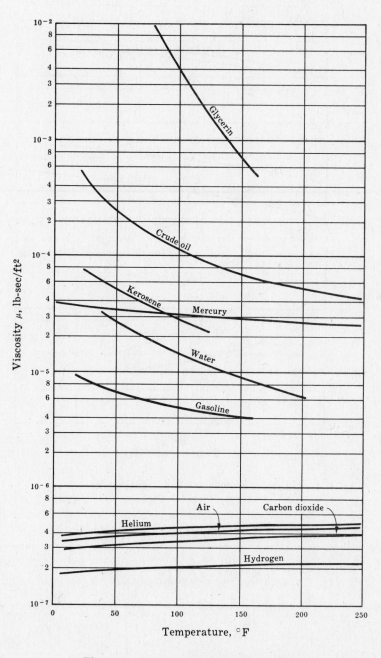

Fig. A-1. Absolute viscosity of some fluids.

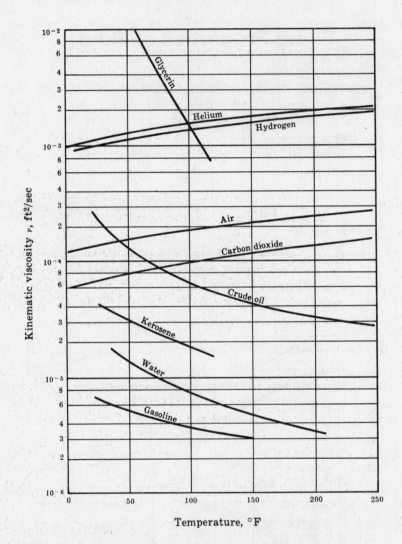

Fig. A-2. Kinematic viscosity of some fluids.

Table A.1 Properties of Water at Atmospheric Pressure

Temperature		Density			Viscosity			Kinematic viscosity		
°C	°F	g/cm^3	kg/m^3	$slugs/ft^3$	$dyn \cdot s/cm^2$ (poise)	$Pa \cdot s$ $(N \cdot s/m^2)$	$lb_f \cdot s/ft^2$	cm^2/s (stoke)	m^2/s	ft^2/s
0	32	0.99987	999.87	1.940	1.794×10^{-2}	1.794×10^{-3}	3.746×10^{-5}	1.794×10^{-2}	1.794×10^{-6}	1.930×10^{-5}
4	39	1.00000	1000.00	1.941	1.568	1.568	3.274	1.568	1.568	1.687
5	41	0.99999	999.99	1.941	1.519	1.519	3.172	1.519	1.519	1.634
10	50	0.99973	999.73	1.940	1.310	1.310	2.735	1.310	1.310	1.407
15	59	0.99913	999.13	1.940	1.145	1.145	2.391	1.146	1.146	1.233
20	68	0.998	998.00	1.937	1.009	1.009	2.107	1.011	1.011	1.088
30	86	0.996	996.00	1.932	0.800	0.800	1.670	0.803	0.803	0.864
40	104	0.992	992.00	1.925	0.654	0.654	1.366	0.659	0.659	0.709
50	122	0.988	988.00	1.917	0.549	0.549	1.146	0.556	0.556	0.598
60	140	0.983	983.00	1.907	0.470	0.470	0.981	0.478	0.478	0.514
70	158	0.978	978.00	1.897	0.407	0.407	0.850	0.416	0.416	0.448
80	176	0.972	972.00	1.885	0.357	0.357	0.745	0.367	0.367	0.395
90	194	0.965	965.00	1.872	0.317	0.317	0.662	0.328	0.328	0.353
100	212	0.958	958.00	1.858	0.284	0.284	0.593	0.296	0.296	0.318

Table A.2 Properties of Air at Atmospheric Pressure

Temperature		Density			Viscosity			Kinematic viscosity		
°C	°F	g/cm^3	kg/m^3	$slugs/ft^3$	$dyn \cdot s/cm^2$ (poise)	$Pa \cdot s$ $(N \cdot s/m^2)$	$lb_f \cdot s/ft^2$	cm^2/s (stoke)	m^2/s	ft^2/s
0	32	1.293×10^{-3}	1.293	2.510×10^{-3}	1.709×10^{-4}	1.709×10^{-5}	3.568×10^{-7}	0.1322	1.322×10^{-5}	1.427×10^{-4}
50	122	1.093	1.093	2.122	1.951	1.951	4.074	0.1785	1.785	1.921
100	212	0.946	0.946	1.836	2.175	2.175	4.541	0.2299	2.299	2.474
150	302	0.834	0.834	1.619	2.385	2.385	4.980	0.2860	2.860	3.077
200	392	0.746	0.746	1.448	2.582	2.582	5.391	0.3461	3.461	3.724
250	482	0.675	0.675	1.310	2.770	2.770	5.784	0.4104	4.104	4.416
300	572	0.616	0.616	1.196	2.946	2.946	6.151	0.4782	4.782	5.145
350	662	0.567	0.567	1.101	3.113	3.113	6.500	0.5490	5.490	5.907
400	752	0.525	0.525	1.019	3.277	3.277	6.842	0.6246	6.246	6.721
450	842	0.488	0.488	0.947	3.433	3.433	7.168	0.7035	7.035	7.570
500	932	0.457	0.457	0.887	3.583	3.583	7.481	0.7840	7.840	8.436

Table A.3 Properties of Air at Standard Conditions*

Molecular weight	R	c_p	c_v	k
28.97	$53.34 \dfrac{\text{ft-lb}_f}{\text{lb}_m\ ^\circ\text{R}}$	$0.240 \dfrac{\text{Btu}}{\text{lb}_m\ ^\circ\text{R}}$	$0.171 \dfrac{\text{Btu}}{\text{lb}_m\ ^\circ\text{R}}$	1.4
	$1716 \dfrac{\text{ft-lb}_f}{\text{slug}\ ^\circ\text{R}}$	$7.72 \dfrac{\text{Btu}}{\text{slug}\ ^\circ\text{R}}$	$5.50 \dfrac{\text{Btu}}{\text{slug}\ ^\circ\text{R}}$	
	$0.287 \dfrac{\text{kJ}}{\text{kg K}}$	$1.0035 \dfrac{\text{kJ}}{\text{kg K}}$	$0.7165 \dfrac{\text{kJ}}{\text{kg K}}$	

* Values in SI units are given at 300 K.

Appendix B

Units and Dimensions

In fluid mechanics we must deal with measurable quantities such as pressure, velocity, density and viscosity. These quantities are related through equations derived from laws or definitions. Each contains some or all of the basic dimensions of force (F), mass (M), length (L), time (T) and temperature (θ). For quantitative purposes a set of units must be established for these basic dimensions. (In electromagnetic theory there is one additional basic dimension (which is arbitrary). It is often convenient to take this dimension as charge which is measured, in mks units, in coulombs. Thus in magnetohydrodynamics there are five basic dimensions.)

Equations expressing relationships among physical quantities must be dimensionally homogeneous. That is, each term of an equation must have the same dimensions.

A difficulty arises when the units of mass and force are established independently. The different disciplines related to fluid mechanics, such as aerodynamics, thermodynamics and heat transfer, all have their own well developed system for handling this problem—the one which best serves their specific needs.

From Newton's law we find that force and mass are not independent dimensions. Force is proportional to the product of mass and acceleration; that is,

$$\mathbf{F} \propto \mathbf{ma}$$

And by requiring dimensional homogeneity, we find that

$$(F) = \left(\frac{ML}{T^2}\right) \quad \text{or} \quad (M) = \left(\frac{FT^2}{L}\right)$$

The actual choice of basic dimensions then is somewhat arbitrary. One may use an F, L, T, θ system or an M, L, T, θ system. Then all other dimensions may be expressed in terms of the chosen independent basic dimensions by means of laws and definitions.

Now we may use Newton's law to define the unit of mass in terms of force and acceleration. We write

$$\mathbf{F} = \mathbf{ma} \quad \text{or} \quad m = \frac{F}{a}$$

If the unit of force is lb_f and the unit of acceleration is ft/sec^2, then the unit of mass is

$$\frac{lb_f \ sec^2}{ft}$$

This is the amount of mass which is accelerated at the rate of 1 ft/sec^2 when acted upon by 1 lb_f. This unit of mass is called a *slug*.

In the SI system of units (Système International d'Unités) the unit of force is taken as the newton (N) and the unit of acceleration is m/s^2. The unit of mass is then

$$\frac{N \ s^2}{m}$$

This is the amount of mass which is accelerated at the rate of 1 m/s^2 when acted upon by 1 N. The unit of mass is called the kilogram (kg).

Another unit of mass established independently of Newton's law is also used. The pound mass (lb_m) is defined as the amount of mass which would be attracted toward the earth's surface by a force of one pound (at a particular location). When units of force and mass are defined independent of Newton's law, we must write the equation with a conversion factor, k, in order to make the equation dimensionally homogeneous. We have

$$\mathbf{F} = k m \mathbf{a} \qquad \text{or} \qquad \mathbf{F} = \frac{1}{g_c} m \mathbf{a}$$

and

$$g_c = \frac{ma}{F}$$

The quantity g_c has a numerical value and units which depend on the particular units chosen for force, mass and acceleration. Some particular sets of values are

$$g_c = \frac{1(\text{slug})(\text{ft/sec}^2)}{\text{lb}_f} \qquad\qquad g_c = \frac{1(\text{gm})(\text{cm/sec}^2)}{\text{dyne}}$$

$$g_c = \frac{32.2(\text{lb}_m)(\text{ft/sec}^2)}{\text{lb}_f} \qquad\qquad g_c = \frac{1(\text{kg})(\text{m/s}^2)}{\text{N}}$$

$$g_c = \frac{9.8(\text{kg}_m)(\text{m/s}^2)}{\text{kg}_f}$$

where we see that

$$1 \text{ slug} = 32.2 \text{ lb}_m$$

and we note that 1 lb_m will be accelerated at the rate of 32.2 ft/sec² when acted upon by 1 lb_f. The slug or kg are the units of mass implied throughout most of this book and thus g_c does not appear in the equations.

Some physical quantities and their dimensions are listed in Table B.1. Some conversion factors are given in Table B.2.

The method of making the proper conversions is illustrated in the following example. Solve numerically for h_0 in ft lb_f/lb_m where the numerical value of each quantity of the right side is given in terms of the units shown.

$$h_0 = u + pv + \frac{V^2}{2}$$

$$u = A\,\frac{\text{Btu}}{\text{lb}_m} \qquad p = B\,\frac{\text{lb}_f}{\text{in}^2} \qquad v = C\,\frac{\text{ft}^3}{\text{lb}_m} \qquad V = D\,\frac{\text{ft}}{\text{sec}}$$

We substitute the numerical values into the equation along with the units and conversions.

$$h_0 = A\,\frac{\text{Btu}}{\text{lb}_m} \times \frac{778 \text{ ft lb}_f}{\text{Btu}} + B\,\frac{\text{lb}_f}{\text{in}^2}\,C\,\frac{\text{ft}^3}{\text{lb}_m} \times \frac{144 \text{ in}^2}{\text{ft}^2} + \frac{D^2}{2}\,\frac{\text{ft}^2}{\text{sec}^2} \times \frac{\text{sec}^2 \text{ lb}_f}{32.2 \text{ lb}_m \text{ ft}}$$

$$= \left(778\,A + 144\,BC + \frac{D^2}{2(32.2)} \right) \frac{\text{ft lb}_f}{\text{lb}_m}$$

Thus we see that g_c is treated as a conversion factor.

In summary, it is useful to write the equation $F = ma$ in various systems of units. The following equations indicate the relationships among the dimensions and units (when g_c is not used).

$$\begin{array}{llll}
\text{CGS System:} & F \text{ (dynes)} & = m \text{ (grams)} & \times a \text{ (cm/sec}^2) \\
\text{SI System:} & F \text{ (newtons)} & = m \text{ (kilograms)} & \times a \text{ (m/sec}^2) \\
\text{FPS System:} & F \text{ (poundals)} & = m \text{ (lb}_m) & \times a \text{ (ft/sec}^2) \\
\text{FSS System:} & F \text{ (lb}_f) & = m \text{ (slugs)} & \times a \text{ (ft/sec}^2)
\end{array}$$

Table B.1 Dimensions and Units

Physical Quantity	Dimensions		Units			
			Metric		English	
	MLT System	FLT System	CGS System	SI System	FPS System	Engineering System
Length	L	L	cm	m	ft	ft
Mass	M	$FL^{-1}T^2$	gm	kg	lb_m	slug
Time	T	T	sec	s	sec	sec
Velocity	LT^{-1}	LT^{-1}	cm/sec	m/s	ft/sec	ft/sec
Acceleration	LT^{-2}	LT^{-2}	cm/sec²	m/s²	ft/sec²	ft/sec²
Force	MLT^{-2}	F	gm cm/sec² = dyne	kg m/s² = N	lb_m ft/sec² = poundal	slug ft/sec² = lb_f
Momentum, Impulse	MLT^{-1}	FT	gm cm/sec = dyne sec	kg m/s = N s	lb_m ft/sec = pdl sec	slug ft/sec = lb_f sec
Energy, Work	ML^2T^{-2}	FL	gm cm²/sec² = dyne cm = erg	kg m²/sec² = N m = J	lb_m ft²/sec² = ft pdl	slug ft²/sec² = ft lb_f
Power	ML^2T^{-3}	FLT^{-1}	gm cm²/sec³ = dyne cm/sec = erg/sec	kg m²/s³ = J/s = watt	lb_m ft²/sec³ = ft pdl/sec	slug ft²/sec³ = ft lb_f/sec
Density	ML^{-3}	$FL^{-4}T^2$	gm/cm³	kg/m³	lb_m/ft³	slug/ft³
Angular velocity	T^{-1}	T^{-1}	rad/sec	rad/s	rad/sec	rad/sec
Angular acceleration	T^{-2}	T^{-2}	rad/sec²	rad/s²	rad/sec²	rad/sec²
Torque	ML^2T^{-2}	FL	gm cm²/sec² = dyne cm	kg m²/s² = N m	lb_m ft²/sec² = ft pdl	slug ft²/sec² = ft lb_f
Angular momentum	ML^2T^{-1}	FLT	gm cm²/sec	kg m²/s	lb_m ft²/sec	slug ft²/sec
Moment of inertia	ML^2	FLT^2	gm cm²	kg m²	lb_m ft²	slug ft²
Pressure, Stress	$ML^{-1}T^{-2}$	FL^{-2}	gm/(cm sec²) = dyne/cm²	kg/(m s²) = N/m²	pdl/ft²	lb_f/ft²
Viscosity (μ)	$ML^{-1}T^{-1}$	$FL^{-2}T$	gm/(cm sec) = dyne sec/cm²	kg/(m s) = N s/m²	lb_m/ft sec) = pdl sec/ft²	slug/(ft sec) = lb_f sec/ft²
Kinematic viscosity (v)	L^2T^{-1}	L^2T^{-1}	cm²/sec	m²/s	ft²/sec	ft²/sec
Surface tension	MT^{-2}	FL^{-1}	gm/sec² = dyne/cm	kg/s² = N/m	lb_m/sec² = pdl/ft	slug/sec² = lb_f/ft

Table B.2   Conversion of Units

Length	
1 kilometer (km) $= 1000$ meters	1 inch (in.) $= 2.540$ cm
1 meter (m) $= 100$ centimeters	1 foot (ft) $= 30.48$ cm
1 centimeter (cm) $= 10^{-2}$ m	1 mile (mi) $= 1.609$ km
1 millimeter (mm) $= 10^{-3}$ m	1 mil $= 10^{-3}$ in.
1 micron (μ) $= 10^{-6}$ m	1 centimeter $= 0.3937$ in.
1 millimicron (mμ) $= 10^{-9}$ m	1 meter $= 39.37$ in.
1 angstrom (A) $= 10^{-10}$ m	1 kilometer $= 0.6214$ mile

Area

1 square meter (m^2) $= 10.76$ ft^2
1 square foot (ft^2) $= 929$ cm^2

Volume

1 liter (l) $= 1000$ cm$^3 = 1.057$ quart (qt) $= 61.02$ in$^3 = 0.03532$ ft^3
1 cubic meter (m^3) $= 1000\ l = 35.32$ ft^3
1 cubic foot (ft^3) $= 7.481$ U.S. gal $= 0.02832$ m$^3 = 28.32\ l$
1 U.S. gallon (gal) $= 231$ in$^3 = 3.785\ l$; 1 British gallon $= 1.201$ U.S. gallon $= 277.4$ in^3

Mass

1 kilogram (kg) $= 2.2046$ lb$_m = 0.06852$ slug; 1 lb$_m = 453.6$ gm $= 0.03108$ slug
1 slug $= 32.174$ lb$_m = 14.59$ kg

Speed

1 km/hr $= 0.2778$ m/sec $= 0.6214$ mi/hr $= 0.9113$ ft/sec
1 mi/hr $= 1.467$ ft/sec $= 1.609$ km/hr $= 0.4470$ m/sec

Density

1 gm/cm$^3 = 10^3$ kg/m$^3 = 62.43$ lb$_m$/ft$^3 = 1.940$ slug/ft^3
1 lb$_m$/ft$^3 = 0.01602$ gm/cm^3;   1 slug/ft$^3 = 0.5154$ gm/cm^3

Force

1 newton (N) $= 10^5$ dynes $= 0.1020$ kg$_f = 0.2248$ lb$_f$
1 pound force (lb$_f$) $= 4.448$ N $= 0.4536$ kg$_f = 32.17$ poundals
1 kilogram force (kg$_f$) $= 2.205$ lb$_f = 9.807$ N
1 U.S. short ton $= 2000$ lb$_f$;   1 long ton $= 2240$ lb$_f$;   1 metric ton $= 2205$ lb$_f$

Energy

1 joule (J) $= 1$ N m $= 10^7$ ergs $= 0.7376$ ft lb$_f = 0.2389$ cal $= 9.481 \times 10^{-4}$ Btu
1 ft lb$_f = 1.356$ joules (J) $= 0.3239$ cal $= 1.285 \times 10^{-3}$ Btu
1 calorie (cal) $= 4.186$ joules (J) $= 3.087$ ft lb$_f \leqslant 3.968 \times 10^{-3}$ Btu
1 Btu $= 778$ ft lb$_f = 1055$ joules (J) $= 0.293$ watt hr
1 kilowatt hour (kw hr) $= 3.60 \times 10^6$ joules (J) $= 860.0$ kcal $= 3413$ Btu
1 electron volt (ev) $= 1.602 \times 10^{-19}$ joule (J)

Power

1 watt $= 1$ joule (J)/s $= 10^7$ ergs/sec $= 0.2389$ cal/sec
1 horsepower (hp) $= 550$ ft lb$_f$/sec $= 33,000$ ft lb$_f$/min $= 745.7$ watts
1 kilowatt (kw) $= 1.341$ hp $= 737.6$ ft lb$_f$/sec $= 0.9483$ Btu/sec

Pressure

1 N/m$^2 = 10$ dynes/cm$^2 = 9.869 \times 10^{-6}$ atmosphere $= 2.089 \times 10^{-2}$ lb$_f$/ft$^2 = 1$ Pascal (Pa)
1 lb$_f$/in$^2 = 6895$ N/m$^2 = 5.171$ cm mercury $= 27.68$ in. water
1 atmosphere (atm) $= 1.013 \times 10^5$ N/m^2 (Pa) $= 1.013 \times 10^6$ dynes/cm$^2 = 14.70$ lb$_f$/in^2
$\qquad\qquad = 76$ cm mercury $= 406.8$ in. water

Angle

1 radian (rad) $= 57.296°$; $1° = 0.017453$ rad

Appendix C

Some Basic Equations
in Various Coordinate Systems[1]

1. NAVIER-STOKES EQUATIONS OF MOTION FOR AN INCOMPRESSIBLE FLUID WITH VISCOSITY CONSTANT

These equations may be used with a high degree of accuracy for problems involving viscosity variations if the viscosity gradient is not too large. In most physical problems this assumption is adequate and the Navier-Stokes equations may be used in most incompressible flow problems. The following symbols are used:

$$p = \text{Pressure}$$

$$\mathbf{F} = \text{Body force density}$$

$$\mu = \text{Viscosity}$$

$$\frac{D}{Dt} = \text{Material derivative (not the same on a component as on a vector).}$$

Vector

$\mathbf{V}$ is the velocity vector.

$$\rho \frac{D\mathbf{V}}{Dt} = \rho \left[\frac{\partial \mathbf{V}}{\partial t} + (\mathbf{V} \cdot \nabla)\mathbf{V} \right] = -\nabla p + \mathbf{F} + \mu \nabla^2 \mathbf{V} \tag{A.1}$$

The term $(\mathbf{V} \cdot \nabla)\mathbf{V}$ is actually a pseudo-vector expression and care must be used in its expansion in other than Cartesian coordinates. It is convenient to express this acceleration term in true vector form, and the equation of motion may be written in the alternative form:

$$\rho \left[\frac{\partial \mathbf{V}}{\partial t} + \nabla \left(\frac{V^2}{2} \right) - \mathbf{V} \times (\nabla \times \mathbf{V}) \right] = -\nabla p + \mathbf{F} + \mu \nabla^2 \mathbf{V} \tag{A.2}$$

Care must be taken in expanding $\nabla^2 \mathbf{V}$ and $D\mathbf{V}/Dt$, since the operation on a vector is not the same as the operation on a scalar component. The following vector identity is useful:

$$\nabla^2 \mathbf{V} = \nabla(\nabla \cdot \mathbf{V}) - \nabla \times (\nabla \times \mathbf{V})$$

Cartesian Tensor

w_i is the velocity in the x_i direction.

$$\rho \left[\frac{\partial w_i}{\partial t} + w_j \frac{\partial w_i}{\partial x_j} \right] = -\frac{\partial p}{\partial x_i} + F_i + \mu \frac{\partial^2 w_i}{\partial x_j \, \partial x_j} \tag{A.3}$$

---

[1] For a more complete listing of useful equations see: Hughes, W. F., and Gaylor, E. W., *Basic Equations of Engineering Science*, Schaum Publishing Co., New York, 1964.

Cartesian

u, v, and w are the velocities in the x, y, and z directions respectively. In the following section:

$$\frac{D}{Dt} = \frac{\partial}{\partial t} + u\frac{\partial}{\partial x} + v\frac{\partial}{\partial y} + w\frac{\partial}{\partial z}, \qquad \nabla^2 = \frac{\partial^2}{\partial x^2} + \frac{\partial^2}{\partial y^2} + \frac{\partial^2}{\partial z^2}$$

$$\rho\frac{Du}{Dt} = F_x - \frac{\partial p}{\partial x} + \mu\nabla^2 u$$

$$\rho\frac{Dv}{Dt} = F_y - \frac{\partial p}{\partial y} + \mu\nabla^2 v \tag{A.4}$$

$$\rho\frac{Dw}{Dt} = F_z - \frac{\partial p}{\partial z} + \mu\nabla^2 w$$

Written out in full, these become:

$$\rho\left(\frac{\partial u}{\partial t} + u\frac{\partial u}{\partial x} + v\frac{\partial u}{\partial y} + w\frac{\partial u}{\partial z}\right) = -\frac{\partial p}{\partial x} + F_x + \mu\left(\frac{\partial^2 u}{\partial x^2} + \frac{\partial^2 u}{\partial y^2} + \frac{\partial^2 u}{\partial z^2}\right)$$

$$\rho\left(\frac{\partial v}{\partial t} + u\frac{\partial v}{\partial x} + v\frac{\partial v}{\partial y} + w\frac{\partial v}{\partial z}\right) = -\frac{\partial p}{\partial y} + F_y + \mu\left(\frac{\partial^2 v}{\partial x^2} + \frac{\partial^2 v}{\partial y^2} + \frac{\partial^2 v}{\partial z^2}\right) \tag{A.5}$$

$$\rho\left(\frac{\partial w}{\partial t} + u\frac{\partial w}{\partial x} + v\frac{\partial w}{\partial y} + w\frac{\partial w}{\partial z}\right) = -\frac{\partial p}{\partial z} + F_z + \mu\left(\frac{\partial^2 w}{\partial x^2} + \frac{\partial^2 w}{\partial y^2} + \frac{\partial^2 w}{\partial z^2}\right)$$

Cylindrical

v_r, v_θ, and v_z are the velocities in the r, θ, and z directions respectively. In the following section:

$$\frac{D}{Dt} = \frac{\partial}{\partial t} + v_r\frac{\partial}{\partial r} + \frac{v_\theta}{r}\frac{\partial}{\partial \theta} + v_z\frac{\partial}{\partial z}$$

$$\nabla^2 = \frac{\partial^2}{\partial r^2} + \frac{1}{r}\frac{\partial}{\partial r} + \frac{1}{r^2}\frac{\partial^2}{\partial \theta^2} + \frac{\partial^2}{\partial z^2}$$

$$\rho\left[\frac{Dv_r}{Dt} - \frac{v_\theta^2}{r}\right] = F_r - \frac{\partial p}{\partial r} + \mu\left[\nabla^2 v_r - \frac{v_r}{r^2} - \frac{2}{r^2}\frac{\partial v_\theta}{\partial \theta}\right] \tag{A.6}$$

$$\rho\left[\frac{Dv_\theta}{Dt} + \frac{v_r v_\theta}{r}\right] = F_\theta - \frac{1}{r}\frac{\partial p}{\partial \theta} + \mu\left[\nabla^2 v_\theta + \frac{2}{r^2}\frac{\partial v_r}{\partial \theta} - \frac{v_\theta}{r^2}\right]$$

$$\rho\frac{Dv_z}{Dt} = F_z - \frac{\partial p}{\partial z} + \mu\nabla^2 v_z$$

Written out in full, these become:

$$\rho\left[\frac{\partial v_r}{\partial t} + v_r\frac{\partial v_r}{\partial r} + \frac{v_\theta}{r}\frac{\partial v_r}{\partial \theta} + v_z\frac{\partial v_r}{\partial z} - \frac{v_\theta^2}{r}\right]$$

$$= F_r - \frac{\partial p}{\partial r} + \mu\left[\frac{\partial^2 v_r}{\partial r^2} + \frac{1}{r}\frac{\partial v_r}{\partial r} + \frac{1}{r^2}\frac{\partial^2 v_r}{\partial \theta^2} + \frac{\partial^2 v_r}{\partial z^2} - \frac{v_r}{r^2} - \frac{2}{r^2}\frac{\partial v_\theta}{\partial \theta}\right]$$

$$\rho\left[\frac{\partial v_\theta}{\partial t} + v_r\frac{\partial v_\theta}{\partial r} + \frac{v_\theta}{r}\frac{\partial v_\theta}{\partial \theta} + v_z\frac{\partial v_\theta}{\partial z} + \frac{v_r v_\theta}{r}\right] \tag{A.7}$$

$$= F_\theta - \frac{1}{r}\frac{\partial p}{\partial \theta} + \mu\left[\frac{\partial^2 v_\theta}{\partial r^2} + \frac{1}{r}\frac{\partial v_\theta}{\partial r} + \frac{1}{r^2}\frac{\partial^2 v_\theta}{\partial \theta^2} + \frac{\partial^2 v_\theta}{\partial z^2} + \frac{2}{r^2}\frac{\partial v_r}{\partial \theta} - \frac{v_\theta}{r^2}\right]$$

$$\rho\left[\frac{\partial v_z}{\partial t} + v_r\frac{\partial v_z}{\partial r} + \frac{v_\theta}{r}\frac{\partial v_z}{\partial \theta} + v_z\frac{\partial v_z}{\partial z}\right]$$

$$= F_z - \frac{\partial p}{\partial z} + \mu\left[\frac{\partial^2 v_z}{\partial r^2} + \frac{1}{r}\frac{\partial v_z}{\partial r} + \frac{1}{r^2}\frac{\partial^2 v_z}{\partial \theta^2} + \frac{\partial^2 v_z}{\partial z^2}\right]$$

Spherical

v_r, v_θ, and v_ϕ are the velocities in the r, θ, and ϕ directions respectively. In the following section:

$$\frac{D}{Dt} = \frac{\partial}{\partial t} + v_r\frac{\partial}{\partial r} + \frac{v_\theta}{r}\frac{\partial}{\partial \theta} + \frac{v_\phi}{r\sin\theta}\frac{\partial}{\partial \phi}$$

$$\nabla^2 = \frac{1}{r^2}\frac{\partial}{\partial r}\left(r^2\frac{\partial}{\partial r}\right) + \frac{1}{r^2\sin\theta}\frac{\partial}{\partial \theta}\left(\sin\theta\frac{\partial}{\partial \theta}\right) + \frac{1}{r^2\sin^2\theta}\frac{\partial^2}{\partial \phi^2}$$

$$\rho\left[\frac{Dv_r}{Dt} - \frac{v_\theta^2 + v_\phi^2}{r}\right]$$

$$= F_r - \frac{\partial p}{\partial r} + \mu\left[\nabla^2 v_r - \frac{2v_r}{r^2} - \frac{2}{r^2}\frac{\partial v_\theta}{\partial \theta} - \frac{2v_\theta\cot\theta}{r^2} - \frac{2}{r^2\sin\theta}\frac{\partial v_\phi}{\partial \phi}\right]$$

$$\rho\left[\frac{Dv_\theta}{Dt} + \frac{v_r v_\theta - v_\phi^2\cot\theta}{r}\right]$$

$$= F_\theta - \frac{1}{r}\frac{\partial p}{\partial \theta} + \mu\left[\nabla^2 v_\theta + \frac{2}{r^2}\frac{\partial v_r}{\partial \theta} - \frac{v_\theta}{r^2\sin^2\theta} - \frac{2\cos\theta}{r^2\sin^2\theta}\frac{\partial v_\phi}{\partial \phi}\right] \qquad (A.8)$$

$$\rho\left[\frac{Dv_\phi}{Dt} + \frac{v_\phi v_r}{r} + \frac{v_\theta v_\phi\cot\theta}{r}\right]$$

$$= F_\phi - \frac{1}{r\sin\theta}\frac{\partial p}{\partial \phi} + \mu\left[\nabla^2 v_\phi - \frac{v_\phi}{r^2\sin^2\theta} + \frac{2}{r^2\sin^2\theta}\frac{\partial v_r}{\partial \phi} + \frac{2\cos\theta}{r^2\sin^2\theta}\frac{\partial v_\theta}{\partial \phi}\right]$$

Written out in full, these become:

$$\rho\left[\frac{\partial v_r}{\partial t} + v_r\frac{\partial v_r}{\partial r} + \frac{v_\theta}{r}\frac{\partial v_r}{\partial \theta} + \frac{v_\phi}{r\sin\theta}\frac{\partial v_r}{\partial \phi} - \frac{v_\theta^2 + v_\phi^2}{r}\right]$$

$$= F_r - \frac{\partial p}{\partial r} + \mu\left[\frac{1}{r^2}\frac{\partial}{\partial r}\left(r^2\frac{\partial v_r}{\partial r}\right) + \frac{1}{r^2\sin\theta}\frac{\partial}{\partial \theta}\left(\sin\theta\frac{\partial v_r}{\partial \theta}\right)\right.$$

$$\left. + \frac{1}{r^2\sin^2\theta}\frac{\partial^2 v_r}{\partial \phi^2} - \frac{2v_r}{r^2} - \frac{2}{r^2}\frac{\partial v_\theta}{\partial \theta} - \frac{2v_\theta\cot\theta}{r^2} - \frac{2}{r^2\sin\theta}\frac{\partial v_\phi}{\partial \phi}\right]$$

$$\rho\left[\frac{\partial v_\theta}{\partial t} + v_r\frac{\partial v_\theta}{\partial r} + \frac{v_\theta}{r}\frac{\partial v_\theta}{\partial \theta} + \frac{v_\phi}{r\sin\theta}\frac{\partial v_\theta}{\partial \phi} + \frac{v_r v_\theta}{r} - \frac{v_\phi^2\cot\theta}{r}\right] \qquad (A.9)$$

$$= F_\theta - \frac{1}{r}\frac{\partial p}{\partial \theta} + \mu\left[\frac{1}{r^2}\frac{\partial}{\partial r}\left(r^2\frac{\partial v_\theta}{\partial r}\right) + \frac{1}{r^2\sin\theta}\frac{\partial}{\partial \theta}\left(\sin\theta\frac{\partial v_\theta}{\partial \theta}\right)\right.$$

$$\left. + \frac{1}{r^2\sin^2\theta}\frac{\partial^2 v_\theta}{\partial \phi^2} + \frac{2}{r^2}\frac{\partial v_r}{\partial \theta} - \frac{v_\theta}{r^2\sin^2\theta} - \frac{2\cos\theta}{r^2\sin^2\theta}\frac{\partial v_\phi}{\partial \phi}\right]$$

$$\rho\left[\frac{\partial v_\phi}{\partial t} + v_r \frac{\partial v_\phi}{\partial r} + \frac{v_\theta}{r}\frac{\partial v_\phi}{\partial \theta} + \frac{v_\phi}{r \sin \theta}\frac{\partial v_\phi}{\partial \phi} + \frac{v_\phi v_r}{r} + \frac{v_\theta v_\phi \cot \theta}{r}\right]$$

$$= F_\phi - \frac{1}{r \sin \theta}\frac{\partial p}{\partial \phi} + \mu\left[\frac{1}{r^2}\frac{\partial}{\partial r}\left(r^2 \frac{\partial v_\phi}{\partial r}\right) + \frac{1}{r^2 \sin \theta}\frac{\partial}{\partial \theta}\left(\sin \theta \frac{\partial v_\phi}{\partial \theta}\right)\right.$$

$$\left. + \frac{1}{r^2 \sin^2 \theta}\frac{\partial^2 v_\phi}{\partial \phi^2} - \frac{v_\phi}{r^2 \sin^2 \theta} + \frac{2}{r^2 \sin^2 \theta}\frac{\partial v_r}{\partial \phi} + \frac{2 \cos \theta}{r^2 \sin^2 \theta}\frac{\partial v_\theta}{\partial \phi}\right]$$

2. STRESS-STRAIN RATE RELATIONSHIPS

Cartesian Tensor

w_i is the velocity in the x_i direction. ϕ is dilatation, $\nabla \cdot \mathbf{V}$.

$$\sigma_{ij} = -P\delta_{ij} + \sigma'_{ij} = -P\delta_{ij} + 2\mu e_{ij} + \delta_{ij}\lambda\phi \tag{A.10}$$

$$= -P\delta_{ij} + \mu\left(\frac{\partial w_i}{\partial x_j} + \frac{\partial w_j}{\partial x_i}\right) + \lambda\delta_{ij}\frac{\partial w_k}{\partial x_k}$$

$$= -P\delta_{ij} + \mu\left(\frac{\partial w_i}{\partial x_j} + \frac{\partial w_j}{\partial x_i} - \tfrac{2}{3}\delta_{ij}\frac{\partial w_k}{\partial x_k}\right) + \zeta\delta_{ij}\frac{\partial w_k}{\partial x_k}$$

Cartesian

u, v, and w are the velocities in the x, y, and z directions respectively.

$$\sigma_{xx} = -P + \sigma'_{xx} = -P + 2\mu e_{xx} + \lambda\nabla \cdot \mathbf{V} \tag{A.11}$$

$$= -P + 2\mu\frac{\partial u}{\partial x} + \lambda\left(\frac{\partial u}{\partial x} + \frac{\partial v}{\partial y} + \frac{\partial w}{\partial z}\right)$$

$$\sigma_{yy} = -P + \sigma'_{yy} = -P + 2\mu e_{yy} + \lambda\nabla \cdot \mathbf{V}$$

$$= -P + 2\mu\frac{\partial v}{\partial y} + \lambda\left(\frac{\partial u}{\partial x} + \frac{\partial v}{\partial y} + \frac{\partial w}{\partial z}\right)$$

$$\sigma_{zz} = -P + \sigma'_{zz} = -P + 2\mu e_{zz} + \lambda\nabla \cdot \mathbf{V}$$

$$= -P + 2\mu\frac{\partial w}{\partial z} + \lambda\left(\frac{\partial u}{\partial x} + \frac{\partial v}{\partial y} + \frac{\partial w}{\partial z}\right)$$

$$\sigma_{xy} = \sigma_{yx} = 2\mu e_{xy} = \mu\left(\frac{\partial u}{\partial y} + \frac{\partial v}{\partial x}\right)$$

$$\sigma_{xz} = \sigma_{zx} = 2\mu e_{xz} = \mu\left(\frac{\partial w}{\partial x} + \frac{\partial u}{\partial z}\right)$$

$$\sigma_{yz} = \sigma_{zy} = 2\mu e_{yz} = \mu\left(\frac{\partial v}{\partial z} + \frac{\partial w}{\partial y}\right)$$

Cylindrical

v_r, v_θ, and v_z are the velocities in the r, θ, and z directions respectively.

$$\sigma_{rr} = -P + \sigma'_{rr} = -P + 2\mu e_{rr} + \lambda \nabla \cdot \mathbf{V} \tag{A.12}$$

$$= -P + 2\mu \frac{\partial v_r}{\partial r} + \lambda \left\{ \frac{1}{r} \frac{\partial}{\partial r}(rv_r) + \frac{1}{r} \frac{\partial v_\theta}{\partial \theta} + \frac{\partial v_z}{\partial z} \right\}$$

$$\sigma_{\theta\theta} = -P + \sigma'_{\theta\theta} = -P + 2\mu e_{\theta\theta} + \lambda \nabla \cdot \mathbf{V}$$

$$= -P + 2\mu \left(\frac{1}{r} \frac{\partial v_\theta}{\partial \theta} + \frac{v_r}{r} \right) + \lambda \left\{ \frac{1}{r} \frac{\partial}{\partial r}(rv_r) + \frac{1}{r} \frac{\partial v_\theta}{\partial \theta} + \frac{\partial v_z}{\partial z} \right\}$$

$$\sigma_{zz} = -P + \sigma'_{zz} = -P + 2\mu e_{zz} + \lambda \nabla \cdot \mathbf{V}$$

$$= -P + 2\mu \frac{\partial v_z}{\partial z} + \lambda \left\{ \frac{1}{r} \frac{\partial}{\partial r}(rv_r) + \frac{1}{r} \frac{\partial v_\theta}{\partial \theta} + \frac{\partial v_z}{\partial z} \right\}$$

$$\sigma_{r\theta} = \sigma_{\theta r} = 2\mu e_{r\theta} = \mu \left(\frac{1}{r} \frac{\partial v_r}{\partial \theta} + \frac{\partial v_\theta}{\partial r} - \frac{v_\theta}{r} \right)$$

$$\sigma_{rz} = \sigma_{zr} = 2\mu e_{rz} = \mu \left(\frac{\partial v_r}{\partial z} + \frac{\partial v_z}{\partial r} \right)$$

$$\sigma_{\theta z} = \sigma_{z\theta} = 2\mu e_{\theta z} = \mu \left(\frac{1}{r} \frac{\partial v_z}{\partial \theta} + \frac{\partial v_\theta}{\partial z} \right)$$

Spherical

v_r, v_θ, and v_ϕ are the velocities in the r, θ, and ϕ directions respectively.

$$\sigma_{rr} = -P + \sigma'_{rr} = -P + 2\mu e_{rr} + \lambda \nabla \cdot \mathbf{V} \tag{A.13}$$

$$= -P + 2\mu \frac{\partial v_r}{\partial r} + \lambda \left\{ \frac{1}{r^2} \frac{\partial}{\partial r}(r^2 v_r) + \frac{1}{r \sin \theta} \frac{\partial}{\partial \theta}(v_\theta \sin \theta) + \frac{1}{r \sin \theta} \frac{\partial v_\phi}{\partial \phi} \right\}$$

$$\sigma_{\theta\theta} = -P + \sigma'_{\theta\theta} = -P + 2\mu e_{\theta\theta} + \lambda \nabla \cdot \mathbf{V}$$

$$= -P + 2\mu \left(\frac{1}{r} \frac{\partial v_\theta}{\partial \theta} + \frac{v_r}{r} \right) + \lambda \left\{ \frac{1}{r^2} \frac{\partial}{\partial r}(r^2 v_r) \right.$$

$$\left. + \frac{1}{r \sin \theta} \frac{\partial}{\partial \theta}(v_\theta \sin \theta) + \frac{1}{r \sin \theta} \frac{\partial v_\phi}{\partial \phi} \right\}$$

$$\sigma_{\phi\phi} = -P + \sigma'_{\phi\phi} = -P + 2\mu e_{\phi\phi} + \lambda \nabla \cdot \mathbf{V}$$

$$= -P + 2\mu \left(\frac{1}{r \sin \theta} \frac{\partial v_\phi}{\partial \phi} + \frac{v_r}{r} + \frac{v_\theta \cot \theta}{r} \right)$$

$$+ \lambda \left\{ \frac{1}{r^2} \frac{\partial}{\partial r}(r^2 v_r) + \frac{1}{r \sin \theta} \frac{\partial}{\partial \theta}(v_\theta \sin \theta) + \frac{1}{r \sin \theta} \frac{\partial v_\phi}{\partial \phi} \right\}$$

$$\sigma_{r\theta} = \sigma_{\theta r} = 2\mu e_{r\theta} = \mu \left\{ r \frac{\partial}{\partial r}\left(\frac{v_\theta}{r} \right) + \frac{1}{r} \frac{\partial v_r}{\partial \theta} \right\}$$

$$\sigma_{r\phi} = \sigma_{\phi r} = 2\mu e_{r\phi} = \mu \left\{ \frac{1}{r \sin \theta} \frac{\partial v_r}{\partial \phi} + r \frac{\partial}{\partial r}\left(\frac{v_\phi}{r} \right) \right\}$$

$$\sigma_{\theta\phi} = \sigma_{\phi\theta} = 2\mu e_{\theta\phi} = \mu \left\{ \frac{\sin \theta}{r} \frac{\partial}{\partial \theta}\left(\frac{v_\phi}{\sin \theta} \right) + \frac{1}{r \sin \theta} \frac{\partial v_\theta}{\partial \phi} \right\}$$

3. SOME USEFUL VECTOR OPERATIONS

The operations $\dfrac{D}{Dt}$ and ∇^2 listed below are for operations on a scalar. These are not the same as operations on a vector except in cartesian coordinates.

Cartesian

$$\frac{D}{Dt} = \frac{\partial}{\partial t} + u\frac{\partial}{\partial x} + v\frac{\partial}{\partial y} + w\frac{\partial}{\partial z}$$

$$\nabla \cdot \mathbf{V} = \frac{\partial u}{\partial x} + \frac{\partial v}{\partial y} + \frac{\partial w}{\partial z} \tag{A.14}$$

$$\nabla^2 = \frac{\partial^2}{\partial x^2} + \frac{\partial^2}{\partial y^2} + \frac{\partial^2}{\partial z^2}$$

Cylindrical

$$\frac{D}{Dt} = \frac{\partial}{\partial t} + v_r\frac{\partial}{\partial r} + \frac{v_\theta}{r}\frac{\partial}{\partial \theta} + v_z\frac{\partial}{\partial z}$$

$$\nabla \cdot \mathbf{V} = \frac{1}{r}\frac{\partial}{\partial r}(rv_r) + \frac{1}{r}\frac{\partial v_\theta}{\partial \theta} + \frac{\partial v_z}{\partial z} \tag{A.15}$$

$$\nabla^2 = \frac{\partial^2}{\partial r^2} + \frac{1}{r}\frac{\partial}{\partial r} + \frac{1}{r^2}\frac{\partial^2}{\partial \theta^2} + \frac{\partial^2}{\partial z^2}$$

Spherical

$$\frac{D}{Dt} = \frac{\partial}{\partial t} + v_r\frac{\partial}{\partial r} + \frac{v_\theta}{r}\frac{\partial}{\partial \theta} + \frac{v_\phi}{r\sin\theta}\frac{\partial}{\partial \phi}$$

$$\nabla \cdot \mathbf{V} = \frac{1}{r^2}\frac{\partial}{\partial r}(r^2 v_r) + \frac{1}{r\sin\theta}\frac{\partial}{\partial \theta}(v_\theta\sin\theta) + \frac{1}{r\sin\theta}\frac{\partial v_\phi}{\partial \phi} \tag{A.16}$$

$$\nabla^2 = \frac{1}{r^2}\frac{\partial}{\partial r}\left(r^2\frac{\partial}{\partial r}\right) + \frac{1}{r^2\sin\theta}\frac{\partial}{\partial \theta}\left(\sin\theta\frac{\partial}{\partial \theta}\right) + \frac{1}{r^2\sin^2\theta}\frac{\partial^2}{\partial \phi^2}$$

The ∇^2 operation and D/Dt operation on a vector are listed below. (^) indicates a unit vector.

Cartesian

$$\mathbf{V} = \hat{x}u + \hat{y}v + \hat{z}w$$

$$(\nabla^2\mathbf{V})_x = \frac{\partial^2 u}{\partial x^2} + \frac{\partial^2 u}{\partial y^2} + \frac{\partial^2 u}{\partial z^2} \tag{A.17}$$

$$(\nabla^2\mathbf{V})_y = \frac{\partial^2 v}{\partial x^2} + \frac{\partial^2 v}{\partial y^2} + \frac{\partial^2 v}{\partial z^2}$$

$$(\nabla^2\mathbf{V})_z = \frac{\partial^2 w}{\partial x^2} + \frac{\partial^2 w}{\partial y^2} + \frac{\partial^2 w}{\partial z^2}$$

$$\left(\frac{D\mathbf{V}}{Dt}\right)_x = \frac{\partial u}{\partial t} + u\frac{\partial u}{\partial x} + v\frac{\partial u}{\partial y} + w\frac{\partial u}{\partial z}$$

$$\left(\frac{D\mathbf{V}}{Dt}\right)_y = \frac{\partial v}{\partial t} + u\frac{\partial v}{\partial x} + v\frac{\partial v}{\partial y} + w\frac{\partial v}{\partial z}$$

$$\left(\frac{D\mathbf{V}}{Dt}\right)_z = \frac{\partial w}{\partial t} + u\frac{\partial w}{\partial x} + v\frac{\partial w}{\partial y} + w\frac{\partial w}{\partial z}$$

Cylindrical

$$\mathbf{V} = \hat{r}v_r + \hat{\theta}v_\theta + \hat{z}v_z$$

$$(\nabla^2\mathbf{V})_r = \nabla^2 v_r - \frac{v_r}{r^2} - \frac{2}{r^2}\frac{\partial v_\theta}{\partial\theta} = \left[\frac{\partial^2 v_r}{\partial r^2} + \frac{1}{r}\frac{\partial v_r}{\partial r} + \frac{1}{r^2}\frac{\partial^2 v_r}{\partial\theta^2} + \frac{\partial^2 v_r}{\partial z^2} - \frac{v_r}{r^2} - \frac{2}{r^2}\frac{\partial v_\theta}{\partial\theta}\right]$$

$$(\nabla^2\mathbf{V})_\theta = \nabla^2 v_\theta + \frac{2}{r^2}\frac{\partial v_r}{\partial_\theta} - \frac{v_\theta}{r^2} = \left[\frac{\partial^2 v_\theta}{\partial r^2} + \frac{1}{r}\frac{\partial v_\theta}{\partial r} + \frac{1}{r^2}\frac{\partial^2 v_\theta}{\partial\theta^2} + \frac{\partial^2 v_\theta}{\partial z^2} + \frac{2}{r^2}\frac{\partial v_r}{\partial\theta} - \frac{v_\theta}{r^2}\right] \quad \textbf{(A.18)}$$

$$(\nabla^2\mathbf{V})_z = \nabla^2 v_z = \frac{\partial^2 v_z}{\partial r^2} + \frac{1}{r}\frac{\partial v_z}{\partial r} + \frac{1}{r^2}\frac{\partial^2 v_z}{\partial\theta^2} + \frac{\partial^2 v_z}{\partial z^2}$$

$$\left(\frac{D\mathbf{V}}{Dt}\right)_r = \frac{Dv_r}{Dt} - \frac{v_\theta^2}{r} = \frac{\partial v_r}{\partial t} + v_r\frac{\partial v_r}{\partial r} + \frac{v_\theta}{r}\frac{\partial v_r}{\partial\theta} + v_z\frac{\partial v_r}{\partial z} - \frac{v_\theta^2}{r}$$

$$\left(\frac{D\mathbf{V}}{Dt}\right)_\theta = \frac{Dv_\theta}{Dt} + \frac{v_r v_\theta}{r} = \frac{\partial v_\theta}{\partial t} + v_r\frac{\partial v_\theta}{\partial r} + \frac{v_\theta}{r}\frac{\partial v_\theta}{\partial\theta} + v_z\frac{\partial v_\theta}{\partial z} + \frac{v_r v_\theta}{r}$$

$$\left(\frac{D\mathbf{V}}{Dt}\right)_z = \frac{Dv_z}{Dt} = \frac{\partial v_z}{\partial t} + v_r\frac{\partial v_z}{\partial r} + \frac{v_\theta}{r}\frac{\partial v_z}{\partial\theta} + v_z\frac{\partial v_z}{\partial z}$$

Spherical

$$\mathbf{V} = \hat{r}v_r + \hat{\theta}v_\theta + \hat{\phi}v_\phi \quad \textbf{(A.19)}$$

$$(\nabla^2\mathbf{V})_r = \nabla^2 v_r - \frac{2v_r}{r^2} - \frac{2}{r^2}\frac{\partial v_\theta}{\partial\theta} - \frac{2v_\theta\cot\theta}{r^2} - \frac{2}{r^2\sin\theta}\frac{\partial v_\phi}{\partial\phi}$$

$$(\nabla^2\mathbf{V})_\theta = \nabla^2 v_\theta + \frac{2}{r^2}\frac{\partial v_r}{\partial\theta} - \frac{v_\theta}{r^2\sin^2\theta} - \frac{2}{r^2\sin\theta}\frac{\partial v_\phi}{\partial\phi}$$

$$(\nabla^2\mathbf{V})_\phi = \nabla^2 v_\phi - \frac{v_\phi}{r^2\sin^2\theta} + \frac{2}{r^2\sin^2\theta}\frac{\partial v_r}{\partial\phi} + \frac{2\cos\theta}{r^2\sin^2\theta}\frac{\partial v_\theta}{\partial\phi}$$

$$\left(\frac{D\mathbf{V}}{Dt}\right)_r = \frac{Dv_r}{Dt} - \frac{v_\theta^2 + v_\phi^2}{r} = \frac{\partial v_r}{\partial t} + v_r\frac{\partial v_r}{\partial r} + \frac{v_\theta}{r}\frac{\partial v_r}{\partial\theta} + \frac{v_\phi}{r\sin\theta}\frac{\partial v_r}{\partial\phi} - \frac{v_\theta^2 + v_\phi^2}{r}$$

$$\left(\frac{D\mathbf{V}}{Dt}\right)_\theta = \frac{Dv_\theta}{Dt} + \frac{v_r v_\theta - v_\phi^2\cot\theta}{r}$$

$$= \frac{\partial v_\theta}{\partial t} + v_r\frac{\partial v_\theta}{\partial r} + \frac{v_\theta}{r}\frac{\partial v_\theta}{\partial\theta} + \frac{v_\phi}{r\sin\theta}\frac{\partial v_\theta}{\partial\phi} + \frac{v_r v_\theta - v_\phi^2\cot\theta}{r}$$

$$\left(\frac{D\mathbf{V}}{Dt}\right)_\phi = \frac{Dv_\phi}{Dt} + \frac{v_\phi v_r + v_\theta v_\phi\cot\theta}{r}$$

$$= \frac{\partial v_\phi}{\partial t} + v_r\frac{\partial v_\phi}{\partial r} + \frac{v_\theta}{r}\frac{\partial v_\phi}{\partial\phi} + \frac{v_\phi}{r\sin\theta}\frac{\partial v_\phi}{\partial\phi} + \frac{v_\phi v_r + v_\theta v_\phi\cot\theta}{r}$$

The expression for $\nabla^2 v_r$, $\nabla^2 v_\theta$, and $\nabla^2 v_\phi$ in various coordinate systems are given by the operator ∇^2 on a scalar by equations (A.14), (A.15), and (A.16).

4. THE DISSIPATION FUNCTION

The mechanical or viscous dissipation function Φ is defined in generalized orthogonal coordinates as (in terms of the strain rate tensor e_{ij})

$$\Phi = \mu[2(e_{11}^2 + e_{22}^2 + e_{33}^2) + (2e_{23})^2 + (2e_{31})^2 + (2e_{12})^2] + \lambda(e_{11} + e_{22} + e_{33})^2 \quad \textbf{(A.20)}$$

(In some texts, the definition of Φ differs by a factor of μ from the one defined here.) λ is the second coefficient of viscosity defined in Chapter 3.

Cartesian Tensor in Terms of the Stress Tensor

$$\Phi = \sigma'_{ij} \frac{\partial w_i}{\partial x_j} \tag{A.21}$$

Cartesian

$$\Phi = 2\mu\left[\left(\frac{\partial u}{\partial x}\right)^2 + \left(\frac{\partial v}{\partial y}\right)^2 + \left(\frac{\partial w}{\partial z}\right)^2 + \frac{1}{2}\left(\frac{\partial u}{\partial y} + \frac{\partial v}{\partial x}\right)^2 + \frac{1}{2}\left(\frac{\partial v}{\partial z} + \frac{\partial w}{\partial y}\right)^2\right.$$
$$\left. + \frac{1}{2}\left(\frac{\partial w}{\partial x} + \frac{\partial u}{\partial z}\right)^2\right] + \lambda\left[\frac{\partial u}{\partial x} + \frac{\partial v}{\partial y} + \frac{\partial w}{\partial z}\right]^2 \tag{A.22}$$

Cylindrical

$$\Phi = \mu\left[2\left\{\left(\frac{\partial v_r}{\partial r}\right)^2 + \left(\frac{1}{r}\frac{\partial v_\theta}{\partial \theta} + \frac{v_r}{r}\right)^2 + \left(\frac{\partial v_z}{\partial z}\right)^2\right\} + \left(\frac{1}{r}\frac{\partial v_z}{\partial \theta} + \frac{\partial v_\theta}{\partial z}\right)^2\right.$$
$$\left. + \left(\frac{\partial v_r}{\partial z} + \frac{\partial v_z}{\partial r}\right)^2 + \left(\frac{1}{r}\frac{\partial v_r}{\partial \theta} + \frac{\partial v_\theta}{\partial r} - \frac{v_\theta}{r}\right)^2\right]$$
$$+ \lambda\left[\frac{\partial v_r}{\partial r} + \frac{1}{r}\frac{\partial v_\theta}{\partial \theta} + \frac{v_r}{r} + \frac{\partial v_z}{\partial z}\right]^2 \tag{A.23}$$

Spherical

$$\Phi = \mu\left[2\left\{\left(\frac{\partial v_r}{\partial r}\right)^2 + \left(\frac{1}{r}\frac{\partial v_\theta}{\partial \theta} + \frac{v_r}{r}\right)^2 + \left(\frac{1}{r\sin\theta}\frac{\partial v_\phi}{\partial \phi} + \frac{v_r}{r} + \frac{v_\theta\cot\theta}{r}\right)^2\right\}\right.$$
$$+ \left\{\frac{1}{r\sin\theta}\frac{\partial v_\theta}{\partial \phi} + \frac{\sin\theta}{r}\frac{\partial}{\partial \theta}\left(\frac{v_\phi}{\sin\theta}\right)\right\}^2$$
$$\left. + \left\{\frac{1}{r\sin\theta}\frac{\partial v_r}{\partial \phi} + r\frac{\partial}{\partial r}\left(\frac{v_\phi}{r}\right)\right\}^2 + \left\{r\frac{\partial}{\partial r}\left(\frac{v_\theta}{r}\right) + \frac{1}{r}\frac{\partial v_r}{\partial \theta}\right\}^2\right]$$
$$+ \lambda\left[\frac{\partial v_r}{\partial r} + \frac{1}{r}\frac{\partial v_\theta}{\partial \theta} + \frac{2v_r}{r} + \frac{1}{r\sin\theta}\frac{\partial v_\phi}{\partial \phi} + \frac{v_\theta\cot\theta}{r}\right]^2 \tag{A.24}$$

Appendix D

Tables for Compressible Flow

Table D.1   Flow Parameters versus M for Isentropic Flow (Perfect Gas, $k = 1.4$)

M	p/p_0	ρ/ρ_0	T/T_0	a/a_0	A^*/A
0.00	1.0000	1.0000	1.0000	1.0000	0.00000
0.10	0.9930	0.9950	0.9980	0.9990	0.1718
0.20	0.9725	0.9803	0.9921	0.9960	0.3374
0.30	0.9395	0.9564	0.9823	0.9911	0.4914
0.40	0.8956	0.9243	0.9690	0.9844	0.6289
0.50	0.8430	0.8852	0.9524	0.9759	0.7464
0.60	0.7840	0.8405	0.9328	0.9658	0.8416
0.70	0.7209	0.7916	0.9107	0.9543	0.9138
0.80	0.6560	0.7400	0.8865	0.9416	0.9632
0.90	0.5913	0.6870	0.8606	0.9277	0.9912
1.00	0.5283	0.6339	0.8333	0.9129	1.0000
1.10	0.4684	0.5817	0.8052	0.8973	0.9921
1.20	0.4124	0.5311	0.7764	0.8811	0.9705
1.30	0.3609	0.4829	0.7474	0.8645	0.9378
1.40	0.3142	0.4374	0.7184	0.8476	0.8969
1.50	0.2724	0.3950	0.6897	0.8305	0.8502
1.60	0.2353	0.3557	0.6614	0.8133	0.7998
1.70	0.2026	0.3197	0.6337	0.7961	0.7476
1.80	0.1740	0.2868	0.6068	0.7790	0.6949
1.90	0.1492	0.2570	0.5807	0.7620	0.6430
2.00	0.1278	0.2300	0.5556	0.7454	0.5926
2.10	0.1094	0.2058	0.5313	0.7289	0.5444
2.20	0.09352	0.1841	0.5081	0.7128	0.4988
2.30	0.07997	0.1646	0.4859	0.6971	0.4560
2.40	0.06840	0.1472	0.4647	0.6817	0.4161
2.50	0.05853	0.1317	0.4444	0.6667	0.3793
2.60	0.05012	0.1179	0.4252	0.6521	0.3453
2.70	0.04295	0.1056	0.4068	0.6378	0.3142
2.80	0.03685	0.09463	0.3894	0.6240	0.2857
2.90	0.03165	0.08489	0.3729	0.6106	0.2598
3.00	0.02722	0.07623	0.3571	0.5976	0.2362
3.10	0.02345	0.06852	0.3422	0.5850	0.2147
3.20	0.02023	0.06165	0.3281	0.5728	0.1953
3.30	0.01748	0.05554	0.3147	0.5609	0.1777
3.40	0.01513	0.05009	0.3019	0.5495	0.1617
3.50	0.01311	0.04523	0.2899	0.5384	0.1473
3.60	0.01138	0.04089	0.2784	0.5276	0.1342
3.70	9.903×10^{-3}	0.03702	0.2675	0.5172	0.1224
3.80	8.629×10^{-3}	0.03355	0.2572	0.5072	0.1117
3.90	7.532×10^{-3}	0.03044	0.2474	0.4974	0.1021

Table D.1—(Continued)

M	p/p_0	ρ/ρ_0	T/T_0	a/a_0	A^*/A
4.00	6.586×10^{-3}	0.02766	0.2381	0.4880	0.09329
4.10	5.769×10^{-3}	0.02516	0.2293	0.4788	0.08536
4.20	5.062×10^{-3}	0.02292	0.2208	0.4699	0.07818
4.30	4.449×10^{-3}	0.02090	0.2129	0.4614	0.07166
4.40	3.918×10^{-3}	0.01909	0.2053	0.4531	0.06575
4.50	3.455×10^{-3}	0.01745	0.1980	0.4450	0.06038
4.60	3.053×10^{-3}	0.01597	0.1911	0.4372	0.05550
4.70	2.701×10^{-3}	0.01464	0.1846	0.4296	0.05107
4.80	2.394×10^{-3}	0.01343	0.1783	0.4223	0.04703
4.90	2.126×10^{-3}	0.01233	0.1724	0.4152	0.04335
5.00	1.890×10^{-3}	0.01134	0.1667	0.4082	0.04000
6.00	6.334×10^{-4}	5.194×10^{-3}	0.1220	0.3492	0.01880
7.00	2.416×10^{-4}	2.609×10^{-3}	0.09259	0.3043	9.602×10^{-3}
8.00	1.024×10^{-4}	1.414×10^{-3}	0.07246	0.2692	5.260×10^{-3}
9.00	4.739×10^{-5}	8.150×10^{-4}	0.05814	0.2411	3.056×10^{-3}
10.00	2.356×10^{-5}	4.948×10^{-4}	0.04762	0.2182	1.866×10^{-3}
100.00	2.790×10^{-12}	5.583×10^{-9}	4.998×10^{-4}	0.02236	2.157×10^{-8}
∞	0	0	0	0	0

Table D.2   Mach Number and Mach Angle versus Prandtl-Meyer Function

ν (deg)	M	μ (deg)	ν (deg)	M	μ (deg)	ν (deg)	M	μ (deg)
0.0	1.000	90.000	5.0	1.256	52.738	10.0	1.435	44.177
0.5	1.051	72.099	5.5	1.275	51.642	10.5	1.452	43.523
1.0	1.082	67.574	6.0	1.294	50.619	11.0	1.469	42.894
1.5	1.108	64.451	6.5	1.312	49.658	11.5	1.486	42.287
2.0	1.133	61.997	7.0	1.330	48.753	12.0	1.503	41.701
2.5	1.155	59.950	7.5	1.348	47.896	12.5	1.520	41.134
3.0	1.177	58.180	8.0	1.366	47.082	13.0	1.537	40.585
3.5	1.198	56.614	8.5	1.383	46.306	13.5	1.554	40.053
4.0	1.218	55.205	9.0	1.400	45.566	14.0	1.571	39.537
4.5	1.237	53.920	9.5	1.418	44.857	14.5	1.588	39.035

Table D.2—(Continued)

v (deg)	M	μ (deg)	v (deg)	M	μ (deg)	v (deg)	M	μ (deg)
15.0	1.605	38.547	37.5	2.431	24.287	60.0	3.594	16.155
15.5	1.622	38.073	38.0	2.452	24.066	60.5	3.627	16.005
16.0	1.639	37.611	38.5	2.473	23.847	61.0	3.660	15.856
16.5	1.655	37.160	39.0	2.495	23.631	61.5	3.694	15.708
17.0	1.672	36.721	39.5	2.516	23.418	62.0	3.728	15.561
17.5	1.689	36.293	40.0	2.538	23.206	62.5	3.762	15.415
18.0	1.706	35.874	40.5	2.560	22.997	63.0	3.797	15.270
18.5	1.724	35.465	41.0	2.582	22.790	63.5	3.832	15.126
19.0	1.741	35.065	41.5	2.604	22.585	64.0	3.868	14.983
19.5	1.758	34.673	42.0	2.626	22.382	64.5	3.904	14.840
20.0	1.775	34.290	42.5	2.649	22.182	65.0	3.941	14.698
20.5	1.792	33.915	43.0	2.671	21.983	65.5	3.979	14.557
21.0	1.810	33.548	43.5	2.694	21.786	66.0	4.016	14.417
21.5	1.827	33.188	44.0	2.718	21.591	66.5	4.055	14.278
22.0	1.844	32.834	44.5	2.741	21.398	67.0	4.094	14.140
22.5	1.862	32.488	45.0	2.764	21.207	67.5	4.133	14.002
23.0	1.879	32.148	45.5	2.788	21.017	68.0	4.173	13.865
23.5	1.897	31.814	46.0	2.812	20.830	68.5	4.214	13.729
24.0	1.915	31.486	46.5	2.836	20.644	69.0	4.255	13.593
24.5	1.932	31.164	47.0	2.861	20.459	69.5	4.297	13.459
25.0	1.950	30.847	47.5	2.886	20.277	70.0	4.339	13.325
25.5	1.968	30.536	48.0	2.910	20.096	70.5	4.382	13.191
26.0	1.986	30.229	48.5	2.936	19.916	71.0	4.426	13.059
26.5	2.004	29.928	49.0	2.961	19.738	71.5	4.470	12.927
27.0	2.023	29.632	49.5	2.987	19.561	72.0	4.515	12.795
27.5	2.041	29.340	50.0	3.013	19.386	72.5	4.561	12.665
28.0	2.059	29.052	50.5	3.039	19.213	73.0	4.608	12.535
28.5	2.078	28.769	51.0	3.065	19.041	73.5	4.655	12.406
29.0	2.096	28.491	51.5	3.092	18.870	74.0	4.703	12.277
29.5	2.115	28.216	52.0	3.119	18.701	74.5	4.752	12.149
30.0	2.134	27.945	52.5	3.146	18.532	75.0	4.801	12.021
30.5	2.153	27.678	53.0	3.174	18.366	75.5	4.852	11.894
31.0	2.172	27.415	53.5	3.202	18.200	76.0	4.903	11.768
31.5	2.191	27.155	54.0	3.230	18.036	76.5	4.955	11.642
32.0	2.210	26.899	54.5	3.258	17.873	77.0	5.009	11.517
32.5	2.230	26.646	55.0	3.287	17.711	77.5	5.063	11.392
33.0	2.249	26.397	55.5	3.316	17.551	78.0	5.118	11.268
33.5	2.269	26.151	56.0	3.346	17.391	78.5	5.174	11.145
34.0	2.289	25.908	56.5	3.375	17.233	79.0	5.231	11.022
34.5	2.309	25.668	57.0	3.406	17.076	79.5	5.289	10.899
35.0	2.329	25.430	57.5	3.436	16.920	80.0	5.348	10.777
35.5	2.349	25.196	58.0	3.467	16.765	80.5	5.408	10.656
36.0	2.369	24.965	58.5	3.498	16.611	81.0	5.470	10.535
36.5	2.390	24.736	59.0	3.530	16.458	81.5	5.532	10.414
37.0	2.410	24.510	59.5	3.562	16.306	82.0	5.596	10.294

Table D.2—(Continued)

v (deg)	M	μ (deg)	v (deg)	M	μ (deg)	v (deg)	M	μ (deg)
82.5	5.661	10.175	90.0	6.819	8.433	97.5	8.480	6.772
83.0	5.727	10.056	90.5	6.911	8.320	98.0	8.618	6.664
83.5	5.795	9.937	91.0	7.005	8.207	98.5	8.759	6.556
84.0	5.864	9.819	91.5	7.102	8.095	99.0	8.905	6.448
84.5	5.935	9.701	92.0	7.201	7.983	99.5	9.055	6.340
85.0	6.006	9.584	92.5	7.302	7.871	100.0	9.210	6.233
85.5	6.080	9.467	93.0	7.406	7.760	100.5	9.371	6.126
86.0	6.155	9.350	93.5	7.513	7.649	101.0	9.536	6.019
86.5	6.232	9.234	94.0	7.623	7.538	101.5	9.708	5.913
87.0	6.310	9.119	94.5	7.735	7.428	102.0	9.885	5.806
87.5	6.390	9.003	95.0	7.851	7.318			
88.0	6.472	8.888	95.5	7.970	7.208			
88.5	6.556	8.774	96.0	8.092	7.099			
89.0	6.642	8.660	96.5	8.218	6.989			
89.5	6.729	8.546	97.0	8.347	6.881			

Table D.3   Parameters for Shock Flow (Perfect Gas, $k = 1.4$)

M_{1n}	p_2/p_1	ρ_2/ρ_1	T_2/T_1	a_2/a_1	p_2^0/p_1^0	M_2 for normal shocks only
1.00	1.000	1.000	1.000	1.000	1.0000	1.0000
1.10	1.245	1.169	1.065	1.032	0.9989	0.9118
1.20	1.513	1.342	1.128	1.062	0.9928	0.8422
1.30	1.805	1.516	1.191	1.091	0.9794	0.7860
1.40	2.120	1.690	1.255	1.120	0.9582	0.7397
1.50	2.458	1.862	1.320	1.149	0.9298	0.7011
1.60	2.820	2.032	1.388	1.178	0.8952	0.6684
1.70	3.205	2.198	1.458	1.208	0.8557	0.6405
1.80	3.613	2.359	1.532	1.238	0.8127	0.6165
1.90	4.045	2.516	1.608	1.268	0.7674	0.5956
2.00	4.500	2.667	1.688	1.299	0.7209	0.5773
2.10	4.978	2.812	1.770	1.331	0.6742	0.5613
2.20	5.480	2.951	1.857	1.363	0.6281	0.5471
2.30	6.005	3.085	1.947	1.395	0.5833	0.5344
2.40	6.553	3.212	2.040	1.428	0.5401	0.5231
2.50	7.125	3.333	2.138	1.462	0.4990	0.5130
2.60	7.720	3.449	2.238	1.496	0.4601	0.5039
2.70	8.338	3.559	2.343	1.531	0.4236	0.4956
2.80	8.980	3.664	2.451	1.566	0.3895	0.4882
2.90	9.645	3.763	2.563	1.601	0.3577	0.4814
3.00	10.33	3.857	2.679	1.637	0.3283	0.4752
4.00	18.50	4.571	4.047	2.012	0.1388	0.4350
5.00	29.00	5.000	5.800	2.408	0.06172	0.4152
6.00	41.83	5.268	7.941	2.818	0.02965	0.4042
7.00	57.00	5.444	10.47	3.236	0.01535	0.3974
8.00	74.50	5.565	13.39	3.659	8.488×10^{-3}	0.3929
9.00	94.33	5.651	16.69	4.086	4.964×10^{-3}	0.3898
10.00	116.5	5.714	20.39	4.515	3.045×10^{-3}	0.3876
100.00	11,666.5	5.997	1945.4	44.11	3.593×10^{-8}	0.3781
∞	∞	6	∞	∞	0	0.3780

Table D.4　Frictional, Adiabatic, Constant-Area Flow (Fanno Line) Perfect Gas, $k = 1.4$

M	T/T^*	p/p^*	p_0/p_0^*	V/V^* and ρ^*/ρ	$4fL_{max}/D$
0.00	1.2000	∞	∞	0.00000	∞
0.20	1.1905	5.4555	2.9635	0.21822	14.533
0.40	1.1628	2.6958	1.5901	0.43133	2.3085
0.60	1.1194	1.7634	1.1882	0.63481	0.49081
0.80	1.06383	1.2892	1.03823	0.82514	0.07229
1.00	1.00000	1.00000	1.00000	1.00000	0
1.20	0.93168	0.80436	1.03044	1.1583	0.03364
1.40	0.86207	0.66320	1.1149	1.2999	0.09974
1.60	0.79365	0.55679	1.2502	1.4254	0.17236
1.80	0.72816	0.47407	1.4390	1.5360	0.24189
2.00	0.66667	0.40825	1.6875	1.6330	0.30499
2.50	0.53333	0.29212	2.6367	1.8257	0.43197
3.00	0.42857	0.21822	4.2346	1.9640	0.52216
3.50	0.34783	0.16850	6.7896	2.0642	0.58643
4.00	0.28571	0.13363	10.719	2.1381	0.63306
4.50	0.23762	0.10833	16.562	2.1936	0.66764
5.00	0.20000	0.08944	25.000	2.2361	0.69381
6.00	0.14634	0.06376	53.180	2.2953	0.72987
7.00	0.11111	0.04762	104.14	2.3333	0.75281
8.00	0.08696	0.03686	190.11	2.3591	0.76820
9.00	0.06977	0.02935	327.19	2.3772	0.77898
10.00	0.05714	0.02390	535.94	2.3905	0.78683
∞	0	0	∞	2.4495	0.82153

Appendix E

Cartesian Tensors

Cartesian tensor notation is a shorthand way of writing equations in cartesian coordinates. It should be emphasized that this notation is not really tensor notation and is not applicable to any coordinate system other than cartesian. From a cartesian tensor equation it is not possible to infer the proper form of an equation in any other arbitrary coordinate system as it is from a generalized tensor or vector form.

Consider any vector, such as velocity $\mathbf{V}$, with cartesian components $w_1 = V_x$, $w_2 = V_y$, and $w_3 = V_z$. We denote the components of a vector by a subscript i, j, or any other dummy suffix which may take on a value of 1, 2, or 3. The 1, 2, and 3 represent the x, y, and z components, respectively.

The velocity components may be represented then as

$$w_i$$

where the i takes on the value 1, 2, or 3, corresponding to x, y, or z.

A scalar is represented without a suffix, as, for example, the velocity potential ϕ.

The cartesian coordinates x, y, and z are represented as x_i (or x_j, say).

When a suffix appears twice in a term, a summation is usually implied (over the three dimensions). For example, the expression

$$\frac{\partial w_i}{\partial x_j}$$

is an array of nine terms where both i and j may take on values of 1, 2, 3. The expression

$$\frac{\partial w_i}{\partial x_i}$$

is a scalar which represents (by the summation convention)

$$\frac{\partial w_i}{\partial x_i} = \frac{\partial w_1}{\partial x_1} + \frac{\partial w_2}{\partial x_2} + \frac{\partial w_3}{\partial x_3} = \nabla \cdot \mathbf{V}$$

Similarly, σ_{ij} are the nine components of the stress tensor (which is a 3 by 3 matrix). The symmetry condition is expressed as $\sigma_{ij} = \sigma_{ji}$.

A comma is sometimes used to denote differentiation with respect to a spatial coordinate.

$$\frac{\partial w_i}{\partial x_j} = w_{i,j}$$

so that $w_{i,i} = \nabla \cdot \mathbf{V}$.

The Kronecker delta δ_{ij} is an operator that has the following values:

$$\delta_{ij} = 0; \quad i \neq j$$

$$\delta_{ij} = 1; \quad i = j$$

For example,

$$\sigma_{ij}\delta_{ij}$$

represents the diagonal terms of the stress tensor, σ_{11}, σ_{22}, and σ_{33}. Summation is generally not implied when the Kronecker delta is used.

Some vector operations in cartesian tensor notation are as follows (ϕ is a scalar, $\mathbf{V}$ is a vector with components w_i): These expressions are *valid only in cartesian coordinates.*

$$\nabla\phi = \frac{\partial\phi}{\partial x_i}$$

$$\nabla \cdot \mathbf{V} = \frac{\partial w_i}{\partial x_i} = \frac{\partial w_1}{\partial x_1} + \frac{\partial w_2}{\partial x_2} + \frac{\partial w_3}{\partial x_3}$$

$$\nabla^2\phi = \frac{\partial^2\phi}{\partial x_i\,\partial x_i} = \frac{\partial^2\phi}{\partial x_1^2} + \frac{\partial^2\phi}{\partial x_2^2} + \frac{\partial^2\phi}{\partial x_3^2}$$

$$\nabla^2\mathbf{V} = \frac{\partial^2 w_j}{\partial x_i\,\partial x_i} = \frac{\partial^2 w_j}{\partial x_1^2} + \frac{\partial^2 w_j}{\partial x_2^2} + \frac{\partial^2 w_j}{\partial x_3^2}$$

(This is a vector with components given by j, the summation being carried out over i.)

For further information about tensor notation in general the following references are recommended:

Borg, S. F., *Matrix-Tensor Methods in Continuum Mechanics*, Van Nostrand, 1963.

Myklestad, N. O., *Cartesian Tensors*, Van Nostrand, 1967.

Synge, J. L., and Schild, A., *Tensor Calculus*, University of Toronto Press, 1949.

Vector Identities

A, **B**, and **C** are vectors; ϕ is a scalar.

$$\mathbf{A} \cdot \mathbf{B} = \mathbf{B} \cdot \mathbf{A}$$

$$\mathbf{A} \cdot (\mathbf{B} + \mathbf{C}) = \mathbf{A} \cdot \mathbf{B} + \mathbf{A} \cdot \mathbf{C}$$

$$\mathbf{A} \times \mathbf{B} = -\mathbf{B} \times \mathbf{A} = \begin{vmatrix} e_1 & e_2 & e_3 \\ A_1 & A_2 & A_3 \\ B_1 & B_2 & B_3 \end{vmatrix}$$

$$(\mathbf{A} + \mathbf{B}) \times \mathbf{C} = (\mathbf{A} \times \mathbf{C}) + (\mathbf{B} \times \mathbf{C})$$

$$\mathbf{A} \times (\mathbf{B} + \mathbf{C}) = (\mathbf{A} \times \mathbf{B}) + (\mathbf{A} \times \mathbf{C})$$

$$\mathbf{A} \times (\mathbf{B} \times \mathbf{C}) = \mathbf{B}(\mathbf{A} \cdot \mathbf{C}) - \mathbf{C}(\mathbf{A} \cdot \mathbf{B})$$

$$\mathbf{A} \cdot (\mathbf{B} \times \mathbf{C}) = (\mathbf{A} \times \mathbf{B}) \cdot \mathbf{C} = \mathbf{B} \cdot (\mathbf{C} \times \mathbf{A}) = \begin{vmatrix} A_1 & A_2 & A_3 \\ B_1 & B_2 & B_3 \\ C_1 & C_2 & C_3 \end{vmatrix}$$

$$(\mathbf{A} \times \mathbf{B}) \cdot (\mathbf{C} \times \mathbf{D}) = (\mathbf{A} \cdot \mathbf{C})(\mathbf{B} \cdot \mathbf{D}) - (\mathbf{A} \cdot \mathbf{D})(\mathbf{B} \cdot \mathbf{C})$$

$$(\mathbf{A} \times \mathbf{B}) \times (\mathbf{C} \times \mathbf{D}) = \mathbf{B}[\mathbf{A} \cdot (\mathbf{C} \times \mathbf{D})] - \mathbf{A}[\mathbf{B} \cdot (\mathbf{C} \times \mathbf{D})]$$

$$= \mathbf{C}[\mathbf{A} \cdot (\mathbf{B} \times \mathbf{D})] - \mathbf{D}[\mathbf{A} \cdot (\mathbf{B} \times \mathbf{C})]$$

$$\nabla^2 \Phi = \nabla \cdot \nabla \Phi$$

$$\nabla^2 \mathbf{A} = (\nabla \cdot \nabla)\mathbf{A}$$

$$\nabla \cdot \nabla \times \mathbf{A} = 0$$

$$\nabla \times \nabla \Phi = 0$$

$$\nabla \times (\nabla \times \mathbf{A}) = \nabla(\nabla \cdot \mathbf{A}) - \nabla^2 \mathbf{A}$$

$$(\mathbf{A} \cdot \nabla)\mathbf{A} = \nabla\left(\frac{|\mathbf{A}|^2}{2}\right) - \mathbf{A} \times (\nabla \times \mathbf{A})$$

$$\nabla \times (\mathbf{A} \times \mathbf{B}) = (\mathbf{B} \cdot \nabla)\mathbf{A} - \mathbf{B}(\nabla \cdot \mathbf{A}) - (\mathbf{A} \cdot \nabla)\mathbf{B} + \mathbf{A}(\nabla \cdot \mathbf{B})$$

$$\nabla \cdot (\mathbf{A} \times \mathbf{B}) = \mathbf{B} \cdot \nabla \times \mathbf{A} - \mathbf{A} \cdot \nabla \times \mathbf{B}$$

$$\nabla(\mathbf{A} \cdot \mathbf{B}) = (\mathbf{B} \cdot \nabla)\mathbf{A} + (\mathbf{A} \cdot \nabla)\mathbf{B} + \mathbf{B} \times (\nabla \times \mathbf{A}) + \mathbf{A} \times (\nabla \times \mathbf{B})$$

Appendix G

Flow Measuring Techniques

Several simple devices are used extensively to measure flow rates in pipes and ducts. The application of the extended Bernoulli equation allows a simple relationship to be established between a measured pressure drop and flow rate. We will discuss a few practical devices.

FLOW THROUGH A SHARP EDGED ORIFICE

First, consider the flow of an incompressible liquid through a sharp edged opening in a tank as shown in Fig. G-1. If the cross sectional area of the tank is large compared to the cross sectional area of the orifice so that the velocity of the top surface is negligible, we can write the Bernoulli equation between points 1 and 2 (with $p_1 = p_2 = 0$ gage) to yield

$$V = \sqrt{2gh}$$

Now the area of the issuing jet will neck down because of the curvature of the stream lines and will actually have a minimum where the velocity is aligned at the point 2. This minimum section is called the *vena contracta*. The flow rate Q through the orifice may be written

$$Q = C_v C_c A_0 \sqrt{2gh} = C_d A_0 \sqrt{2gh}$$

where C_c is the area correction between A_0 and A_v

$$A_v = C_c A_0$$

and C_v is a coefficient that corrects for frictional losses in the Bernoulli equation

$$V = C_v \sqrt{2gh}$$

The product $C_v C_c$ is known as the discharge coefficient C_d and is usually determined experimentally for a given orifice design. Values of C_c vary between 0.6 and 1.00 depending on how rounded the orifice is, and C_v usually varies between 0.8 and 0.99.

In practice, a sharp edged orifice is often inserted in a pipe and the pressure drop across the orifice is measured with a manometer or other type of pressure gage as shown in Fig. G-2. A_0 is the area of the

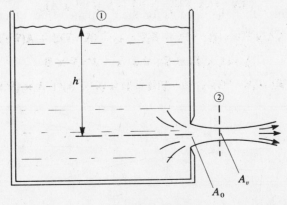

Fig. G-1

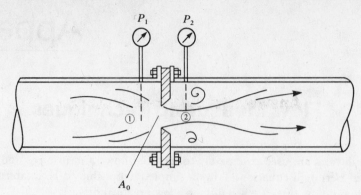

Fig. G-2 The flat plate orifice.

flat plate orifice opening. Writing the Bernoulli equation between 1 (upstream) and 2 and assuming an incompressible fluid

$$\frac{p_1}{\rho} + \frac{V_1^2}{2} = \frac{p_2}{\rho} + \frac{V_2^2}{2}$$

and the continuity equation

$$Q = A_1 V_1 = A_2 V_2 = A_0 V_0$$

where A_2 is the area of the downstream jet (which may or may not be at the vena contracta) we obtain

$$Q = \frac{C_v A_2 \sqrt{(2/\rho)(p_1 - p_2)}}{\sqrt{1 - (A_2/A_1)^2}} = \frac{C_d A_0 \sqrt{(2/\rho)(p_1 - p_2)}}{\sqrt{1 - (A_0/A_1)^2}}$$

The overall discharge coefficient C_d depends on the exact orifice shape and on the precise locations of the pressure taps at points 1 and 2 and takes into account the corrections for the area difference between A_0 and A_2 and all losses. Typical values of C_d are in the range of 0.6 to 0.8. They are determined experimentally for each individual flat plate orifice.

The equation above is still valid for compressible flow if the pressure drop $(p_1 - p_2)$ is small compared to p_1 so that ρ is nearly constant. In gas pipe lines $\Delta p = (p_1 - p_2)$ might be the equivalent of a few inches of water and p_1 might be several hundred psi. If compressibility effects are important, the Bernoulli equation may be used in the compressible form to obtain more accurate results.

FLOW THROUGH A VENTURI METER

The flat plate orifice discussed above is easy to install at a pipe flange and is used extensively in practice. Another type of meter is the venturi (Fig. G-3), which has a low loss but is more difficult to install. The flow contraction is defined by the shape and the overall loss coefficient C_d is usually of the order of 0.98 to 0.99.

The expression for flow rate is the same as for the flat plate orifice where A_0 is the minimum area where the pressure p_2 is measured.

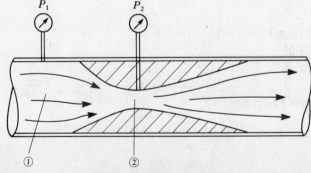

Fig. G-3 The venturi meter.

Answers to Selected Problems

CHAPTER 2

2.15. (a) 111 in., (b) 8.14 in.

2.16. 3.78 psi

2.17. 0.747 psi

2.18. 0.0205 psi

2.19. 0.00419 psi

2.20. 585 lb_f

2.22. $8.47(10)^4$ lb_f; $F_N = 0$

2.23. The same as if the container were at rest.

2.24. 6.94 ft

2.25. $p_B - p_A = \gamma$ psf; $p_D - p_A = 1.61\gamma$ psf; $p_C - p_A = 0.61\gamma$ psf

2.26. $\dfrac{\Delta a}{\Delta p} = \dfrac{D}{4t\gamma}$ where D is the mean diameter of the cup, and t is the wall thickness of the cup.

2.30. Water surface is parallel to the inclined plane.

2.31. Same as for rotation of an open container with its minimum depth equal to the height of the container with the lid.

2.32. $p = \frac{1}{2}\rho r^2 \omega^2$, independent of vertical position.

2.33. $p_{top} + \rho a(H - h)$

2.34. $p - p_0 = \frac{1}{2}\rho_1 \omega^2 r^2 + \rho_1 g(z_{01} - z)$ for $z_{01} > z - \dfrac{\omega^2 r^2}{2g} > z_{02}$

$p - p_0 = \frac{1}{2}\rho_2 \omega^2 r^2 + \rho_2 g(z_{02} - z) + \rho_1 g(z_{01} - z_{02})$ for $z_{02} > z - \dfrac{\omega^2 r^2}{2g}$

Free surface: $z = z_{01} + \frac{1}{2}\rho\omega^2 r^2$

Interface: $z = z_{02} + \frac{1}{2}\rho\omega^2 r^2$

2.35. Shape is unimportant. L and h are relevant parameters.

2.37. If the balloon is filled with helium it will slant forward as the car accelerates.

2.38. $\sigma = \dfrac{\pi D^2}{4NA}(p_c + \rho\omega^2 D^2/16)$ where ρ = density of oil, σ = tensile stress, N = number of bolts, A = cross-sectional area of a bolt, p_c = pressure from charging pump, ω is measured in radians per sec.

CHAPTER 3

3.17. 0.0119 ft^3/sec

3.18. $F = \frac{1}{3}\pi\rho a^2 V_0^2 + \pi a^2(p_2 - p_1)$

3.19. $p_1 = p_2 + 2\tau_0 L/a + \rho V_0^2/3$

3.21. 10.6 fps

3.22. 16.21 psi

3.23. (a) $Q = \pi a^2 V_c/2$, (b) $\frac{1}{3}\pi\rho a^2 V_c^2$, (c) $\frac{1}{8}\pi\rho a^2 V_c^3$

3.24. 47.0 horsepower

3.25. 18.0 ft^3/sec; 4760 lb_f

3.26. 437 ft/sec

3.27. 8.66 sec

3.29. 27.4 psi/ft

3.33. 2.32 ft/sec^2

3.34. $Q = A_2\sqrt{2gh_1}$ if $A_2 \ll A_1$ and $h_2 < \dfrac{p_{atm}}{\rho g}$

3.35. $-\rho U^2 h/6$

3.36. $D = -0.63\rho U^2 D$

3.37. 76.5 rpm

3.38. $\omega = \sqrt{p_0/x_0 \rho H}$ where p_0 is atmospheric pressure.

CHAPTER 4

4.11. Lift = 2150 lb

4.12. $f = 0.028$. Pressure drop per mile = 168 psi = 484 ft oil = 387 ft water, assuming sp. gr. 0.8.

4.14. $C_D = 0.18$. $D = 20$ lb, the same as for the model test.

4.18. Π's are $\rho DV/\mu$ and $\omega D/V$

4.19. $\omega D/V$ modeling can be achieved but not $\rho DV/\mu$ which is less important.

4.20. Froude modeling requires $V_m = \sqrt{20}$ mph and is independent of the fluid viscosity. Reynolds number modeling is impractical here since for $V_m = 5$ mph, say, $v = 1.36 \times 10^{-7}$ and such a liquid is not available.

4.22. Efficiency = 57%, $Q = 135$ ft^3/sec, $H = 90$ ft, power = 6075 hp

4.24. 11.25°F

CHAPTER 5

5.15. 0.305 in.

5.16. 0.00151 lb$_f$/ft^2

5.17. 0.125 lb$_f$; 0.000705

5.21. $1.29(UL/v)^{-1/2}$

5.24. $\dfrac{0.075\, h^2 V_1^2}{[h - 0.1\sqrt{x/8}]^3 \sqrt{x}}$; 32.5 ft/sec^2

5.25. $\rho U^2/12$

5.26. $p - p_{entrance} = -1.03(10)^{-3}$ psi

5.28. 0.338 lb$_f$/ft

5.29. $p - p_{entrance} = -6.67(10)^{-3}$ psi

5.30. 3.43 lb$_f$

5.33. 0.460 lb$_f$

5.34. 0.0735 horsepower

5.35. $\mu = 1.16(10)^{-4}$ lb$_f$ sec/in^2

5.36. 7270 rpm

5.37. $\pi\mu(\omega_1 - \omega_2)(R_2^4 - R_1^4)/2T_{in}$

5.38. 1.12 ft/sec

5.39. 23.0 ft

5.40. 23.6 ft/sec

5.42. 12.6 ft/sec

5.43. 0.0923 psi

5.44. 0.216 psi

5.45. $T = \dfrac{A}{a}\sqrt{\dfrac{2}{g}}\left[\sqrt{h_0 + h_1} - \sqrt{h_0}\right]$

5.47. $F = \dfrac{\pi DL\mu V}{h}\left[1 + \dfrac{3}{h^2}\left(\dfrac{d^2}{4} - h^2\right)\right] \approx \tfrac{3}{4}(d/h)^3 \pi L\mu V$ where $h = \dfrac{D_c - D}{2}$

ANSWERS TO SELECTED PROBLEMS

339

CHAPTER 6

6.13. 26.6 lb$_f$/ft

6.16. Force $= 2m^2\rho \int_0^\infty \dfrac{(x^2 - a^2)^2}{(x^2 + a^2)^4}\, dx = \dfrac{3\pi m^2 \rho}{16a^3}$

6.17. Force $= \dfrac{m^2\rho}{a^3}\left(\dfrac{3\pi}{16}\cos^2\alpha + \dfrac{\pi}{4}\sin^2\alpha - \tfrac{1}{3}\sin 2\alpha\right)$

6.18. No difference since F depends on m^2.

6.20. $F = -m\ln(\sinh \pi z/2d)$

6.21. $\phi = A(x^2 - y^2)$, $\psi = (A)(2xy)$

6.23. $F = -\dfrac{Q}{2\pi}\ln(\sinh \pi z/d) - U_0 z$

6.24. $F = \dfrac{i\Gamma}{2\pi}\ln\left(\dfrac{z - id}{z + id}\right)$, streamlines are nested circles.

6.31. $u - iv = -\dfrac{Q}{2\pi}\left[\dfrac{z - (b + ia)}{z - (b - ia)} + \dfrac{z + (b - ia)}{z - (b + ia)} + \dfrac{z + (b + ia)}{z + (b - ia)} + \dfrac{z + (b - ia)}{z + (b + ia)}\right]$

6.32. $F = -\dfrac{Q}{2\pi}\ln\left\{\displaystyle\prod_{n=0}^{7}\left[z - ae^{i(\pi/4)(1/2 + n)}\right]\right\}$

6.34. $ry = \Gamma^2/4\pi^2 g$, where y is measured downward from the level of the free surface at infinity to the free surface at r.

CHAPTER 7

7.28. 4880 ft/sec

7.29. 0.313; 1.11

7.30 0.182

7.31 16.1 psi

7.32. 72.5 psi; 179.2 Btu/lb; 822°R

7.33. 0.00256 ft^2

7.34. 18.1 lb$_m$/sec

7.35. (a) 0.00806 ft^2; 0.01385 ft^2. (b) 2.01

7.36. 0.222 lb/sec; 1590 ft/sec

7.37. 1.40

7.38. 1198 ft/sec

CHAPTER 8

8.8. $M_2 = 2.6$; $T_2 = 381°R$; $p_2 = 5.5$ psia

8.9. $M_2 = 1.45$; $T_2 = 638°R$; $p_2 = 31$ psia

8.10. β (bottom) $= 15°$; β (top) $= 19.5°$

8.11. $M_2 = 2.6$; $T_2 = 437°R$; $p_2 = 3.93$ psia

8.12. $D_2 = 0.68D_1$

8.13. $D_2 = 1.75D_1$

8.15. For the nth half wavelength, $\dfrac{\Delta p}{p_1} = \dfrac{(-1)^n \epsilon}{\lambda\sqrt{M_1^2 - 1}}$

8.16. $L = 1.03cp_1$ per unit length (shock expansion theory).

8.17. $G = \dfrac{c}{2\sqrt{M_1^2 - 1}}$

8.19. $C_L = \dfrac{4\bar\alpha}{\sqrt{M_1^2 - 1}}$, $C_D = \dfrac{4}{\sqrt{M_1^2 - 1}}\left[\dfrac{\pi^2}{2}\left(\dfrac{H}{c}\right)^2 + \bar\alpha^2 + 2\pi^2\left(\dfrac{Y}{c}\right)^2\right]$

8.22. $|\theta| = 53°$

8.24. Velocity of shock before reflection $= 1160$ ft/sec.

Index

Dihedral, 230
Dilatant fluid, 4, 298
Dilatation, 50, 52
Dimensional analysis, 84
Dimensions, 315
   table of, 317
Dipole, 157
Dispersion equation, in MHD, 283
Displacement thickness, 108
Dissipation friction, 57
Dissipation function,
   in various coordinate systems, 325
Downwash, 167
Drag, 91, 113
   coefficients of, 114
Dynamics, particle vs. fluid, 2

Eckert number, 87, 123
Eddy viscosity, 245
Electrodynamics, 270
Electromagnetic body force, 274
Ellipse, potential flow over, 170
Elliptic equations, 223
Energy equation, 40, 56
   differential form, 58
   in MHD, 276
   integral form, 42
   in viscous flow, 122
   relation to Bernoulli equation, 61
Entrance flow, 116
Entropy production, 58
Equivalent circuit, 280
Euler equation, 48, 49
Eulerian coordinate system, 9, 44
Expansion waves, centered, 194
Expansion, supersonic, 194, 213
External flow, 6
Eyring model, 299

Fan, modeling of, 93
Fanning friction factor, 188
Fanno line flow, 187
   tables of, 331
Fanno line, 186
Faraday disk, 291
Faraday's law, 271
Fick's law, 258
Flat plate, flow over, 101
   flow transition on, 102
Flow measurement, 335
Flow work, 42
Flow,
   classification of, 6
   compressible, 5

Flow—*continued*
   external, 6
   ideal, 3, 144
   incompressible, 5
   internal, 9
   potential, 144
   steady, 6
   subsonic, 5
   supersonic, 5, 208
   turbulent, 105, 108, 240
Fluctuations, in turbulent flow, 241
Fluid, definition of, 1
Forces on submerged bodies, 16, 20
Fourier's law, 57
Fracturing, of oil wells, 297
Friction coefficient, 105
Friction factor, 92, 120
   pseudoplastic fluid, 305
   relation to shear stress, 120
Friction velocity, 110
Friction, 3
Frictional flow in ducts, 187, 192
Friedrichs diagram, 285
Froude number, 85
"Frozen in" fields, 278
"Frozen" properties, 259
"Frozen" vorticity, 55

Gas wells, 297
Gas, definition of, 1
Generator, MHD, 291
   vortex, MHD, 293
Grashof number, 88
Gravity, 14

Hartmann flow, 278
Hartmann number, 277
Head loss, 43, 60, 91, 120
   coefficients of for fittings, 121
Heat of formation, 259
Heating and cooling in ducts, 191
Homentropic flow, 209
Horseshoe vortex, 166
Hydraulic jump, 75, 89
Hydrostatic bearing, 133, 138, 140
Hyperbolic equations, 223
Hypersonic flow, 257
Hysteresis in fluid, 301

Ideal gas, 178
Images, method of, 153
Incompressible approximation for low Mach
   number, 193